WINDFALL

How the New Energy Abundance
Upends Global Politics and
Strengthens America's Power

Meghan L. O'Sullivan

SIMON & SCHUSTER
New York   London   Toronto   Sydney   New Delhi

Simon & Schuster
1230 Avenue of the Americas
New York, NY 10020

First Simon & Schuster hardcover edition September 2017

SIMON & SCHUSTER and colophon are registered trademarks
of Simon & Schuster, Inc.

For information about special discounts for bulk purchases,
please contact Simon & Schuster Special Sales at 1-866-506-1949
or business@simonandschuster.com.

The Simon & Schuster Speakers Bureau can bring authors to
your live event. For more information or to book an event, contact
the Simon & Schuster Speakers Bureau at 1-866-248-3049
or visit our website at www.simonspeakers.com.

Interior design by Lewelin Polanco

Manufactured in the United States of America

1 3 5 7 9 10 8 6 4 2

Library of Congress Cataloging-in-Publication Data
Names: O'Sullivan, Meghan L., author.
Title: Windfall: How the new energy abundance upends global politics and
strengthens America's power / Meghan L. O'Sullivan.
Description: New York : Simon & Schuster, [2017] | Includes bibliographical
references and index.
Identifiers: LCCN 2016057135 (print) | LCCN 2017032073 (ebook) |
ISBN 9781501107955 (ebook) | ISBN 9781501107931 (hardcover : alk. paper) |
ISBN 9781501107948 (trade pbk. : alk. paper)
Subjects: LCSH: World politics—21st century. | Power resources.
Classification: LCC D863 (ebook) | LCC D863 .O85 2017 (print) |
DDC 327—dc23 LC record available at https://lccn.loc.gov/2016057135

ISBN 978-1-5011-0793-1
ISBN 978-1-5011-0795-5 (ebook)

To my parents,
Michael and Kathi O'Sullivan

Contents

Preface

On January 17, 2017, the leader of one of the world's largest economies took the stage at a gathering of global elites in Davos, Switzerland. He made a forceful and compelling case in favor of globalization and free trade, evoking Abraham Lincoln's Gettysburg Address in the process. "Say no to protectionism," the speaker implored. "It is like locking yourself in a dark room. Wind and rain are kept out, but so are light and air." What was unusual was not the message—particularly at this meeting of 1,250 global leaders and CEOs—but the messenger.

The speaker was Xi Jinping, the first Chinese president to ever address this annual assembly of the World Economic Forum. Although China's embrace of free trade was far from complete, Xi was clearly maneuvering to position China as the new vanguard of globalization. To further confuse those surprised by China becoming the global advocate for free trade, only three days later, a newly sworn-in President Donald Trump stood in light rain on the steps of the Capitol in Washington, D.C., and delivered a fiery inaugural address extolling the virtues of guarding borders and promising the American people that "protection will lead to great prosperity and strength."

This pairing of speeches, in which the American and Chinese presidents seemed to have swapped texts, roles, and global orientations, is only one of the indications that the world in 2017 is in tumult and in the throes of historical change. A growing tide of populism, the rise of once-marginal powers, and real questions about continued American global leadership are shaping the geopolitical landscape. To make sense of the changes, we would normally look to historical, cultural, economic, or political trends. Such matters will continue to provide insights into foreign affairs. But new variables—such as technological and social change—need more of our attention, as do old drivers that have been consistently underappreciated. In a quest to better understand the world unfolding around us, understanding energy is critical.

Often overlooked as a determinant of global politics, energy has long been a driver of international affairs. The shift from wood to coal allowed for the making of steel, helping usher in the Industrial Revolution in Britain and elsewhere in the late-eighteenth and early-nineteenth centuries. By the mid-1900s, however, oil had overtaken coal, bringing with it a surge of game-changing innovations, including the internal combustion engine and the tank, which ended the stalemate of trench warfare in favor of Britain in World War I.

As discussed so vividly in Daniel Yergin's Pulitzer Prize–winning book, *The Prize*, for much of the twentieth century, the economics of oil and gas in particular have permeated geopolitics and vice versa. The history of grand strategy during that era was often the history of efforts to gain or deny access to energy. For instance, many pivotal moments in World War II—from Hitler's drive to the Caucasus to Japan's quest for Borneo to the failed drive of Germany's Afrika Korps across North Africa—were shaped decisively by oil. Decades later, perceptions that the Soviet invasion of Afghanistan could be the first step in a push to control energy resources in the Gulf informed U.S. and Saudi efforts to support the Afghan mujahideen.

More recently, oil and gas have funded the rise of separatist groups in Nigeria and played a critical role in the surprising rapprochement between Turkey and the Kurds of Iraqi Kurdistan. The need for reliable energy supplies has also underpinned unlikely partnerships, such as those between Washington and Riyadh or between Europe and Russia. And for the bulk of the last thirty years, nervousness over energy scarcity was one of the most important animators of Chinese foreign policy.

Today, the impact of energy on international affairs is as pronounced as ever. Yet energy's bearing on geopolitics has arguably never been less understood. Why is this the case? One possibility is the rate of change. In the last decade alone, developments in the world of energy have unfolded at breakneck speed. Technology has brought large quantities of oil and gas once thought too expensive to produce to global markets. The declining costs of some renewable energies are making them competitive in some locations without government support. Digitization is introducing the possibility of once-unimagined efficiencies. And concerns over environment and climate change are spurring new forms of global action. All of these changes, moreover, are part of dynamic systems, which will continue to evolve, injecting new incentives and obstacles into the political domain as they do.

Another reason why energy has not figured more prominently in the analysis and making of foreign policy may be explained in the work of Robert Jervis, a professor now at Columbia University who has applied psychology to policy and decision-making. Jervis writes of the tendency of all people, when seeking to explain complex phenomena, to unconsciously discount the importance of factors they do not understand. The workings of energy markets are often complicated and technical, possibly leading many to gloss over their critical role in shaping international affairs and to focus instead on more intuitive explanations such as politics and history.

This book intends to remove that obstacle for nonexpert readers seeking to appreciate one of the most longstanding and consequential drivers of global politics: energy. It demystifies energy markets and powerfully and tangibly relates them to the most basic and fundamental drivers of foreign affairs. It demonstrates how the energy revolution that has taken the world by surprise in the last decade is creating both new opportunities and new challenges at a global level, altering the balance of power between countries, and shaping their actions and attitudes toward one another.

Understanding this interaction between energy and international politics is and will continue to be essential to appreciate the unfolding global landscape. It will arguably be more important in driving foreign policy outcomes than many of the other issues consuming the calories of policymakers and the airtime of pundits.

The twelve months of 2016 were sobering for those who believed they understood the underlying dynamics of many global trends. The frequency with which conventional wisdoms and established understandings were proven wrong should spur us to look for new lenses through which to comprehend the world. This book offers its readers just that. Many readers will be surprised at how powerful the prism of energy is in making sense of global events. While surely not determinative on its own, energy is and will continue to be a major driver of how the world works.

WINDFALL

Introduction

I sat waiting a bit self-consciously on the sofa in a large, tidy office—something of a cross between a workspace and a *diwan*, which in the Arab world is the section of a house that is always open to guests. I wore the customary long, loose black abaya, but I had let my headscarf fall and rest draped around my shoulders. In previous meetings in Saudi Arabia, I had opted for what I called the "Benazir Bhutto look," where I covered my head but let more than a few wisps of red hair escape. Yet the assistant in the outer office had insisted that the senior ministry official I was waiting to see "was a modern man" and there was no need to cover my head in a private meeting. Still, I was uncertain whether I was transgressing what was considered appropriate in this highly conservative—and, to me, still mysterious—society.

I had slipped out of a large conference of analysts and diplomats I was attending in Riyadh at the invitation of the Saudi Ministry of Foreign Affairs to hold this private meeting at the Ministry of Petroleum and Mineral Resources. It was September of 2014 and the Middle East was smoldering. Earlier hopes that the removal of rulers from Tunisia to Egypt would lead to more participatory governments now seemed shockingly naive. The civil war to upend another autocratic ruler, President Bashar al-Assad in Syria, was raging. The United Nations had estimated a month earlier that 191,000 people had been killed there—with little hope for an end to the violence.

Two months previously, the Islamic State for Iraq and al-Sham (ISIS) had commandeered international attention with its brutal tactics and the shocking ease with which it had wrenched control of nearly a third of Iraqi territory away from the government in Baghdad. The United States had just begun limited air strikes. Perhaps in response, ISIS had beheaded yet another American, journalist Steven Sotloff, only days before I arrived in Riyadh. Tensions within Yemen were also simmering. Several days later, they would boil over when Iranian-supported Houthi insurgents stormed the Yemeni capital of Sana'a and forced the resignation of the country's prime minister.

In the eyes of the Saudis, one factor tied nearly all these developments together: the nefarious efforts of Iran to destabilize the Arab world and assert dominance in the region. Most Arabs I met that September believed the United States was foolishly abetting the Iranians through its pursuit of a nuclear accord and remaining aloof from the region as one fire after another lit up the Middle Eastern sky.

I had sought private meetings at the ministry, not to speak about the regional political and security meltdown under way at the time, but to talk of the economic crisis I saw on the horizon. I had been hearing from oil producers from North Dakota to Texas to Pennsylvania about the remarkable transformation of the U.S. oil industry as American entrepreneurs tapped new resources. Since 2010, U.S. crude oil production had surged, consistently surpassing even the most bullish expectations— sometimes exceeding the previous year's forecast by more than a million barrels. It was clear to me that, almost one for one, this surge in American crude was substituting for barrels coming *off* the global market as Middle East producers wobbled. The result was that the price of oil remained remarkably stable despite the dramatic events unfolding. It was an amazing pairing of developments, and one I suspected would not last long.

Barring some grave unanticipated event, global oil production would soon outstrip global demand. Thanks to what many were calling an "energy revolution," American oil production was nearing all-time highs, while Russia was producing record amounts and Saudi Arabia very close to it. What's more, oil demand growth seemed to be stalling, reflecting slowing economies and rising efficiency; global oil demand for the first half of 2014 flatlined from the end of 2013.

In my writings and speeches I had suggested that, as a result, the price of oil was in for a significant decline. And I wanted to get a sense from Saudis themselves whether the kingdom was poised to take action to stem a weakening price, as they had done so often in the past.

My view of the future was not widely held at the time. Even as the price of oil dipped just below $100 a barrel for the first time in more than two years, producers remained sanguine. Just before I had boarded the plane to Saudi Arabia, in fact, I had defended my views at a workshop in New York. Several participants, pointing to the high costs in the oil industry, adamantly disputed my assertion that the world was moving into an age of "energy abundance." My Harvard colleague Leonardo Maugeri, who had been even earlier and bolder in his predictions of an oil glut, elicited snickering from an audience when he predicted in 2012 that oil could fall to as low as $50.

At the time, the Paris-based International Energy Agency (IEA) was advancing the more conventional view that new American oil production, rather than undercutting global prices, would continue to contribute to global oil price stability. Indeed, it predicted that over the next couple of years, prices would remain slightly above $100. Abdalla Salem El-Badri, the secretary general of the Organization of the Petroleum Exporting Countries (OPEC) at the time, took the same position during a May 15, 2014, speech in Moscow. El-Badri interpreted the steady oil price over the previous several years as evidence of consumer and producer satisfaction with the price level, which clocked in at $110 a barrel on the day of his speech. El-Badri focused on steady demand growth and sufficient supply, calling the market balanced and predicting it would stay that way for the rest of the year.

By the time I had reached Riyadh, oil prices had just begun to soften, but people in positions of authority expressed confidence that prices would remain stable. Prince Abdul Aziz bin Salman, the then–assistant minister of petroleum and an influential and highly competent royal family member, had made a forceful case to the conference I came to Saudi Arabia to attend. He argued oil markets would remain balanced and the high costs of producing some resources "help put a floor [under] the long-term oil price." Dismissing gyrations in the price of oil as temporary, the prince focused on expectations of a young, burgeoning global middle class driving up oil consumption and "a shrinking pool of cheap and easy oil."

A Meeting at the Ministry

My meeting began with a handshake and quickly turned into a wide-ranging and fast-paced conversation covering energy trends, U.S. military power, and American allegations that Saudi Arabia was the root of Islamic terrorism. With a look that was either playful or mischievous, or both, the official told me that Saudi Arabia welcomed America's growing oil production. "We should be happy for our friends for this good fortune," he suggested.

The two countries, he added, were destined to become even stronger partners as their interests in oil became more closely aligned. Like his colleague at the conference, this official also seemed unfazed by the possibility of a significant drop in the price of oil. When I pressed the issue, he pointed to how the world had absorbed millions of barrels of additional oil from the Caspian and Angola in the previous decade, all without a major dip in the price of the commodity. He took a long-term view. While some in the United States saw the new energy abundance as a path to energy independence, freeing the nation from reliance on Saudi oil, he saw, instead, the basis for cooperation. Moreover, he posited that the United States would no longer be interested in lower prices, as they would undermine America's newfound "strategic advantage."

But what made the meeting so memorable was the official's response to one of my questions in particular.

"Will Saudi Arabia continue to produce today's large volumes of oil even in the face of a falling price?" I asked.

Without a moment's pause, he replied, "You can bet on it."

He then referred back to earlier times when Saudi Arabia cut its production in a failed effort to boost oil price. Revenues plunged as the price remained low, lurching the kingdom into economic crisis and political uncertainty. "We remember 1985 and 1998 and how we can't hold people's hands while our feet are to the fire. The price will be what it is."

Wanting to make sure there was no misunderstanding, I then diplomatically inquired if this approach effectively made OPEC "less relevant."

"Are you asking me if OPEC is dead?" The official quipped, "I never like to say OPEC is dead, but . . ."

Our chat ended shortly thereafter, and I left the ministry. Repositioning my headscarf before stepping out into the Saudi heat, I felt a sort of excitement, the sort of rush one feels when one has gotten the final piece of a complex intellectual puzzle. But I also felt a foreboding. If what the official told me proved to be true, the global oil market—and the world—was in for a dramatic shock if a falling price would elicit no action from Saudi Arabia or OPEC to stabilize it. There would be huge winners and losers, and the process of reshuffling would be both jarring and destabilizing for governments and people around the world.

The Price Plunge

The coming months demonstrated that my interlocutor had indeed told me exactly what would happen—or, rather, not happen. The oil price began to drop sharply, but Saudi Arabia and OPEC sat on the sidelines. The extent and duration of the resulting price plunge far exceeded my expectations or, almost certainly, those of individuals with whom I had met in Riyadh. The price of oil didn't just dip—it took a nosedive, declining more than a fifth in the two months after I left Riyadh.

The economic effect of low prices rippled around the world—awakening it to the new energy abundance that had been building over previous years. For some oil importers, seemingly rock-bottom prices were an economic stimulus. They helped keep Europe's growth modestly positive when the fundamentals might have pulled it in another direction. They injected a boost into the Chinese economy when the government might have not otherwise been willing to lift demand. In contrast, for some oil exporters, from Venezuela to Angola, low oil prices created immediate fiscal crises and doubts about the ability of governments to fund commitments.

For others, low oil prices were a mixed blessing. In Japan, consumers welcomed relief from high energy prices, even while such prices frustrated the government's efforts to combat persistent deflation. In the United States, consumers did not respond by spending more as they had in other periods of low oil prices. After the deepest recession since the

Great Depression and years of slow economic recovery, many Americans preferred to save their dollars rather than splurge their savings from the pump. Similarly, the boost that stock markets traditionally received from low oil prices did not materialize. Energy companies worldwide almost uniformly swooned under the pressure of low prices, with the large oil corporations turning in their worst financial performances since the 2008 financial crisis. Stock markets around the world dipped, weighed down by poorly performing energy shares.

As great as this immediate economic tumult was, the impact of the new energy abundance goes far beyond balance sheets and stock markets. In fact, changes in oil and gas markets have provoked massive global changes. As the world has lurched unexpectedly from energy scarcity to energy abundance in recent years, geopolitical mainstays have been upended. The low price of oil itself halted one of the biggest transfers of wealth in history, allowing consumers to save an estimated $3 trillion a year that they would have otherwise paid to producers.

Low prices also changed the strategic orientation and priorities of countries around the globe. The United States, for example, has moved from being the world's thirstiest consumer of overseas oil to a position of greater self-sufficiency. Among other impacts, this dynamic has helped temper predictions and perceptions of American decline. In Asia, bountiful American natural gas inadvertently and indirectly helped Japan manage the aftermath of the nuclear disaster at Fukushima, which had led to a suspension of the nuclear power generating a third of Japan's electricity before the 2011 earthquake and tsunami. Plentiful oil both eased Chinese anxieties about meeting domestic needs and reduced predictions of inevitable conflicts over the pursuit of energy resources. This abundance even enabled China to broaden its foreign policy focus to embrace new priorities, such as exporting excess capacity and promoting "the Chinese dream."

In contrast to consumers, major oil producers who depend on oil as a primary source of revenue had their geopolitical wings clipped by the persistently low prices. Venezuela can no longer readily supply neighboring countries with cheap fuel oil, diminishing its ability to wield influence over regional politics and pushing the country to the brink of collapse. Low prices exposed the unsustainability of many socialist policies in

Latin America, accelerating the end of a period of leftist politics throughout the continent. Another massive producer, Russia, is also finding it more difficult to translate its vast energy reserves into geopolitical influence in a low-energy-price environment. Pronouncements of Russia as "an energy superpower"—made just a decade ago—now sound absurd.

More changes are undoubtedly to come. For one, Russia's economic troubles could eventually deepen to a point where Moscow loses effective control of its autonomous republics, particularly those in the crisis-prone Caucasus, with consequences for Russia's internal stability and security. Abundant energy is also complicating the seemingly historic rapprochement between Russia and China. More positively—if not derailed by politics—energy abundance could drive further integration between the U.S., Canadian, and Mexican economies, leading to the most competitive manufacturing zone in the world.

The internal politics of countries are also being transformed. Continued low oil prices are straining, and could perhaps ultimately break, the social contract between the Saudi people and their rulers that has for so long underpinned Saudi stability and prosperity. This would only further stoke current Middle Eastern fires. At a minimum, these low prices are providing the leadership in Riyadh with a real impetus for serious economic reform. In Africa, countries from Mozambique to Uganda to Sierra Leone will be far harder pressed to capitalize on recent natural resource finds as sagging prices dash hopes of propelling their populations out of poverty. Iraq's prospects are also even more sobering in the face of low oil prices. Petroleum revenue is necessary not only to keep ISIS at bay, but also to rebuild destroyed cities and help Baghdad keep provinces bound to the center.

The likelihood that some of these geopolitical developments—and perhaps many more—will come to fruition increases the longer energy prices stay well below the level needed to keep producer budgets afloat. Yet, as important as price is, it is not the only way in which today's new energy abundance is shaping geopolitics. We are seeing big changes in the structure of energy markets that will have their own geopolitical ramifications. For instance, the gradual but distinct movement away from regional natural gas markets toward a global one will make trade in natural gas harder to utilize as a geopolitical tool. Patterns of trade are shifting as the United States, the largest consumer of both oil and natural gas, becomes more self-sufficient in the first and nearly independent in the second, affecting the national conversation about U.S. global engagement. Old institutions, such as OPEC, have lost their vigor. As more countries discover and develop energy sources of their own, diminished dependencies will transform bilateral relations. In short, the new energy abundance shifts the world from a seller's market to a buyer's one, empowering consumers and wrenching geopolitical influence from producers.

The new energy abundance is erasing the long-held vulnerabilities of some countries, creating leverage for weak states over strong, and offering new opportunities to address persistent challenges to the international order. It is both advancing and deterring efforts to combat climate change around the world. On the whole, the new energy abundance is a boon to American power—and a bane to Russian brawn. On balance, China is already a winner from this energy revolution, both from the lower energy prices it has brought and through the geopolitical opportunities that it now offers to Beijing. These new energy realities have presented unforeseen avenues of cooperation between the United States and China, while creating strains on long-standing partnerships between Washington and the capitals of the Gulf in the Middle East.

The impact that energy has on geopolitics is no game at the margins. It will, in fact, be a major determinant of the international order or, rather, how the world works. It will alternatively hasten and help arrest the major trends now discernible to any global strategist: the corrosion of the rules and norms that have shaped the liberal international order since World War II, the shift of power and wealth from West to East, the push by Russia and China to establish spheres of influence, the rise of nonstate actors at the expense of sovereign governments, and the retrenchment of the United States and Europe from the global stage.

Energy—its abundance, scarcity, price, method of production, et cetera—will not be the only factor shaping geopolitics in the years ahead. The future always has many engines. The pace of technological progress, the balance of power between countries, the durability of political alliances, the robustness of the global economy and its institutions, the vulnerability of fragile regimes, the distribution of natural endowments, the military strength of great powers, and the decisions of certain individuals will all play a role in charting the course of the next decade and beyond.

But the vicissitudes of energy can and will influence each of these factors. And, in turn, energy will shape the conduct of foreign policy and national security and the contours of global affairs. While this interaction itself is not novel, the energy dynamics at work have changed dramatically in the past decade. They are therefore sending new and different signals throughout the international system. How this new energy abundance unfolds will have a greater—if more diffuse—bearing on international affairs than many of the current issues that dominate headline news.

This book concentrates primarily—although not exclusively—on the impact of energy changes in the oil and gas sector on global politics. This focus is not to imply that renewable sources of energy lack importance. To the contrary. We have begun to see renewables make real inroads into the world's energy mix, particularly in the power sector where they are the fastest-growing source of electricity generation, albeit from a low base.

Every major change in the global energy mix or in the energy system brings with it its own geopolitical ramifications. We should therefore expect the widespread deployment of renewable energy eventually to have major repercussions for global politics. These changes may take familiar forms, such as the formation of cartels not around oil, but around lithium and other critical resources. Or they could spur the need to manage state collapse among some oil producers, if renewable energy penetrates the transportation sector on a large scale. The energy poverty that currently keeps so many people from enjoying the fruits of growth could also be addressed more quickly than imagined. Yet at the same time, countries powered primarily by renewable energy may find themselves subject to new vulnerabilities as economies become heavily electrified. And to those who have battled the politics of pipelines, the politics of supergrids may become familiar. While renewable energy itself is unlikely to cross borders too often, the electricity it generates might, as will the technologies and know-how that give a country a competitive edge.

These intriguing possibilities notwithstanding, for the time being, global politics are shaped far more by fossil fuels than by any other energy source. There are several reasons for this dominance. To begin with, fossil fuels still account for more than four-fifths of all the world's energy, and will continue to be the main source for the foreseeable future. Even many scenarios that envision the world as successful in making the changes required to avert "catastrophic" climate change still posit that the majority of energy used globally will come from fossil fuels. Moreover, virtually all cross-border trade in energy is in fossil fuels; renewables are generally consumed in the country in which they are generated. As a result, a pipeline snaking across the Caspian Sea has many more geopolitical implications than a field of solar panels in Nevada's desert. While the potential is large, cross-border electricity trade generated from renewable energy is still limited.

Moreover, the exact geopolitical contours that this energy transition will take remain essentially unknown; they will depend in large part on which technologies and energy sources replace fossil fuels. In a 2014 book, *Game Changers: Energy on the Move*, Stanford and MIT faculty explore energy innovations in natural gas, solar photovoltaics, grid-scale storage, electric cars, and LED lighting. The big takeaways from that book are the sheer number of energy innovations bearing fruit or holding promise, and the wide variety of outcomes that could emerge over the coming years and decades. Given this, efforts to attribute broad geopolitical shifts to more sustainable energies in a systematic way necessitate some speculation, whereas the impact of oil and gas on geopolitics is clear and in the present.

Roadmap

This book is divided into three parts that, collectively, explain the new energy landscape and its impact on the world of foreign affairs and international security. The first section is devoted to illuminating the new energy abundance. Chapter One explains the forces of technology and politics that were behind the big price plunge beginning in 2014. Chapters Two and Three delve deeper into oil and gas, respectively, revealing how the new energy abundance shapes not just price, but also the structure of markets in ways that will be lasting and have geopolitical consequences.

The second section of the book pertains to the new energy landscape and America, the genesis of many of the energy developments transforming global markets and geopolitics. Chapter Four looks at America's misguided pursuit of energy independence, while the following two chapters examine how the new energy dynamics are reinforcing American sources of strength. Chapter Five looks at how energy is bolstering American hard power; Chapter Six focuses on the energy boom's impact on American soft power. Chapter Seven examines the U.S. experience when it comes to the complex relationship between the energy boom and the environment and climate.

The third section of the book focuses on the international arena beyond the Americas. Even though the boom in oil and gas production has been thus far largely limited to the United States, its geopolitical impacts are much broader. Because energy markets are global or regional in nature, and because of the huge footprint of the United States as a consumer and producer, the new oil and gas coming from North America reverberates beyond its borders. It is felt on every continent, in every country, to some degree. While Africa and Latin America are also affected by this new energy landscape, this book concentrates on the regions most likely to be the main drivers of global politics in the years ahead. Chapters Eight, Nine, Ten, and Eleven examine how the new energy abundance is transforming politics and international affairs in the important power centers of Europe, Russia, China, and the Middle East.

Finally, the Conclusion takes a step back and considers the entirety of this complex landscape and offers thoughts for policymakers who are looking to do what great powers have done for centuries: use energy as either a means or an end to their grand strategies. In particular, it urges the United States to seize the good fortune of the energy boom not only by focusing on the economic benefits it brings at home, but also on the strategic advantages that can accrue to it in many parts of the world as a result of the new energy realities.

SECTION ONE

The New Energy Abundance

ONE

Behind the Price Plunge

I n 2005, *New York Times* columnist John Tierney made a $10,000 bet with Matt Simmons, an outspoken figure in the world of energy finance, over the future price of oil. Tierney had gotten in touch with Simmons after reading about his assessment that global oil production had hit a "peak" level, and his prediction that as a result, shortages would soon wreak havoc on the oil-dependent world. Peak oil, Simmons claimed, was a looming calamity for the global economy. Tierney was unconvinced. He had a strong belief in human ingenuity and the ability of technology to solve problems like decreasing oil reserves. So Tierney challenged Simmons to put a price on his prediction. After some friendly negotiation, each put $5,000 in escrow. The two agreed to focus on the price of oil at the end of 2010. If the average, inflation-adjusted price of oil for that year exceeded $200, Simmons would collect the full $10,000 plus interest. Otherwise, Tierney would be declared the winner and reap the gains.

Simmons's anticipation of a world where demand outpaced supply put him in good company in 2005. A wide spectrum of energy experts was also predicting rapidly increasing competition for oil and pursuant ruinous consequences. Among them was the U.S. intelligence community, which produces a report called *Global Trends* roughly every five years in order to give the president or president-elect its best assessment of what the world could look like fifteen to twenty years in the future. The report is the product of exhaustive and intense consultations with experts inside and outside the U.S. government, as well as from many countries around the world. As a result, *Global Trends* gives the best possible window into how the broadest number of internationally recognized experts saw dominant geostrategic trends and their implications for the future less than a decade ago.

Energy—and increasing competition for dwindling resources—is a recurring theme in the *Global Trends* report produced in 2008. The expert consensus at that time was that emerging economies would be ever more thirsty for energy, severely taxing the ability of supply growth to keep pace. The situation would be exacerbated by the declining oil output of many traditional, non-OPEC producers such as Norway, the U.K., Colombia, Argentina, and Indonesia. As a result, oil and natural gas production would be concentrated in fewer and fewer countries, the majority of them located in the increasingly politically volatile Middle East. Yet experts also anticipated that conflict driven by nervousness over energy security would extend well beyond the borders of that region. This growing competition for secure energy sources from the Arabian/Persian Gulf would drive naval competition and tension over sea lanes between the world's greatest economic and military powers, including China, India, and the United States. New alliances would develop as countries sought to guarantee their access to resources in ways other than relying on the market.

Other prominent policymakers and analysts saw similar trends at the time. In 2008, Nobuo Tanaka, then head of the Paris-based IEA, wrote that, if unaltered, "the course on which we are now set . . . would lead to possible energy-related conflict and social disruption." In a similar vein, U.K. defense minister John Reid, on the eve of a 2006 summit between Prime Minister Tony Blair and environmentalist activists, predicted that the British military would need to deal with conflicts related to scarce energy and water resources in the years ahead.

The future, however, did not arrive as predicted. In the years since then–newly elected President Barack Obama received his *Global Trends* briefing, the world has indeed seen an energy transformation. But it is an entirely different one than that anticipated by the hundreds of experts consulted for the *Global Trends* report. Rather than being dragged into an energy-scarce landscape, the world finds itself—very unexpectedly—in a situation of global energy abundance.

Even if analysts had been tipped off that the world was about to lurch into a supply-driven state of energy surfeit, they likely would still have been unable to predict the source of this energy largesse. The authors of the *Global Trends* report produced in 2008 did in fact grapple with the question of whether technological revolutions would change the supply side of the energy equation. Specifically, they, like most analysts at the time, were focused on the growth potential of new sources of cleaner energy. Would the world realize the full fruits of innovations in energy storage, clean water, and biofuels? Would solar, wind, and hydropower allow the world to transition away from fossil fuels to alternative low-carbon energies?

Certainly, there have been important advances in the world of renewable energy in recent years. In the first decade of this century, renewable electricity generation from nonhydro sources nearly quadrupled. Generous subsidies and policy incentives from European, Chinese, and American governments contributed to the maturation of solar and wind energy. This support encouraged expansion and helped drive down costs to the point where, in some sites, both wind and solar are cost competitive with fossil fuels without subsidies. Such growth will continue in the years ahead, with renewables accounting for more than a quarter of electricity generation by 2021. But the actual and anticipated dramatic growth in renewable energy—coming from such a small base—cannot itself explain the sudden and remarkable global shift from energy scarcity to energy abundance. After all, *nonhydro* renewable energy still only accounted for 3 percent of overall global energy usage in 2016.

The energy revolution that analysts overlooked in 2008 was not in renewables, but in oil and gas. The experts did not foresee the emergence of new resources that would fundamentally change the balance between the supply and demand of oil and gas, as well as the structure of energy markets. These new, unanticipated oil and gas resources are known as "unconventional," but not because their molecular structure is any different from those considered "conventional." Instead, the term refers to the method required to extract the oil and gas. Unlike more familiar oil and gas resources, unconventional ones do not reside in large reservoirs that can be tapped and drained with a small number of wells. The challenge of unconventional resources is to liberate the oil or gas from millions of small pockets in shale rock, where they reside, like bubbles in champagne.

It is the type of extraction process that provides the distinction between conventional and unconventional oil and gas, and over time what is considered to be unconventional may change. The next important distinctions are *within* the category of what is broadly labeled unconventional. Think of "unconventional oil" as an umbrella term with many different unconventional oils grouped under it, including kerogen oil, gas converted to liquids, and tight oil (also known as shale oil). Likewise, the term "unconventional gas" is also an umbrella term that encompasses several different types of natural gas, one of which is shale gas. There is also more than one unconventional extraction technique. The most well-known process is "fracking," which is used for extracting resources such as shale gas and tight oil. But oil sands, also considered an unconventional oil, is generally extracted by very different means: the intense heating of the resource while it is underground before extracting it.

The diagram below helps explain the relationship between unconventional and conventional energy, and the different forms of energy that fall under each category. For the purposes of this book, and for most conversations in the energy world today, the "unconventional boom" refers to dramatic changes in the production of a wide variety of oil and natural gas resources. Two specific unconventional resources, shale gas and tight oil, are also frequently referred to in this book. They warrant special mention because they have turned out to be the most prolific new unconventional energy sources in the United States and, to a lesser extent, in Canada, and have prospects of being produced in significant quantities outside North America.

The roots of today's surfeit of oil and, to a lesser extent, natural gas lie in two related stories, half a world apart. The first is a tale of technology, innovation, and human persistence in America's heartland. The second is a saga of politics, markets, and power in the deserts of the Middle East. The two stories came together to deliver a major shock to the world, spurring us to look at the globe through a lens of plenty, not paucity.

Figure 1.1: Unconventional Resources is an Umbrella Term

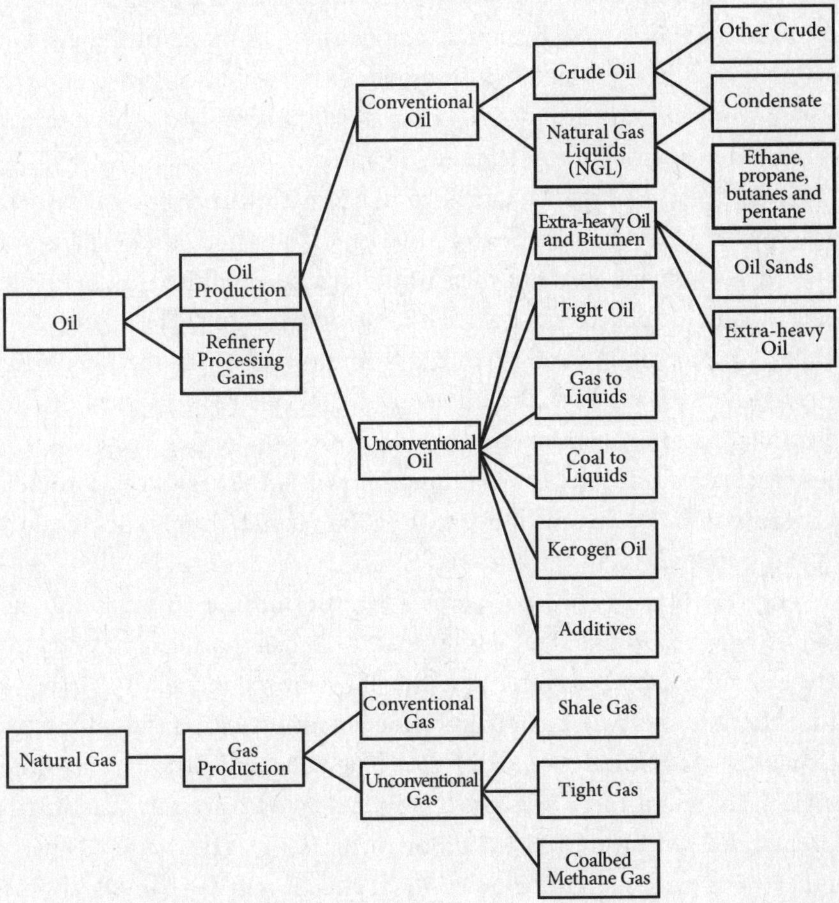

Source: International Energy Agency, *World Energy Outlook, 2015*, 678.

The Man Who Squeezed Oil from Stone

In 1952, George Mitchell, a struggling petroleum engineer running a small company with his brother, received a tip from an unlikely source.

A bookie who usually took horse racing bets insisted that a fortune could be made drilling on a piece of land north of Dallas nicknamed "wildcatters' graveyard." Neither the name of the land nor the source of the tip was particularly encouraging, but Mitchell checked out the ranch. After his initial survey, he saw some possibility and arranged for a small lease. The first ten wells produced gas, but no oil—explaining why so many others had declined to make a similar investment. But Mitchell was encouraged and convinced he could turn natural gas into a profit, even at the low prices of the time. Within ninety days, he had collected enough money from his investors to purchase 300,000 acres—or a little under five hundred square miles—at about $3 apiece.

Thus began George Mitchell's romance with the geological formation known as the Barnett Shale. Mitchell's fixation on that expanse of shale rock south and west of Dallas would nearly drive him to ruin before it eventually propelled him to great fame and fortune. The natural gas Mitchell initially found and produced from the shallow formations of his property soon tapered off. But he was aware that there was a great deal of additional natural gas trapped in the fissures of shale rock deep beneath these formations. Geologists had long known of the resources confined in these rocks. The first reference to them had come from French explorers and missionaries in the mid-1600s.

The difference between George Mitchell and the countless others who had dismissed this shale gas as uncommercial mostly came down to determination. Mitchell would spend half a century and nearly a quarter of a billion dollars researching and experimenting with different modes of coaxing the natural gas out of the shale rock and bringing it to the surface. His main focus was on a practice called hydraulic fracturing. Mitchell did not invent the technique. It had in fact been used as early as the 1860s, with its first commercial application in 1949 in Oklahoma and Texas. The essential idea was—and still is—to pump large volumes of fluid down a well at high pressure to create cracks in the source rock. Those cracks are then prised open by introducing a porous substance, often sand, that allows gas to flow out from the rock when the well is depressurized.

Mitchell and his engineers sought to find the right combination of materials, apply them in the correct fashion, and create fractures of the right size. It was a lonely endeavor, but Mitchell's team had help. They were the beneficiaries of U.S. government programs created in the 1970s to reduce dependence on Middle Eastern oil by backing initiatives to exploit domestic resources. Government-funded projects mapped and assessed shale formations throughout the country and provided tax breaks and other incentives to entice companies to explore nascent technologies and drilling techniques. As the U.S. government has done in so many other areas from supercomputers to advanced prosthetics, it hoped to catalyze innovation in areas that were not, on their own, significantly attractive to private investors and researchers. Well aware of the resources trapped within the shale rock, the government sought to motivate and support Mitchell in his pursuit to liberate them.

Mitchell was pushing to succeed where much larger oil and gas companies had given up. But it wasn't only executives from other companies who thought Mitchell was on a fruitless pursuit. His heir apparent, Bill Stevens—and even his own son, Todd, a geologist who sat on the board of Mitchell Energy—argued ardently against Mitchell's plans in professional and private settings. George Mitchell, they insisted, was driving his company into the ground with his obsession to tap into shale gas.

Through it all, Mitchell Energy made incremental progress. The most important advance was dramatically bringing down the costs of hydraulic fracturing by continually innovating and readjusting the formula. Mitchell decreased costs by relying on a relatively inexpensive fracking fluid—a combination of sand, water, and other chemicals—after flirtations with nitrogen foam, nitrogen gel, and other substances. Mitchell's team also increased its overall efficiency, reducing the average time it took to drill a well by nearly a half. These advances helped coax more natural gas out of the wells, and people in the industry began to note a small, but distinct, uptick in Barnett natural gas production. Some began to wonder if that crazy George Mitchell was finally onto something. Devon Energy, an oil and gas company that had earlier declined to buy Mitchell Energy, gave the company a second look and, in 2001, purchased it for $3.1 billion.

Devon combined Mitchell's advances in hydraulic fracturing with its own enhancements in another technology: horizontal drilling. This combination proved powerful. Mitchell's innovation allowed for a cheaper and more effective extraction of gas from the source rock. Devon's knowledge of horizontal drilling, opposed to traditional vertical drilling, enabled the company to maximize exposure to the rock's surface. This drilling was especially beneficial in dealing with shale gas, as shale formations, which run laterally, are often long and thin, and the gas, trapped in very small pockets, requires contact across large faces of the rock. Combined, these two technologies came to be known as "fracking."

Devon added one more ingredient to the mix—the use of 3-D seismic data—to help place the horizontal well in the optimal position vis-à-vis the shale formation. The marriage of these techniques—hydraulic fracturing, horizontal drilling, and 3-D seismic data—began to produce dramatic results. By 2003, initial production rates of a handful of wells were more than three and a half times what Mitchell Energy wells had been producing. Devon filed the production rates of these wells with the Texas Railroad Commission (which had evolved to regulate oil and gas production over time) and they became public in July 2003. In the months that followed, more than two dozen other operators applied for permits to drill more than one hundred horizontal wells in the Barnett. Within a decade, the number of wells in the Barnett had increased 750 percent, and the production of natural gas in the Barnett grew almost sevenfold over the same period. This success was soon replicated in similar basins across the country.

Small companies took the lead in applying Devon's recipe to basins such as the Eagle Ford in South Texas and the Bakken in North Dakota and Montana. At least initially, large oil and gas companies such as ExxonMobil, Chevron, and Shell—often called "the majors"—could not be bothered with such intensive efforts. They were more focused on oil production and, in any case, they had little interest in tending to thousands of wells with such small yields. Instead, they preferred to concentrate on the "big elephant" projects such as the offshore Nigerian field Amenam-Kpono or to tap into the vast natural gas reservoirs in the Arctic shelf. These projects promised enormous returns over time and capitalized on the natural advantages of the majors, which excelled in managing complex projects. In contrast, the smaller oil and gas outfits seemed better suited to develop the shale "plays" due to their comparative flexibility and decentralized decision-making.

A proliferating number of smaller-scale operators seized the opportunity to parlay fortuitous geography into sometimes staggering fortunes. The Wilks brothers—two enterprising bricklayers in Texas—were just two of many, and they entered through a back door. Having learned masonry skills from their father, the two, Farris and Dan, founded Wilks Masonry in 1995. Several years later, new geological imaging technology revealed shale beneath their land. Big oil and gas companies had no interest in fracking their small plot. Being entrepreneurial in spirit, Farris and Dan purchased some equipment, manufactured other tools, and began fracking their own land. Specializing in little jobs, Farris and Dan began to service the wells of their neighbors, eventually founding a company called Frac Tech. The company grew as small landowners—incentivized by U.S. law that grants them rights to what lies below the grass under their feet all the way down to the core of the earth—rushed to exploit their unexpected riches. In 2011, after eight years of building and operating Frac Tech, Farris and Dan sold it to Temasek Holdings, Singapore's sovereign wealth fund, for $3.5 billion. That same year, *Forbes* ranked them among the 400 wealthiest Americans.

The collective efforts of dozens of small and midsized American companies fundamentally altered the U.S. energy landscape. Companies took advantage of readily available financing and private ownership of subsurface minerals, and were kept lean and hungry by America's competitive marketplace. Eventually, larger companies joined in. Natural gas production in the major U.S. shale formations skyrocketed, driving a veritable bonanza in U.S. shale gas production and a stunning turnaround in overall American natural gas output. In 2006, the United States was producing enough shale gas to heat 15 million homes a year; by 2014, this amount had grown sufficient (hypothetically) to heat 200 million homes a year. By 2015, more than half of all natural gas produced in the United States came from shale, compared to just 6 percent a decade earlier.

It didn't take a genius to surmise that what worked in shale formations that held shale gas, might well work in shale formations that held primarily tight oil instead of shale gas. Companies began to apply the same fracking technologies to such basins in the later part of the first decade of the 2000s, again achieving staggering results. Production of tight oil in the Eagle Ford shale in Texas grew *twenty-four-fold* from January 2007 to January 2014. Total production from this one basin surpassed that of OPEC member Algeria and nearly matched the entire production of Kazakhstan. Tight oil production in the Bakken fields increased eightfold during the same period. In 2014, U.S. tight oil production alone surpassed Iraq's overall annual production. In the same year, burgeoning U.S. tight oil production pushed overall American crude output to be 10 percent of the world's total supply. Accounting for nearly half of overall U.S. crude oil production, tight oil was the driving force behind America's oil resurgence.

Figure 1.2: Dry Natural Gas Production by Type (trillion cubic feet)

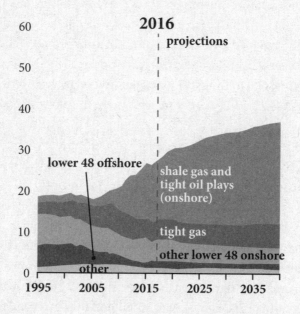

Source: EIA, *Annual Energy Outlook 2017*, 57.

The gush of new American oil was significant. An additional 4.9 million barrels a day (mnb/d) at its first peak in March 2015 might seem like a small addition to a total global market of 96 mnb/d. But the demand for oil is essentially fixed in the short term, so a quick change in supply can have a dramatic impact on price. Consider how a sudden contraction in the supply of oil can lead to a price spike. For instance, in 2005, damage to production, refining, and storage facilities in the Gulf of Mexico by Hurricane Katrina took 1.3 mnb/d of oil offline overnight, leading the price for American oil to rise 7 percent in a matter of days. A huge, relatively sudden increase in global oil supply could have the opposite effect, all other things equal.

Figure 1.3: U.S. Crude Oil Production (million barrels per day)

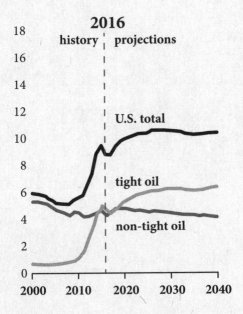

Source: EIA, *Annual Energy Outlook 2017*, 43.

Nevertheless, America's new energy prowess was not *in itself* enough to tip the world into significant oversupply and drive down the price to lows not seen since 2003. In fact, as discussed earlier, from 2011 to 2014, U.S. tight oil inadvertently contributed to price stability. Instead of leading to a glut, it was substituting—nearly one barrel for one barrel—for oil coming offline as a result of political disruption in the Middle East, as seen in the graph on the next page. To fully understand the origins of the new oil abundance and the subsequent price decline, we must supplement the story of American innovation and technological advance with one of the politics and decision-making behind one of the world's frequently derided and long-standing cartels.

Figure 1.4: Supply Disruptions During Arab Upheavals and U.S. Supply
Additions During the Same Time (million barrels per day)

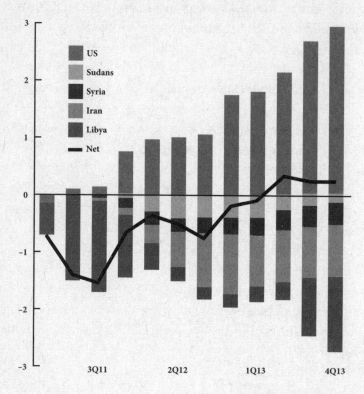

Source: BP, *BP Energy Outlook 2035*, 2014, 34.

The Dethroning of OPEC

Sitting on his couch in Riyadh, Hamad Saud Al-Sayari recounted to me the woes of the 1980s in Saudi Arabia with painful accuracy. In 1983, he had just been appointed governor of the Saudi Arabian Monetary Authority, making him the kingdom's central bank head. The fat years created by the oil price increases in the 1970s—following the 1973 embargo by Arab members of OPEC on countries supporting Israel in the Yom Kippur War and the 1979 Iranian Revolution—had come to an abrupt end. The global economy was slowing. The efforts to reduce fuel usage and oil dependency that consumer countries had initiated following the 1973 embargo were bearing fruit. Where consumers could, they moved away from oil to other sources of energy. In countries such as France and Japan, demand for heavy fuel oil virtually disappeared, as nuclear power, coal, and natural gas moved in as substitutes. Global demand for oil decreased by more than 10 percent. Further exacerbating the situation, new non-OPEC oil was entering the market. In the first half of the 1980s, Mexico, the U.K., Norway, China, Brazil, and India together increased their production by half to bring almost 4 mnb/d to the global market.

Not surprisingly, the combination of growing supply and declining demand forced prices down. In response, OPEC slashed its production by nearly half to bolster the price, but to no avail. As the world's leading oil producer, Saudi Arabia bore the brunt of these cuts; by 1985, the country was producing oil at roughly a third of its 1980 levels. Worse, the increasing non-OPEC oil supply meant that Saudi Arabia was losing market share even as the price of oil continued to fall.

Al-Sayari faced tough choices. As central bank governor, he oversaw the rapid drawdown of fiscal reserves while budget deficits rose. Revenues had plunged, and the government, desperate for income, was forced to rein in expenditures, mostly by suspending construction projects, delaying payments to contractors and suppliers, and even cutting back on military spending. Faced with declining prices and cheating OPEC partners, in December 1985 Saudi Arabia abandoned its restraint and decided to unleash its own production. Prices, which had tumbled by a quarter, sank by another half in the next year alone.

The boom of the 1970s and early 1980s was now a bust. The recession and the Saudi government's response undermined business confidence, deterred private sector investment, and spurred extensive bankruptcies in the country. Perhaps most alarmingly, it stoked discontent among the Saudi population. In late 1984, the government raised heavily subsidized electricity prices, but was forced to backtrack a year later with the aim of "relieving the financial burden of the people." In early 1986, the *Economist* noted:

What should really worry the ruling families is that all over the Gulf people are beginning to complain about the amount of money the rulers are making while everyone [else] is getting worse off. In the good times, no one minded that the rulers took slices of the biggest contracts, or kept the most profitable businesses for themselves. But when the whole cake shrinks, and the amount of it which the rulers are getting stays the same or even increases, people begin to notice. Some of the rulers are being accused of helping to put merchants out of business—behavior that does not endear them to the rest of the trading community.

Figure 1.5: Saudi Arabian Oil Production, Consumption, and Net Exports, 1971–1986 (million barrels per day)

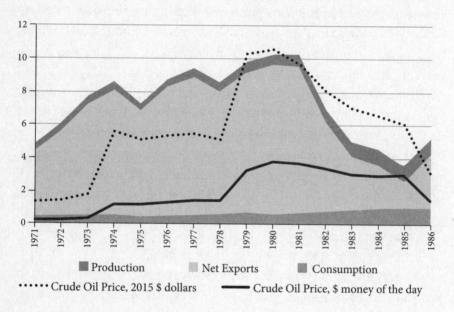

As reported by *BP World Statistical Review 2016*; 1965–1983 prices are Arabian Light prices posted at Ras Tanura; 1984–1986 are Brent prices.

Source: IEA: *World Oil Statistics*, 2016; BP, *BP Statistical Review of World Energy 2016—data workbook*, 2016.

More than two decades later, these negative memories of the 1980s would form the critical backdrop against which Saudi leaders and officials would respond to the 2014 price plunge. As 2014 began, oil prices were still around $100 a barrel and most Saudi leaders didn't regard surging U.S. tight oil production as a challenge to the comfortable status quo.

After meeting with U.S. energy secretary Ernie Moniz in January 2014, then Saudi Arabia's much revered minister of petroleum and mineral resources Ali al-Naimi exuded calm, saying, "The Kingdom welcomes this new source of energy supplies that contribute to meeting rising global energy demand and also contribute to the stability of the oil markets." Certainly, many big oil producers had grown accustomed to high oil prices. Fiscal breakeven prices—the minimum oil prices producers need for their budgets to be in balance—had doubled or more in a few short years, reaching or exceeding the $100 mark. For many Middle Eastern monarchies, high oil prices were a blessing given the volatile politics of the region and the need, in some cases, to buy political quiet within their own borders. Al-Naimi publicly interpreted the price stability as an indication that both consumers and producers were comfortable at this price level, suggesting that it could continue indefinitely.

Figure 1.6: Nominal and Real Oil Prices
Real Prices as of December 2014

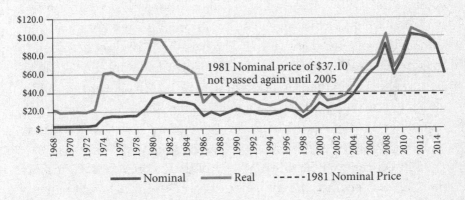

EIA, *Short-Term Energy Outlook Real and Nominal Prices,* December 2014.

While global demand for oil was steady if flat in 2014 before the price began to drop, it was initially weakening demand for the kingdom's own crude throughout the summer of that year that made Saudi officials nervous. As American tight oil production boomed, the United States was importing less and less oil, not just from Saudi Arabia, but also from Venezuela and West African producers, which were also traditional suppliers of U.S. markets. These countries aggressively began to seek other markets for their exports, for example, pushing to secure Chinese customers at the expense of Saudi exports. In the eighteen months following January 2013, Chinese imports of Saudi crude fell by more than a third. As a result, even before global prices faltered, Saudi Arabia was giving discounts to many customers in an effort to win back and hold its market share.

As the global price dipped below $100 in the autumn of 2014, the road ahead looked fairly straightforward to many. The common wisdom was that Saudi Arabia—not to mention other members of OPEC—could not afford to let prices sink too low in the face of bloated budgets and potentially restive populations, especially as the Arab revolutions swept the region. The world expected OPEC, or at least Saudi Arabia, to rein in its oil production as prices began to soften. Global supply and demand would subsequently settle at a price that would—not coincidentally—be close to that needed by OPEC leaders to balance their fiscal budgets, about $100 a barrel in late 2014.

Such a turn of events would have been consistent with what had become a familiar OPEC playbook. Following the 1973 embargo, the Arab countries of OPEC absorbed some tough lessons. That oil embargo, which caused prices to soar, had been catastrophic—not just for oil consumers but for producers as well. The hardship Al-Sayari described to me extended well beyond Saudi Arabia. Not only did the global stagflation resulting from the high prices of the 1970s curb oil demand, but investment into oil-producing economies, on the whole, fared more poorly than other developing countries. The message to members of OPEC was clear: oil embargoes cause economic pain for everyone.

Subsequently, OPEC shifted its approach to more nuanced interventions in the oil markets. Its efforts were generally limited to calibrating its oil production to support prices within a range that its members thought was fair and sustainable. Often, this meant restricting supply and keeping prices higher than they would have been if the market were the only arbiter of price. For instance, OPEC cut millions of barrels of production after oil prices plummeted in the wake of the 2008 financial crisis, quickly boosting the price. On more rare, although significant, occasions, OPEC had also brought additional supply online, dipping into its "spare capacity"—that is, oil from developed fields and infrastructure that some OPEC countries keep offline, but have the ability to bring to market in thirty to ninety days. For example, in 2011, OPEC used its spare capacity to keep prices from spiking when international military action in Libya curtailed global supply on short notice. While OPEC's post-1973 approach did not eliminate volatility in the oil market, its actions often helped shave the peaks and troughs off the changes in the price of oil. Had OPEC not played this familiar role, the only mechanism to bring supply and demand into balance would have been price; given the sluggish response of oil supplies to price, the adjustments would have been longer and more disruptive.

Most analysts had not had the benefit of the conversation I had at the Saudi Ministry earlier that year. As a result, and given OPEC's history, the debate among oil market observers in 2014 was initially more about *when* OPEC, and Saudi Arabia in particular, would step in and cut production to shore up price. Few people were asking *if* OPEC would assume its traditional role of bolstering prices when they flagged. Yet, as described to me by many Saudis, Saudi officials were determined not to repeat the mistakes of the 1980s. The main architect of the Saudi— and therefore OPEC—response was Ali al-Naimi. Widely referred to as "the world's most powerful oilman" or "the closest thing the oil industry has to a celebrity," al-Naimi had seen it all. Born in 1935 to one of Saudi Arabia's last nomadic tribes, al-Naimi spent his early years tending lambs outside the family's tent. He joined Aramco—Saudi Arabia's national oil company—at the age of eleven, working his way up the company over the next thirty years until he was named president in 1984, soon after Minister Zaki Yamani had ratcheted back the country's oil production in response to burgeoning North Sea and Alaskan oil supplies and shortly before the massive price plunge of 1985–1986. Al-Naimi must have tried to distill lessons from this period, when the Saudis lost market share and then revenues as the oil price collapsed. In 2014, when once again faced with a surge in non-OPEC supply, one can see why al-Naimi called into question whether cutting production was the right course of action.

The United States was not the only country putting more oil on the market in the early 2010s. Other non-OPEC producers had also been enticed to increase production by the comparably high prices of the preceding years. Since 2004, Brazil had increased its production by half. Colombia nearly doubled its production, hitting the 1 mnb/d mark for the first time in 2013. Canada's crude production was up by nearly a third. And Russia was producing at a post-Soviet all-time high—indeed, at or above the levels of Saudi Arabia throughout the first half of the decade. Collectively, non-OPEC producers *apart* from the United States added nearly another 5 mnb/d between 2000 and 2014.

However, it was the American tight oil production that was most problematic from al-Naimi's and OPEC's perspective. It was not simply the large volume of tight oil that had come to market. The resource's unusual production characteristics were much more troubling. Up until the entry of millions of barrels of tight oil onto the market, OPEC had a pretty clear picture about what it could expect when it cut production in order to elicit higher global prices for oil. As happened repeatedly over the previous decades, higher prices stimulated new investment and eventually new production, particularly from non-OPEC countries where oil production tended to be more costly.

But this added conventional production was slow to come online, especially as companies needed to go farther and farther afield to look beyond "easy oil" for new resources to tap. The time that elapsed from when high prices spurred the initial investment to first production took years, and the cost often reached billions of dollars. For example, Kazakhstan's Kashagan oil field—sarcastically known to industry insiders as "Cash All Gone"—took thirteen years from discovery to first commercial production, before going offline again for several years. By some estimates, the field required $116 billion to bring online. Less complicated fields still required many years to go from exploration to production. But once these investments were made and oil was flowing, the additional costs to maintain oil production were minimal in comparison. Economists called conventional oil supply "inelastic," meaning it was only very slowly responsive to price. As a result, up until the tight oil phenomenon, OPEC could be confident that if its suppliers cut production, they could enjoy the higher prices and resulting revenues for a long period before new supplies would enter the market and place downward pressure on price.

America's new tight oil was a different animal. Unlike the conventional oil fields that had met the world's needs in the past, tight oil did not entail massive projects or expensive wells with tens or even hundreds of millions in front-loaded costs. Extracting tight oil, in contrast, involved quickly drilling hundreds or thousands of wells into shale rock that were downright cheap—at least up front—compared to their conventional counterparts. The average cost to drill and complete an onshore well in one of the five large American tight oil basins in 2016 was approximately $5 million; it might have been twice that five years earlier. Either way, this sum is seen as paltry when compared to more than $200 or $300 million to drill and complete a deepwater well in the Gulf of Mexico.

Such tight oil investments could also be made incrementally, not only when a company was ready to commit hundreds of millions or more dollars to a project. Ultimately, tight oil costs per barrel were more than conventional ones; because production of a shale gas or tight oil well declined very rapidly after an initial burst, continued production of these unconventional resources required constant investment to maintain a steady or rising level of output. But, tight oil wells took only weeks or months to drill, rather than years, before oil would gush forth and be sped to market. A higher global price of oil would simply—and quickly—encourage more developers onto the plains of North Dakota, Texas, and other states with tight-oil-producing basins. The higher the price, the more fields became commercially viable, even beyond the "sweet spots" sung about by the industry.

For al-Naimi, the implications of this new resource for OPEC seemed ominous. From decades of experience, he knew that if the cartel curtailed production and forced up global prices, others would take it as an invitation to eventually bring even more expensive oil online. But with the new tight oil delivering additional production so quickly, Saudi Arabia and others would have far less time than in the past—perhaps only months—to benefit from higher prices before new supply encroached. Saudi Arabia could end up with a smaller share of the market with prices only marginally higher, if at all. It would, in short, be like the 1980s all over again.

The opposite outcome, however, seemed at least equally plausible. With lower prices, America's tight oil producers might curtail investments quickly. The production growth of the previous years would quickly grind to a halt and reverse, bringing price stability at some lower notional price at which it would not make sense to produce more tight oil.

The responsive, flexible, and plentiful nature of American tight oil had changed the game Saudi Arabia and OPEC could play. Al-Naimi was also skeptical that OPEC countries could agree on and adhere to a program of production cuts. OPEC agreements had always been undermined by a certain amount of cheating; compliance with the 2008 agreement was about 70%, which was good by OPEC standards. But in 2014, political divisions were so high that even reaching an agreement to cut production was going to prove difficult. Both Iraq and Iran argued their historical circumstances warranted that they receive special treatment. Members from Venezuela to Nigeria cautioned that their budgets were already so curtailed that any cut to their own production would create real risks to regime stability. Even Kuwait and the United Arab Emirates (UAE), the countries Saudi Arabia could traditionally count on to join in production cuts, felt more financially vulnerable given the political instability in the region and the need to keep social spending high. Nor could Saudi Arabia rely on non-OPEC producers, such as Russia and Mexico, to join it in a production cut as they had in the past. In November 2014, Venezuela's representative to OPEC, Rafael Ramírez, organized a meeting between al-Naimi and senior energy officials from Russia and Mexico at the upscale Vienna Hyatt. The meeting, designed to explore joint production cuts, ended with no deal. At the time, no country was willing to cut production with revenues already on the decline and so little trust among producers.

These circumstances led al-Naimi to make the case to others in the Saudi government that the right response to the drop in prices was to do nothing. Domestic politics likely strengthened his arguments. Saudi Arabia's King Salman bin Abdulaziz Al Saud had inherited the throne less than a year earlier, and the region was in unprecedented turmoil. Riyadh could little afford an oil strategy that made the kingdom look weak. An effort in which Saudi Arabia assumed its traditional role and cut production, but made no impact on price, would seem to communicate that Riyadh had lost its unique ability to shape the global market for the world's most strategic of commodities.

There were, of course, some significant uncertainties in al-Naimi's calculus. Producers with higher costs, such as those working in Brazilian deepwater fields and Canadian oil sands, would halt new investments when confronted with lower prices. However, as in the past, it would likely take years before these moves would be reflected in the global supply balance. But tight oil was a new and untested phenomenon. Although its responsiveness to price was unquestionably quicker than conventional resources, no one knew exactly how fast a price drop would translate into shuttered rigs and decreased production. Moreover, no one knew exactly at what price America's tight oil producers would pack it in.

The Saudis apparently had spent much of the previous year seeking the advice of analysts from ExxonMobil and elsewhere to help them understand more about U.S. tight oil. Jamie Webster, an oil market analyst who met with Saudi officials dozens of times to discuss the new dynamics, exclaimed to me: "You could make a whole career just briefing the Saudis about new American production." They were eager to comprehend how tight oil producers were financed and what the projections were for future production. But most importantly, the Saudis sought solid answers to the question "What is the breakeven cost for U.S. tight oil producers?" No one inside or outside the desert kingdom knew the answer. But by the time I visited in September 2014, I sensed a consensus had settled at around $80. That same month, Ibrahim al-Muhanna, one of al-Naimi's top advisors, told an audience in Bahrain that he did not expect prices to drop significantly below $90 a barrel, given the high cost of producing U.S. tight oil. Later reports of stress tests done of the Saudi 2015 government budget at an oil price of $80 seemed to suggest similar expectations. After years of high oil prices, the finances of Saudi Arabia were flush. The country had virtually no government debt and boasted more than $700 billion of financial reserves. While others might feel the strain, Riyadh was confident it could endure oil prices of $70 or $80 if that is what it took to force higher-cost producers out of the market and bring back market stability.

Such is the backdrop to the November 2014 OPEC meeting. Pundits excitedly claimed that this meeting would be one of the most significant in OPEC's history. Headlines such as "Cancel Thanksgiving: The Most

Important OPEC Meeting in Years Is Happening on Thursday" were commonplace online and in print media. Hundreds of journalists put on their winter garb, traveled to the organization's headquarters in Vienna, Austria, and waited for hours as OPEC's then–twelve members debated the situation and the correct course of action. Finally, al-Naimi, the most watched man of the event, emerged from the meeting smiling and announced to the press that the group had made "a great decision"—one that, in effect, changed nothing at all.

Although there had been pleas from OPEC's poorer members for a production cut, the view of Saudi Arabia had prevailed. OPEC issued a statement making clear no production cuts would follow. Members had agreed to maintain the current production ceiling of 30 mnb/d, despite the organization's own estimate that this amount would exceed the demand for its oil by more than one million barrels in the first half of the coming year. The meeting adjourned, and representatives of the cartel's members fanned out to speak to the press. Those who spoke on the record emphasized a common message of OPEC unity; many speaking on a not-for-attribution basis grumbled about having no choice given the Saudi stance.

The market gasped in reaction to this news, and then the tumble began. The price of oil dropped $6—or 8 percent—in the twenty-four hours after the meeting adjourned to hit a new yearly low of $71 a barrel. It went south from there. Finally, the abundance of energy resources emerging on the global scene over the past several years was beginning to be reflected in the price. Many expected the drop to be temporary, but the price kept sliding.

Just as surprising to many, the comparatively high-cost tight oil resources that the United States had recently brought online did *not* come to a quick halt as prices plunged. Moreover, Saudi Arabia, as well as Iraq, further increased their already high levels of production, in an effort to gain more market share. Over 2015, the price collapsed even further. The world welcomed the new year of 2016 with the shock of $28 oil, a price not seen since 2003. According to an analyst at one U.S. bank, the downturn had become "deeper and longer than each of the five oil-price crashes since 1970."

Figure 1.7: The Price of Oil,
July 2014–January 2016 (dollars per barrel)

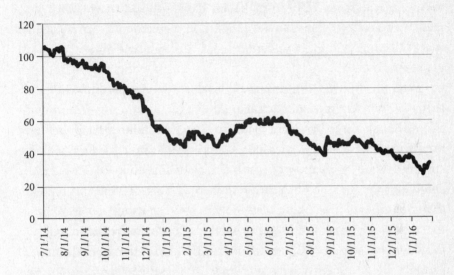

This graph shows the price of oil—specifically the benchmark price called "West Texas Intermediate."

Source: Energy Information Administration, http://www.eia
.gov/dnav/pet/hist/LeafHandler.ashx?n=PET&s=rwtc&f=D.

OPEC's action—or rather inaction—fueled endless conspiracy theories. Russian president Vladimir Putin publicly mused, "We all see the lowering of the oil price. There's lots of talk about what's causing it. Could it be the agreement between the U.S. and Saudi Arabia to punish Iran and affect the economies of Russia and Venezuela? It could." Iranian president Hassan Rouhani was only a bit more circumspect in pinning the blame for the price drop on the United States; speaking to his cabinet in December 2014, he asserted, "The main reason for [the oil price plunge] is [a] political conspiracy by certain countries against the interest of the region and the Islamic world. . . . Iran and people of the region will not forget such . . . treachery against the interests of the Muslim world." Even states on the margins of the fray chimed in. Bolivian president Evo Morales, for example, declared, "The reduction in oil prices was provoked by the U.S. as an attack on the economies of Venezuela and Russia. In the face of such economic and political attacks, the nations must be united."

Such conspiracies also raged in the Western world. Writing for the magazine *Foreign Policy*, Columbia University professor Andrew Scott Cooper mused, "Riyadh's real hope . . . is that escalated production will force [Iranian president] Rouhani's government to implement an austerity budget that will ultimately stoke underlying social unrest and once again push people into the streets."

When analyzing OPEC's and particularly Saudi Arabia's policy shift at the end of 2014, these individuals all made the common error of confusing correlation with cause. Saudi Arabia's oil policy did in fact hurt the finances of its adversaries, but that does not mean such an outcome was the prime motivation of Saudi action at that time. Saudi behavior could also be explained by economic theory. If Saudi Arabia were to act as the "dominant firm" that many analysts have suggested it was in the oil marketplace, economic rationality would lead it *not* to cut production in this energy environment. Instead, it would cause the kingdom to increase production "in order to drive the fringe firms out of the market," which in this case were shale companies, and "later return to monopolist pricing." Geopolitical benefits would be the frosting, not the cake itself. "If in the process, you [shave] 30% off Iran's income, fine. If in the process, you shave 30% off Russia's income, fine," a senior Arab official involved in the negotiations told the *Wall Street Journal*.

Moreover, those wishing to explain the Saudi decision primarily in terms of geopolitics need to account for the timing of the 2014 action. The Saudi government has been locked in a fervent competition with Iran for decades, a rivalry that has reached a crescendo in the last decade. The kingdom has vehemently opposed Iran's efforts to expand its regional influence and its pursuit of nuclear weapons. Riyadh supported international efforts to create economic pressure on Tehran in the hopes of deterring it from its nuclear program. If the Saudi government were willing to endure a price dive in order to put greater pressure on Iran, why would it have waited until the end of 2014, when the contours of a deal it found objectionable between Tehran and the United States and its allies were already clear? Had the Saudis been interested and willing to use an oil weapon against Iran, the optimal moment would have been several years earlier, before the Obama administration had made key concessions to Tehran in order to begin negotiations. When, during a visit to Saudi Arabia, I asked my Saudi interlocutors about why the kingdom did not adopt this approach in 2012, my suggestions were rebuffed with claims that Riyadh could not afford such a strategy. Pointing to a fiscal breakeven price that had risen by 13 percent between just 2010 and 2012, and a more dubious regional and political environment, Saudi officials lamented they could not survive a period of low prices. At the time, with oil prices high and looking sustainable, the Saudis seemed unwilling to risk losing control of the oil market, even in pursuit of a goal so central to their interests.

Al-Naimi's satisfaction after the November 27, 2014, OPEC meeting seemed palpable, but it was not until later that the minister described the logic in plain words: "In a situation like this, it is difficult, if not impossible for the kingdom or OPEC to take any action that may result in lower market share and higher quotas from others, at a time when it is difficult to control prices." Talking to *Middle East Economic Survey* in December 2014, al-Naimi revealed, "It is not in the interest of OPEC producers to cut their production, whatever the price is. . . . If I reduce, what happens to my market share? The price will go up and the Russians, the Brazilians, U.S. shale [tight oil] producers will take my share. . . . Whether [the price] goes down to $20, $40, $50, $60, it is irrelevant."

Al-Naimi's determination was certainly tested. Prices remained historically low for more than two years after this meeting. The enduring financial hardship eventually spurred Saudi Arabia—and the bulk of OPEC countries and eleven non-OPEC countries—to cut production in early 2017 in an effort to rebalance the market and bolster the price.

———

Matt Simmons died before he could see the outcome of the bet he made with John Tierney about the oil price. Had he lived a few months longer, he would have seen his office pay out the $5,000 plus interest he owed Tierney, given that the oil price did not exceed $200 by the end of 2010. Tierney, having placed faith in the driving power of human ingenuity and innovation, won the bet. Although victorious, Tierney—like most others—probably had no idea in 2010 how these two forces were going to affect the oil price and more in a few short years. It seems unlikely that even George Mitchell anticipated the change that he was going to help catalyze.

Although Mitchell had plenty of ambitious goals—such as developing a new resource extraction technique, building his company, and changing America's energy balance—they were relatively close to home. But the unconventional boom he helped launch had even bigger and more widespread impacts. It altered geopolitics in ways Mitchell couldn't have foreseen when he walked around wildcatters' graveyard six decades earlier. These unconventional resources were perhaps the single largest factor in changing the calculations of OPEC and convincing the cartel to shift strategies. In the absence of U.S. tight oil production, OPEC would have almost certainly been discussing how to temper rising prices at the end of 2014, rather than stem declining ones. Not only would the price environment have been different without the added supply of U.S. tight oil, but the tools at OPEC's disposal and the logic behind its actions would have varied as well. The new energy abundance not only shaped OPEC's historical decision in November 2014. As will be discussed, it has more generally challenged OPEC's ability to successfully manage the oil market as it did during its strongest periods over the past decades.

The combination of technological advances and political decisions is not only the key to understanding the recent past, but is also vital to interpreting and anticipating what is happening today and what will happen tomorrow in the world of both energy and geopolitics. Moreover, as explored in the next two chapters, technology and politics have affected much more than simply the price of oil and gas. They have altered the structure of markets in profound and lasting ways. It is these changes, as well as price levels, that we will take into account as we explore the nexus between energy and geopolitics in the United States, Europe, Russia, China, and the Middle East throughout subsequent sections of this book.

TWO

The New Oil Order

The March 6, 1999, cover of the *Economist* portrayed two oil-soaked workers struggling to control a gushing oil well. Superimposed over the dark-brown geyser were three words: "Drowning in Oil." For two years, the price of oil had scraped rock bottom, under $30 in 2017 dollars. Despite the dramatic cover, the *Economist* still cautioned that oil's "abundant flow might be too easily taken for granted today. Normality could last a while; but it is unwise to assume that it will endure for ever." As the magazine predicted, OPEC subsequently got its act together and curtailed production, demand grew, and the market gradually became tighter over the coming years. Once prices rose, investment into building new capacity resumed, and a new cycle was under way.

There is some debate over whether the 2014 price plunge is just a new performance of the same Broadway show, or if—as this book contends—there is something fundamentally different at work this time. For the sake of simplicity, let's consider two groups with different views.

The first group sees today's oil market essentially as resembling a version of earlier boom-and-bust cycles—and expects tomorrow to look similarly familiar. Its members do have some strong facts and arguments on their side. The 2014–2016 price plunge did cause companies to curtail investment in exploration and production on a massive scale. In 2015 and 2016, for the first time in thirty years, investment declined for two consecutive years; in both years investment in oil and natural gas declined by a quarter. The reduced investment will temper oil supply growth in coming years. This anemic new supply growth is particularly problematic in the face of the "decline rates" of existing fields. Although decline rates vary field by field, on average, a conventional oil field loses 5 to 7 percent of production every year. Proponents of this view tend to think that demand for oil will continue to grow positively and close to the recent average. Eventually, growing demand will outstrip lethargic supply growth. Tight oil will respond to higher prices, but will not be sufficient to fill the gap between supply and demand, leading to a price hike in 2018–2020, which will be needed to incentivize further investment in higher cost resources, such as deep water fields and oil sands. Looking out over the longer term to 2040, many of the same observers predict that oil markets then will resemble those of previous decades: the production of tight oil will be exhausted and continued demand for the oil will further boost the reliance of the world on Middle Eastern resources.

The second group believes that more change has occurred in the oil space in the last several years and expects more change to lie ahead. Proponents in this group see the parameters of a new oil order emerging. They do not necessarily rule out a price spike in the years from 2018 to 2020 due to the curtailed investment in the years prior. Yet, they do not see it as inevitable or even probable, as they have greater confidence in tight oil rising to meet the growing demand and also can imagine some of the OPEC countries being in a position to increase capacity in the face of rising prices to meet increased demand. Looking beyond 2020, this group places greater emphasis on tight oil's transformative impact. Advocates of this view see tight oil as seriously diminishing the power of OPEC, leaving the market open to greater fluctuations in prices. They also see tight oil as responding relatively quickly to changes in price, thereby shaving the peaks and troughs off the increased volatility and bringing any disconnects between supply and demand into balance more rapidly. This group also has greater confidence that technology will continue to increase supply, both by making more oil of all kinds available to produce at lower prices and by decreasing decline rates of fields. And it is more likely to perceive that technology—in the form of efficiency or possibly even the greater electrification of transportation—will also chip away at oil demand growth over time. The net result is, looking out to the longer term, this group does not see a return of the oil markets to earlier norms. Except for the occasional geopolitical disruption, the long trend is one of more flush oil markets, with supply finding balance with demand at moderate (although not rock bottom) prices for the foreseeable future.

While appreciating that one cannot entirely dismiss the possibility of an old oil order comeback, this book squarely associates itself with the second group—those who see the emergence of the broad contours of a new order in which the traditional boom-and-bust cycle of the old oil market will be at least greatly muted.

Lower prices are the most noticeable result of this new oil order, but they are not the only dimension of it. Other characteristics are as important as, and arguably more interesting than, the question of price. For the next decade or longer, the world will feel less influence of two phenomena—scarcity and cartel-like organizations—that have long shaped oil markets and have had major geopolitical consequences. In the new oil order, peak oil will no longer drive decisions of governments, businesses, and individuals as it has in the past. Moreover, in this era—despite the OPEC agreement to cut production made at the end of 2016—OPEC is a greatly diminished force. Instead, the market—rather than OPEC or any other institution—will play a greater role as a balancer of oil supply and demand.

A Reprieve from Peak Oil

On March 8, 1956, Marion King Hubbert, a geophysicist with a reputation for stoking controversy, was sitting on the stage at the regional meeting of the American Petroleum Institute, when he was signaled to leave the platform to take an urgent call. A public relations manager at his employer, Shell Oil, wanted to make one last appeal to Hubbert to tone down the "sensational parts" of the speech he planned to deliver to the members of the country's powerful oil industry association. Hubbert was not easily cowed. He hung up, returned to the stage and proceeded to present his paper, "Nuclear Energy and the Fossil Fuels."

Drawing on a paper laden with equations and graphs, Hubbert argued that there was a finite amount of oil and gas in the world. Given that it had taken 500 million years of geological pressure to turn plants and animals into fossil fuels, it stood to reason that these resources would eventually be exhausted long before they could be replenished. He introduced an idea now known as the Hubbert curve, postulating that the production of such a finite resource would resemble a bell curve, exponentially increasing up to a peak and declining afterward. Based on this theory and his own calculations of how much oil remained in the earth, Hubbert predicted that American oil production would reach a "peak" between 1965 and 1970 and then begin to decline until the resource was depleted; the world would follow suit around the year 2000.

Hubbert's thesis was not popular—critics in the oil industry challenged his assumptions, his math, and his methodology. Morgan Davis, the head of Humble Oil, the largest producer of U.S. oil at the time, reportedly tasked people to attend Hubbert's every presentation to publicly refute his theory. Kenneth Deffeyes, a longtime friend and colleague of Hubbert's from the Shell research lab, explained that some of the negative reaction to Hubbert's theory was emotional. He noted, "The oil business was highly profitable, and many did not want to hear the party would soon be over. A deeper reason was that there had been many false prophets before. . . . They had cried 'wolf' . . . and the industry actually grew instead of drying up."

Despite this barrage of criticism, Hubbert's prediction at least *appeared* to come true. Crude oil production in the United States did seem to top off in November 1970 at 10 mnb/d and then embark on a slow steady decline. Even after this apparent validation of Hubbert's peak, however, debates continued over its applicability to the rest of the world. For decades, the salience of "peak oil" would wax and wane. Inevitably, whenever oil prices began to climb, consumers and producers around the world would bite their fingernails, wondering if the world was finally about to face the peak on Hubbert's curve. In 2005, U.S. EIA director Guy Caruso gave a talk in which he chronicled thirty-six times between 1972 and 2004 when various sources had predicted peak oil. When prices declined, the fretting over peak oil would subside but not vanish. In fact, the notion that the world would eventually run out of oil sparked serious inquiries, such as the 2005 U.S. government–funded study *Peaking of World Oil Production: Impacts, Mitigation, and Risk Management* (also known as the Hirsch Report). In 2010, a think tank associated with the German military also conducted a similar analysis, concluding that peak oil could in some instances lead to open conflict and might even jeopardize the foundations of democracy. Peak oil has even made multiple appearances in the realm of popular culture; it has been the inspiration for a bevy of fantasy novels and sci-fi films such as *Blade Runner* and *Mad Max*, movies in which the human race confronts a grim future without oil.

Given the frequent episodes of peak oil hysteria in the past, one might anticipate its return as the subject of editorials and conferences before too long. Yet today's trends suggest otherwise. For the next many years to come, global oil shortages will be far from the minds of governments, corporations, or individuals—the same groups whose behavior was once shaped by the fear of peak oil. Today, the possibility that the world will run out of oil seems so remote that no serious entity will use the notion of peak oil as a driver of its decisions.

At least one, and possibly two, waves of energy abundance will put a stake in the heart of peak oil. The first wave is the supply-induced abundance that has come about thanks to the technological advances and political decisions described in the previous chapter. The second wave involves a reduction in demand. Such a reduction remains speculative, but the mere fact that there is real potential for such a decline has begun to influence the thinking of key actors. If, however, demand growth or even absolute demand begins to taper off, the energy abundance the world is experiencing today will be reinforced. In this scenario, the reprieve from peak oil would be a permanent one.

The first, supply-induced energy abundance will be enduring, notwithstanding many headlines to the contrary. Americans in 2016 were inundated with reports on the plunging number of drill rigs deployed, the shuttering of small businesses that sprang up earlier to support shale producers from North Dakota to Texas, and a raft of bankruptcies of small oil and gas companies. Many declared the end of "the 21st century version of the American gold rush" and wrote obituaries for the tight oil boom.

Unquestionably, the low oil prices of 2014–2016 did temper the energy boom in North America. From 2011 to 2014, while oil prices hovered close to $100 a barrel, U.S. production grew like gangbusters, increasing, on average, by 1 mnb/d per annum. For a year following the price plunge, such production remained remarkably resilient. The combination of advancing technology, the slashing of costs, significant hedging, and plentiful credit enabled companies to continue to produce copious amounts of tight oil even with prices in the doldrums. But eventually, after U.S. crude oil production peaked at 9.6 mnb/d in April 2015, it began a slow decline, exposing the limits to the resilience of tight oil. A year later, instead of being up an additional million barrels—as likely would have been the case had oil prices stayed high—overall American oil production was down by almost that amount, nearly half due to contraction in the tight oil output.

Yet, the poignant stories of loss and financial ruin in this downturn obscured the big picture. Two important points about American tight oil production were easily lost. First, most companies and energy agencies anticipate a return to more hardy levels of U.S. oil production in the coming years, despite the dip of 2014–2016. Despite lower prices, in their main scenarios, both the U.S. EIA and the Paris-based IEA anticipate that overall U.S. oil production (crude *plus* natural gas liquids) will surpass its 1970 record by nearly a quarter by 2020.

Second, even in the less-likely scenario where oil prices returned to early 2016 lows of $30 and stayed there for the foreseeable future, overall American oil production will still be more robust than it was before the boom began. There are shale fields—such as the part of the Eagle Ford formation in DeWitt County, Texas—that will continue to be profitable even at that price. And the continuous advance of technology will make any given field of tight oil profitable to produce at ever-lower prices. Production from these fields will cumulatively be significant. Supporting this view is the "low price case" developed by the U.S. EIA, in which oil prices do not return to $60 until 2025. Even in this low-price scenario, overall U.S. oil production would still be about a million barrels higher than America's all-time production peak in 1970.

This supply-driven wave of oil abundance, however, does not rest solely on the fortunes of the United States. The new energy abundance is a global phenomenon, not just an American one. Today, American tight oil production is a major driver of this new energy picture; in 2015, the United States and Canada were together responsible for virtually all global tight oil production. The dominance of global tight oil supply by these two countries will likely be the case for much of the next decade. But, subsequently, the unconventional boom has prospects for going global, with major implications for energy markets worldwide. The fact is that the world's unconventional oil deposits are both significant and found in several continents. As of the end of 2015, according to the IEA, the world's total technically recoverable resources of tight oil were massive, exceeding the conventional oil resources of all of Africa—which holds roughly 10 percent of the world's proven oil reserves and five of the top thirty oil-producing countries.

It will, however, take time before such unconventional oil production outside North America comes online in significant quantities. While there is already modest production in Argentina, countries from Algeria to Russia to China also possess large resources of unconventional oil and gas but have yet to exploit them anywhere nearly as effectively as the United States and Canada. Why?

First, not all geology is equal. Most countries with significant unconventional resources tried, as the United States did, to begin their exploration with shale gas, rather than tight oil. Hoping to replicate America's experience within their own borders, these countries were frustrated by a number of factors. China, for instance, has vast quantities of shale resources, but the quality of shale and the depth of the deposits make it much more challenging to extract than in North America. Early efforts on the part of countries from Poland to China demonstrate that the leapfrog effect of being able to benefit from the U.S. experience is more limited than originally thought. Success involves not only having access to the needed technology, but knowing how to apply it to the particular shale in question. Fracking, it turns out, is as much an art as a science.

In addition, geology is only *one* of the determinants of how easily a country can produce its unconventional energy. The institutional framework and the incentives or disincentives it creates are equally important. Environmental and regulatory structures also make a difference, with some countries having more centralized governments and higher environmental standards or greater public sensitivity to fracking than in North America. For example, Bulgaria, France, and Scotland are just some of the European economies that have banned fracking altogether. Russia—another country with vast tight oil resources—has focused more of its limited investment dollars on developing its massive conventional oil. Investment in developing Russian unconventional production has also been dampened by international sanctions. Other factors—such as open lands, access to infrastructure, financing, and water, as well as a competitive company landscape—also explain why the unconventional boom that took off in the United States has been difficult to ignite in other countries.

The EIA and other energy outfits anticipate that production of tight oil will overcome these obstacles to become a truly global phenomenon in the decades ahead. Argentina, Russia, Mexico, Colombia, and Australia could account for approximately a fourth of global tight oil production by 2040. At that point, tight oil production may amount to 10 percent of overall global oil production.

The end of peak oil may be, as some academics like to say, overdetermined—meaning that more than one possible factor could deliver it. Technology will help ensure that more resources are produced at lower prices and these new resources will disrupt old markets in ways that cause actors to abandon old strategies to bolster price; demand may falter over the medium to long term as well. For those who have battled with the specter of peak oil for decades, this truly is uncharted territory. For Bob Belfer, a businessman who made much of his fortune in the oil fields of America's heartland, the end of peak oil involves a major revision to how he views the world. "For the last forty years," he told me, "you could wake me up at 3 a.m. and peak oil would be on my mind." Today, Bob can sleep more peacefully.

Uncertainty of Demand

As mentioned earlier, a second, future wave of oil abundance could be generated by the demand side. We know with confidence that the world is moving out of a period of intense growth in *energy* demand (which includes, but is not limited to, oil) to one in which the thirst for energy moderates. The decade from 1997 to 2007 saw tremendous growth in global energy demand, propelled by economic growth, population pressures, urbanization trends, and the rise of a global middle class. But if one driver stood out, it was China, whose focus on heavy industry and government-supported investment pushed GDP growth rates to a peak of 14 percent in 2007; China's demand for energy rose in tandem with these sky-high economic growth rates. Rapid urbanization and the development of megacities with over ten million inhabitants also helped change China's energy profile. In just the few years between 2002 and 2006, China's demand growth for energy was greater than it had been over the entire two previous decades. By 2007, China's needs had transformed the world and accounted for nearly a fifth of all energy demand and almost half of overall global energy demand growth.

Global energy demand growth, while still positive, slowed by more than a third in the years since. Looking forward, various companies and energy agencies agree that demand for energy will rise at slightly higher rates compared to the post-2008 recession period until 2025, at which time energy demand growth will become less robust. In part, these trends are due to weaker, but still positive, rates of economic growth, urbanization, and population expansion. China, in particular, will temper its energy demands as it grows more slowly and as it transforms its economy to be more consumer-oriented and less energy-dependent. Perhaps most significant, the world will continue to become less energy intensive—meaning it will take less and less energy to produce the same amount of economic output, largely due to increases in efficiency. For example, according to one projection, by 2035, the European Union will use the same amount of energy it did half a century earlier, although its economy will be 150 percent larger. This phenomenon will not be limited to the democratic, relatively well-off countries of the Organization for Economic Cooperation and Development (OECD); India and the economies of Africa will also use less energy to produce each unit of output, even while they are industrializing.

Figure 2.1: Energy Consumption by Region
(billion tons of oil equivalent)

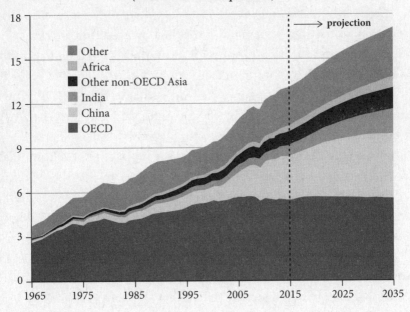

Source: BP, *BP Energy Outlook 2017*, 12.

When it comes to oil—as opposed to the broader concept of energy—many experts and analysts anticipate positive, but declining, demand growth out to 2035 or 2040. One reason to expect positive growth is that nearly all of the world's transportation runs on oil-based fuels. By one account, the need for fuel for transportation will drive as much as two-thirds of the growth in demand for liquid fuels in the decades ahead. With 57 cars per 100 people in OECD countries and only 7 per 100 in developing ones, ExxonMobil, for instance, expects the absolute number of light-duty vehicles to nearly double from the one billion on the road today. This bump-up in global ownership of vehicles and commercial use of heavy-duty vehicles is expected to be so large that it will dwarf the substantial technological improvements that will make cars roughly 60 percent more fuel efficient than they are today. Others—such as Fatih Birol, the executive director of the IEA—see positive demand growth not as a function of transportation, but as the result of increased oil use in other sectors, such as heating, petrochemicals, or industries.

As Lincoln Moses, the first head of the EIA, once told Congress, "there are no facts about the future." This is especially worth bearing in mind when considering projections such as those above, which necessarily rely on a number of huge assumptions. An unrelated number of factors could render demand growth for oil much less robust in the coming years than is currently expected. BP, in fact, identifies two such risks in its *2016 Energy Outlook*. The first is slower global economic growth. Nothing is more tied to energy use than economic activity. BP posits that if China grew at 3.5 percent, compared to the 5 percent in its "base case" 2016 scenario, the world could lurch into a period of the weakest economic growth in recent history. Growth in world energy demand would decrease to just 1 percent a year, leading to slower demand growth in oil and gas, as well as all other fuels. Certainly, there is sufficient uncertainty around China's growth path to make this scenario feasible. Just ask the thirty-five economists, strategists, and Wall Street investors who gathered at a private workshop at the Council on Foreign Relations in April 2016 to discuss China's trajectory. Nearly two-thirds of them anticipated that China's growth would drop from 6.7 percent to between 1 percent and 3 percent for the next decade; almost a tenth were prepared for China's economy to contract.

The second scenario BP identifies that could significantly affect demand for oil is one in which the world accelerates its efforts toward a low-carbon-energy future. Policies such as putting a significant price on each ton of carbon, enacting tougher vehicle standards, and legislating efficiency gains in industry and buildings could decrease global energy and carbon intensity at "unprecedented rates." They would also engineer a steep decline in coal consumption and, over time, cause oil demand to flag. Although these measures would go beyond the policies currently being considered in most economies, it is not impossible to envision their adoption. An extreme natural disaster could galvanize popular opinion and political resolve for more aggressive measures. Less dramatically, countries could revise upward the steps they will take to curb carbon emissions, as much of the world agreed to do periodically at the Paris 2015 climate change conference.

While neither slower global growth nor more aggressive policies to rein in carbon emissions are hard to imagine, the development most likely to eat away at anticipated oil demand growth in the coming decades is technology. By convention, the reference case scenarios—which can be considered "best guess" projections for the future created by BP, Exxon-Mobil, the EIA, and almost all other energy companies and agencies— only take into account existing technologies. As the unconventional boom itself demonstrated, technological advances that may not factor into such projections could dramatically alter the future.

The potential for technology to upend the demand equation—just as it has revolutionized the supply one—is real. Given oil's near monopoly on transportation, the technology that would have the most dramatic effect on global oil demand growth would be one that enables cars, trucks, ships, or airplanes to run on nonoil fuel sources at a large scale. Oil and gas companies have different views on this matter. For example, Exxon-Mobil expects biofuels, natural gas, and other fuels to further encroach on oil in only an additional 5 percent of transportation out to 2040. However, BP's 2017 "most likely" scenario anticipates growth in electric cars; the company also acknowledges that advances in batteries could dramatically affect cost-competitiveness of electric vehicles and their penetration of the automobile markets as early as 2025.

Outside the oil industry, others are making bolder assessments about just how quickly electric cars or other technologies could begin to challenge oil's dominance in the automotive sector. I met Dr. Salman Ghouri on a steamy hot day in Doha in August 2016 during my first trip to the headquarters of Qatar Petroleum. I was looking forward to meeting Dr. Ghouri, the head of its strategic planning department, and a good friend of one of my close colleagues. Once Dr. Ghouri secured a waiver to let me in the gleaming World Trade Center offices despite the knee-length skirt I was wearing, we settled into a surprising conversation. Rather than focusing on the low energy prices then seizing the attention of the rest of the Gulf, Dr. Ghouri spoke energetically about his recent work, which predicts a dramatic absolute decline in oil demand in the transport sector by 2040, mostly due to the penetration of electric vehicles. When I asked about the reaction of his colleagues when he presented his research in Qatar, he said, "People laughed at me. They just laughed. And then they were angry—why am I presenting such ideas?"

While such ideas may still be the source of ridicule in oil- and-gas-producing states, Dr. Ghouri is not alone in his predictions. Months later, I had a similar conversation in Moscow with energy economist Maria Belova about her research on how innovations in batteries will spur electric cars to displace oil demand. She posits a number of scenarios, but none of them look favorable to those banking on growing oil demand; both her "most likely" scenario and her "innovative" scenario anticipate substantial absolute declines in oil demand by 2035. In the United States, Bloomberg New Energy Finance (BNEF), a research group focused on energy trends, told the world in February 2016 to prepare for dramatic changes in the transportation sector, even within the coming decade. Their analysts concluded that, by 2022, new technologies would make electric cars cheaper than the internal combustion ones that most of us drive today. With this advance in mind, BNEF saw plug-in cars going from just one-tenth of 1 percent of the global car market in 2016 to a third of all new cars sold and a quarter of all the cars on the road in 2040. According to BNEF, this would displace 13 million barrels of crude oil a day—significantly more than Saudi Arabia produces every day or nearly one out of every seven barrels of oil consumed globally in 2016. Oil markets, however, would not need to wait until 2040 to see the impact. BNEF made various calculations suggesting that electric vehicles could lessen global demand for oil by 2 million barrels a day as early as 2023, potentially triggering a significant glut in oil markets.

Others see demand for oil clocking in far under conventional expectations for other reasons. Anthony Yuen of Citigroup suggested to me that Chinese consumers may buy many fewer cars than anticipated—perhaps between 300 and 400 per 1,000 people rather than the much larger numbers that others were predicting a decade ago. This change would be largely the result of the Chinese government's aggressive policies to limit driving and gasoline consumption and to promote other forms of transport. Others still attribute slower—or even negative—demand oil growth to growing efficiency, or a "mobility revolution" in which a combination of autonomous vehicles, car sharing, and ride pooling collectively dent overall oil demand.

Sex and Technology: The Market Encroaches on OPEC

Speaking to a large audience in Houston, Texas, in February 2016, then–Saudi oil minister Ali al-Naimi disappointed those who were betting on a Saudi intervention to address low oil prices. "Let markets work," he urged the crowd. There is no need for "meddling," as the market would eventually rebalance supply and demand and lift the price of oil to more sustainable levels, he assured journalists jotting down his every word. Although sounding less sanguine about the future, al-Naimi's successor underscored the same message to his colleagues at an OPEC meeting just four months later. Khalid al-Falih, the man long in charge of Saudi Aramco, the Saudi oil company, was new to an expanded ministerial post that included not just oil, but also other forms of energy, industry, and minerals. He told a small gaggle of reporters visiting him in his penthouse hotel suite in Vienna in 2016 that oil producers should "let the market forces continue to seek and find that equilibrium price between supply and demand."

For nearly two years, Saudi Arabia and OPEC left the rebalancing of their most valuable commodity to market forces, before attempting to reassert themselves through a new OPEC deal in November 2016. An American economics student may see nothing newsworthy about letting the market balance supply and demand. Yet, for almost the entire history of oil markets, there has been a powerful group that held sway over global supply. OPEC is but one institution in an infamous line of mechanisms created to ensure that the market alone did not balance supply and demand. Institutions representing corporate or national interests have almost always exercised some control over the price of oil.

The first institution to wield such power over price was John D. Rockefeller's Standard Oil, which at the turn of the twentieth century had such influence over the distribution infrastructure in the United States that it effectively controlled the price of oil. Certainly, this was the perception of the readers of one of the most popular articles of all time in *The Atlantic*. In 1881, readers gobbled up seven straight printings of the magazine to read "Monopoly on the March," an article calling for the end of Standard Oil.

Twenty years later, the Justice Department broke Standard Oil into thirty-four companies. Yet, before long, a new entity assumed a variant of the same role. The Texas Railroad Commission was essentially the world's first oil cartel. In fact, OPEC later modeled itself on the organization. Initially formed by Texas governor James "Big Jim" Hogg to regulate the much-loathed railroads, the commission eventually assumed duties to regulate oil production in Texas, one of the epicenters of global production at the time. When, near the start of the Great Depression in the early 1930s, the newly discovered East Texas field gushed forth so much new oil that the national price collapsed from $1.27 a barrel to 65 cents a barrel within two years, the commission instituted a quota system limiting the level at which every oil well in Texas could produce in an effort to bolster price. Rampant cheating was so persistent that Governor Ross Sterling declared martial law on August 15, 1931, and deployed the National Guard to shut down wells. Yet, despite these difficulties, the commission held huge sway over the price of oil for decades.

As oil production became increasingly global, a new grouping called the "Seven Sisters" and made up of the world's largest oil companies, became the dominant force influencing the price of oil. The idea of a cartel among them was born in secret on August 28, 1928, at an appropriately secluded castle in the Scottish Highlands. From the 1940s until OPEC was formed in 1960 and the nationalization of oil companies began, these Seven Sisters divided the market among themselves and sustained uncompetitive and high prices by restraining production. In 1953, they controlled almost 90 percent of the world's oil production. Even a dozen years after the establishment of OPEC, the Seven Sisters still held sway over 70 percent of global production. It would take multiple nationalizations to erode their power.

For the last half century or so, OPEC has played this role of managing the market. OPEC has not sought to set the price of oil directly as many American politicians would have their constituents believe. Instead, OPEC—like the Texas Railroad Commission—has influenced price indirectly, by calibrating global supply through its own contributions to the market. In essence, because there is one global market for oil, OPEC has sought to control how much oil is in that market, sometimes adding to supply and sometimes subtracting from it in order to reach the price desired by its members.

The cartel's ability to play this role in the past has relied heavily on two factors. First is cohesion within the cartel and the willingness of its members to cooperate. When this cohesion has been at its highest, OPEC has been most effective in influencing the market. Of even greater importance, however, has been the willingness of OPEC, and Saudi Arabia in particular, to be the "swing producer." Unlike nearly all other producers, the kingdom has maintained significant spare capacity. Despite being expensive to develop and maintain, Saudi Arabia only uses this excess oil production on occasion. But its existence has allowed Riyadh on very short notice to "swing" its production in either direction as necessary to maintain the price of oil at a certain level or within a price band.

There are good reasons to question OPEC's continued ability to play this dominant role in oil markets. The first relates to political solidarity within OPEC. Trust—and therefore cohesion—among OPEC members has been at rock bottom in recent years. Hostility between OPEC members Iran and Saudi Arabia animates the region. The competition between the two regional powers skates just beneath the surface of the Syrian civil war and is even more overt in Yemen. The sectarian conflict between majority Shi'a Iran and Sunni Saudi Arabia burst into full daylight in early 2016 after an Iranian crowd burned the Saudi embassy in Tehran following Saudi Arabia's execution of a dissident Saudi Shi'a cleric. On top of what can generously be called a "trust deficit" are the tenuous economic circumstances of many OPEC members from Venezuela to Iraq.

Nevertheless, the OPEC deal struck at the end of 2016 surprised the naysayers. For the first time since 2008, OPEC agreed to production cuts. Ten of OPEC's members—and eleven non-OPEC countries—collectively agreed to decrease output to the tune of 1.8 mnb/d, or not quite 2 percent of global production at the time. Increasing prices and revenue finally appeared more urgent than advancing narrower political objectives; in 2016, OPEC countries had only earned $341 billion from oil exports, compared to a record $920 billion in 2012. The 20 percent boost that news of the planned supply cuts generated provided much-needed relief to cash-strapped oil producers; equally important, the deal hastened the gradual movement toward balancing supply and demand in the market, after years of burgeoning inventories that held down prices. To many, this agreement augured a new chapter in cooperation among OPEC countries and between them and others outside the cartel, most notably Russia. Oil ministers from Saudi Arabia to Venezuela declared the agreement "historic," appearing almost visibly relieved that OPEC had been resuscitated from its stupor.

However, several realities throw cold water on the idea of a resurgence of OPEC. First, the political compromises required to cement the agreement may end up undermining it. Despite dragging its feet, Iraq agreed to limit production for the first time since the 1991 Gulf War. Yet that achievement was outweighed by the number of exceptions given to others. Iran successfully argued that as it had just been relieved of international sanctions on its energy industry, it deserved the right to increase its production, rather than the opposite. Libya and Nigeria were similarly successful in negotiating exemptions from output decreases, given the civil conflict occurring in each place. At a minimum, any production boosts from these three countries work against the cuts made by other OPEC members hoping to balance the global market and put a floor under prices more quickly. In some scenarios, however, oil supply growth from Iran, Libya, and Nigeria could virtually annul the larger cuts. In the first month of the agreement, these three countries collectively brought supplies online equal to one-third of the cuts made by other OPEC members.

Second, tight oil continues to challenge OPEC's way of doing business. As described in the previous chapter, the new way in which tight oil can respond to price more quickly than conventional resources has undermined the incentive of Saudi Arabia and others to cut production to bolster price. Put succinctly by Abdalla El-Badri in early 2016 when he was still the secretary general of OPEC, "Shale oil in the United States . . . I don't know how we [OPEC and shale] are going to live together." Remember al-Naimi's original logic: it did not make sense for Saudi Arabia to cut production in the face of falling prices in 2014 because tight oil producers would respond to higher prices by increasing production and taking Saudi Arabia's relinquished market share. Why the change?

The OPEC agreement did not annul these tight oil realities. Rather, it suggested two factors were at play. The first was the seriousness of the cash flow crisis experienced by Saudi Arabia and other OPEC producers. They must have valued increased revenues in the very short term more than they feared losses in market share over some unspecified period when tight oil would return to the market. Second, the Saudis and others likely felt that they had gathered important information on how responsive tight oil was to price changes over the two years that had elapsed since the initial price collapse. Tight oil had proved more sluggish in responding to price drops than expected. Rather than quickly curtailing production when faced with falling prices, U.S. tight oil production remained resilient. The Saudis, and others, may have inferred from this track record that, while tight oil production was much more responsive to price than conventional oil production, there was enough of a lag between price changes and production adjustments that OPEC producers could expect to reap the gains of an oil price boost for a half a year or even longer. While not ideal, reaping some gains in the short term may have been viewed as too attractive to resist.

Yet even before OPEC and other producers could translate their agreement into action, tight oil began to respond positively to the tightening of the oil market and slightly firmer prices. From September 2016—when rumors of an OPEC deal first emerged—until June 2017, tight oil production in the United States added more than 800,000 barrels a day. The adherents to the 2016 deal may not like what they see in the subsequent months as tight oil production grows robustly. Technological advances made in the past two years—such as the ability to drill more wells from one platform—mean more tight oil will be produced at lower prices. John B. Hess, the CEO of Hess Corporation, put it well when he told me in the last days of 2016, "the shale you needed a price of $60 to produce two years ago now makes sense to bring on line at $50." Roger Diwan, an energy market expert, made a similar point when in March 2017, at the margins of one of the largest energy conferences in the world, he told me that capital efficiency has increased so much since the price collapse that one dollar of expenditure today will yield two and a half times the oil that it would have at the end of 2014.

In short, while the November 2016 deal offers oil producers some immediate financial relief, it is hardly grounds to claim that OPEC has roared back to center stage. At best, OPEC has proven it is not yet in the grave and has been forced to share the stage with more robust co-stars: tight oil and the market itself. In the closing days of 2014, Nick Butler, a former energy advisor to British prime minister Gordon Brown and longtime energy guru, predicted that "sex and technology" will shape the oil market for the foreseeable future. It is true that sex—as a proxy for population growth—has always had a strong impact on demand, and technology has influenced both supply and demand. This will be more true going forward than it has been for many decades, the efforts of OPEC to reassert itself notwithstanding. In this new world of energy abundance, the fundamentals of supply and demand will have as much or more influence than any cartel in determining the price of oil.

The Next Saudi Arabia?

Given oil's long history of having a mediator between it and the market, one might ask if another entity will emerge now that OPEC is weakened. Some see the United States as a potential successor. Experts from former Federal Reserve Board chairman Alan Greenspan to ConocoPhillips CEO Ryan Lance have suggested that American tight oil makes the United States well positioned to be the swing supplier. The ability of tight oil to adjust relatively quickly when prices rise and fall creates this new possibility. But can U.S. tight oil really substitute for OPEC or Saudi Arabia in the three main roles these entities have played over the past decades?

A great deal of Saudi Arabia's past strategic heft arose from the kingdom's ability to play its first role: stepping in and adding oil to global markets in advance of, or immediately after, a destabilizing geopolitical event. For instance, at the behest of the United States, Saudi Arabia agreed to increase production in advance of both the first Gulf War and the 2003 invasion of Iraq. Washington and Riyadh knew that the military conflict risked the removal of Iraq's oil from the global market, potentially causing a dramatic price spike. More recently, in 2012, when the Obama administration was seeking the cooperation of China, India, and others in curtailing Iran's oil exports, Saudi Arabia's willingness to increase its production was one factor that allayed nervousness about the impact turning on Tehran could have on oil prices. In all these cases, Saudi Arabia's willingness and ability to turn up the taps within days or in advance of geopolitical turmoil helped keep oil markets relatively calm.

American tight oil has no ability to assume this critical role. The expansion or curtailment of American tight oil flows are almost entirely dependent on price signals. There is no U.S. government entity—no "national" oil company—that can decide to unleash or constrain the country's tight oil production. The government can, of course, influence production through taxes, regulation, and permits, but such tools are hardly designed to affect short-term output. As a result, U.S. tight oil will respond to the price jump created by a geopolitical crisis; it is in no position to preempt a price increase or respond to a political imperative.

The second role OPEC has played with varying degrees of success is dampening volatility in the oil markets. Those licking their chops for the end of OPEC have generally overlooked this contribution. Many economists view volatility in the price of energy as more damaging to economies than high prices; three researchers from Oxford University published a study in early 2014 on the volatility of crude oil prices, calling it "a fundamental barrier to stability and hence growth." Surveys of hundreds of senior financial executives found that more than "an astonishing" three-quarters of them would make some sacrifice in overall value of an investment in order to avoid volatile earnings. We might therefore postulate that given a choice between stable oil prices at $70 a barrel versus a constantly fluctuating price between $70 and $100, many oil-producing companies and governments would opt for the first. Volatility complicates decision-making for energy companies looking to make billion-dollar investments, some of which may be profitable at $70 oil, but not at $50 oil. Volatility in energy prices also frustrates the creation of realistic budgets for both producing and consuming governments. And it further complicates the task of governments that rely on energy taxes to fuel their spending, or that provide energy subsidies to their populations. Even individual consumers would allocate their money more sensibly with more certainty about fuel prices and fewer surprises at the pump.

OPEC's calibration of supply has often been geared to temper this price volatility. Rather than every shift in demand or every supply-contracting crisis forcing a change in the price of oil as a way of sending a signal to the market, OPEC—or simply Saudi Arabia—would adjust production at the margins to minimize the ups and downs in price. Although not always successful, the cartel sought to understand and anticipate trends in oil markets enough to keep prices steady.

Again, tight oil cannot be the antidote to oil price volatility because changes in tight oil production will only emerge as a result of a change in price. In the absence of an overseer, each of America's hundreds of individual producers will be making its own calculations about when to stop and start, based on the signals sent by the market. And although tight oil can be brought to market relatively quickly when compared to conventional oil, U.S. tight oil is still a laggard when compared to the speed at which OPEC or Saudi Arabia can tap into their spare capacity. "The big dogs, the Saudis, could snap their fingers and make that [bringing more oil to market] happen by tomorrow," said Mike Wittner, head of oil research at Société Générale in New York. "Here [in the United States], you have a whole sector of a couple hundred companies doing what they do and looking out for their own self-interests, and the whole thing takes a long time." At a minimum, companies working in the United States and looking to ramp up production will first need to recontract for drilling rigs, rehire employees who were let go, and reorder fracking fluids— potentially adding weeks or months to the timeline between the decision to produce and actual output increase. In addition, companies will need to find adequate financing, which may prove to be more difficult than it was in the first frenzy surrounding fracking. If anything, tight oil will increase volatility in price by shortening the response time between price change and production adjustment of conventional oil. Yet, however shorter this response time is, it will inevitably remain longer than that of OPEC, given the ability of several OPEC governments to swiftly increase production.

The third role OPEC has assumed is shaving off the peaks and troughs that might have resulted had the market been the only arbiter of supply and demand. In fact, from 2000 to 2005, OPEC had a formal policy of defending the price within a certain "price band." When prices were edging close to the top of the band, OPEC would bring more oil to market; when they were heading south, it would do the opposite. In this third case, U.S. tight oil *is* likely to take up the mantle of OPEC or Saudi Arabia reasonably well. A relatively small increase in prices will invoke a fairly quick production response, mitigating the need for prices to spike high to curb demand and incentivize investment until supply rebounds. There will be some jaggedness when compared to Saudi Arabia's execution of this task. But in coming on and off the market in response to price in relatively short order, American tight oil may effectively keep global oil prices in a band, where the upper and lower reaches are defined by the varying breakeven costs of producing this resource in different shale formations across the country. The more tight oil produced—in the United States and other countries—the more effective tight oil will be in playing this role.

Nick Butler, in the end, looks to be right that "sex and technology" will take on the most important role in determining the oil markets. Even if OPEC is not out of the game, it is a much diminished player on a more crowded field. Unlike any other time in the history of oil, the market—not a government, institution, or cabal of companies—could be the dominant arbiter between supply and demand. The unique characteristics of U.S. tight oil will help mitigate some of the extremes that could be expected if OPEC were to fade from the scene in other circumstances; it is likely to help keep prices within a band at a moderate price level for some time. But tight oil is unlikely to be able to help the world anticipate or react immediately to geopolitical calamities, as Saudi Arabia and OPEC often have. And it will exacerbate, not ameliorate, the problem of price volatility—creating new challenges for countries and companies looking to navigate the contours of this new energy abundance.

It is tempting to view the tumult in the oil world beginning in 2014 as just another cycle in a notoriously cyclical industry. But in our changing energy landscape, relying on past experience to predict future outcomes is risky. While some elements of the new energy landscape are familiar, there is plenty that is new. Fears of peak oil, and the actions they have motivated, will be defunct for the next decade or longer as tight oil and advancing technology help meet rising global demand. Uncertainty about demand, moreover, at least introduces the possibility that available supply exceeds demand in the future, further reinforcing the new energy abundance. OPEC is wounded, perhaps not mortally, but enough to relinquish the role it has sought to play in the market for the better part of half a century. In this new world, we can expect the occasional price spike, lower prices on the whole, and certainly more volatility. But much of this new oil landscape is uncharted territory—for the energy industry, those who watch it and make money from it, and, most importantly, for the countries that have shaped their foreign policies and international behavior either around securing oil or using it to advance other national security agendas.

THREE

Natural Gas Becomes More Like Oil

N atural gas was long considered a second-class citizen in the world of energy. An old wildcatters' joke says it all. A geologist returns from the field to report on the drilling of a recent well. "I have good news and bad news," he declared. "The bad news is we didn't find oil. The good news is we didn't find natural gas either."

There were a few reasons why natural gas was viewed as the lesser fuel. Compared to oil, natural gas packs less of an energy punch: one gallon of gasoline contains more than three times the energy of a standard cubic meter of natural gas. Put another way, a barrel of oil has nearly a thousand times the energy of a barrel of natural gas.

In addition to being less energy dense than oil, natural gas is also much less convenient to produce, transport, and store. The reason for this difference is self-evident: oil is a liquid; natural gas is . . . a gas. But the ramifications of this simple distinction are enormous. A barrel of oil can move without difficulty by truck, rail, pipeline, boat, or even rickshaw and is easily stored along the way. From the beginning, natural gas has been more complicated. It cannot be barreled right out of the well like oil. Once captured, it requires a massive infrastructure to transport and trade. In some cases, this takes the form of purpose-built pipelines. In other cases, special liquefaction facilities are used to cool the gas to the point where it can be transferred into a tanker. After shipping, the product must then be regasified. Storing natural gas is also much more complicated than husbanding away oil: whereas oil is easily housed in tanks, natural gas storage generally involves underground facilities—be they depleted reservoirs, aquifers, or salt caverns.

Compared to oil, the relatively complicated processes for producing and transporting natural gas create a totally different relationship between the producer and buyer when we look at cross-border trade. A barrel of oil may relatively easily change hands many times, literally or figuratively, from the time it leaves its original producer and arrives at its ultimate consumer. As a result, relationships do not need to be forged between producers and consumers, as long as both put their faith in the market that handles the transaction.

Given huge, up-front infrastructure expenses for natural gas, companies generally have only been willing to make investments with a long-term contract of twenty years or more in place between the producer and the future buyer. Over such a long period of time, both sides also want some agreed mechanism to determine price. Once they make these comparatively complicated arrangements, the producer and buyer are bound together by contract and physical infrastructure. Until roughly the year 2000, this commitment frequently meant a huge network of pipelines for natural gas traded across national boundaries. A long-term relationship, of some nature at least, was essentially inevitable. The difference between buying oil versus natural gas has been somewhat like renting an apartment versus building your own apartment complex in a foreign country. You get a place to live either way, but the second option is far more involved and cumbersome.

The manner in which this expensive infrastructure profoundly tied the consumer and producer together also created conditions for politicizing the natural gas trade in ways that do not exist in oil markets. If, say, Russia refused to sell oil to Ukraine, Ukraine could relatively easily import the oil from elsewhere. While it might pay a small premium owing to higher transportation costs, the fact that the single global market sets the price for oil could give Ukraine some confidence that it could meet its oil needs without being gouged. But when, in 2006, Russia curtailed shipments of natural gas to Ukraine that flowed through its extensive pipeline network, Ukraine was in a tougher spot. It could not easily buy gas from other massive global producers such as Qatar or Australia because the needed infrastructure was not in place.

As a result, natural gas can easily become a tool of foreign policy—and sometimes a more effective one than oil. Compare the 1967 Arab oil embargo (not the better-known 1973 embargo) against countries supporting Israel with Russia's 2006 decision to cut off natural gas to Ukraine. In 1967, given the fluidity of global oil trade, countries and companies were quick to redirect oil shipments to the embargoed countries from other producers; there was little economic damage and no political effect. In contrast, Russia created immediate difficulties in 2006 for Ukraine—and to a lesser extent Europe—when it curtailed natural gas sales to Ukraine as part of an effort to get the new, more pro-Western government in Kiev to pay higher prices. Although the disruptions were minor, Russia was still successful in negotiating a more favorable position for itself.

The realities surrounding the natural gas trade can also empower the buyer in certain circumstances. The subsequent 2009 crisis between Ukraine and Russia provides another vivid example. Although Russia's decision to cut off gas to Ukraine hurt Kiev, it also had—in this case lethal—implications for Russia's customers further downstream. At the time, 80 percent of the natural gas Russia exported to Europe flowed via Ukraine. So not only did Ukraine feel the pinch, but so did Romania, Bulgaria, Macedonia, Greece, and Turkey. This episode in particular called into question the reliability of Russia as a supplier of natural gas to Europe in a way that had not occurred in decades of natural gas trade. Recognizing that the transit of its gas through Ukraine leaves it somewhat at the mercy of the Ukrainians, Russia has prioritized building infrastructure to deliver Russian gas more directly to Western Europe; it has vowed that none of its gas will transit through Ukraine by 2019.

Because natural gas has traditionally been transported via pipelines, gas markets have also been divided by geography. Three largely distinct natural gas markets emerged over time: one in the Americas, one in Europe, and one in Asia. In addition to physical location, these markets also differ in how they have traditionally priced their natural gas. Since deregulation of natural gas markets in the 1980s and 1990s, the United States has had a liberalized and highly competitive market. U.S. prices are established by Henry Hub—not an individual, but a storage and transit location in Louisiana where multiple pipelines meet. Here, supply and demand in effect determine the price of natural gas, much as the price of oil is set. Prices set this way are known as "spot" or prices determined by "gas-on-gas" competition. Asia, in contrast, has customarily set natural gas prices in relationship to oil prices in a practice known as "oil index-ation." The European market has employed a combination of these two mechanisms.

Finally, oil also had the advantage as being viewed as a strategic com-modity, whereas natural gas was not seen in the same way. Oil has been the fuel that has powered militaries, won wars, moved planes, trains, and automobiles, and therefore lubricated trade and the global economy. The fact that, even in 2016, there are few substitutes for oil in the arena in which the majority of the world's oil is used—transportation—adds to oil's strategic dimension. Natural gas, in contrast, was originally used for heating, cooking, and electricity; only later did it became a major indus-trial input. In each area, natural gas is not essential for the task; other fuels can serve the same purpose as natural gas if necessary.

Given these characteristics, it makes sense that our geologist con-sidered natural gas a punch line as much as an energy resource. It also explains why a shocking amount of natural gas is still wasted, flared or burned off, mostly in the process of producing oil. Anita George, former senior director of energy and extractives at the World Bank, estimated that if all the natural gas that was flared around the world in 2015 had been converted into power, it could have met the entire electricity needs of Africa. Even as late as 2016, Fred Julander, a wildcatter and relentless advocate for natural gas, lamented, "Gas is a wonderful fuel and it has been an unappreciated fuel for all of history. I wonder if this is going to be the case until the demise of gas."

Today, however, things are improving for natural gas. This change is in large part because this underdog resource is becoming more like oil. In geopolitics, natural gas has already been elevated from the butt of jokes to the subject of presidential summits. In February 2013, Japanese prime minister Shinzo Abe visited Washington to meet President Obama. The agenda of the two leaders was packed with high-stakes items: Japanese par-ticipation in the Trans-Pacific Partnership trade talks, proposals to counter North Korea's nuclear pursuits with financial sanctions, and heightened tensions between Japan and China in the East China Sea. But amidst a program primarily intended to assure the United States that Japan remained a first-tier power, Abe made certain to raise the issue of natural gas. In an unusual foray into the domestic politics of another country, the Japanese leader asked President Obama to approve applications of American companies wishing to sell natural gas to Japan. At the time of the summit, natural gas was in particularly high demand following the Fukushima Daiichi tragedy. Japan's thirst for natural gas had skyrocketed when the country shut down its nuclear generators; natural gas became more attractive as concerns over the safety of nuclear energy grew worldwide.

The recent trend in favor of natural gas has been due to other factors as well. Technological advances, discussed in detail below, have made natural gas both abundant and much more accessible, even to those outside the reach of pipelines. In addition, natural gas is only half as carbon intensive as coal and a quarter less than oil, giving it a reputation as the "good" fossil fuel—at least in certain parts of the world. President Obama, in his 2014 State of the Union address, declared natural gas to be "the bridge fuel that can power our economy with less of the carbon pollution that causes climate change." Moreover, the ability of natural gas—as a provider of 24/7 power—to easily complement intermittent solar and wind sources has also given the commodity a role to play in boosting renewable power. Scholars Maximilian Kuhn and Frank Umbach have dubbed natural gas "the triple A" fuel, given its widespread availability, its affordability, and its acceptability. Some industry experts, such as Oklahoma oil and gas pioneer Robert Hefner, implore their audiences to think of natural gas as a superior fuel to oil on account of these and other qualities of the resource.

While all this will be of interest to energy buffs, these changes also have major implications for geopolitics. Most notably, the leverage in natural gas trade is shifting from producers to consumers. As explained in this chapter, the new energy abundance and related technological advances have exerted downward pressure not only on oil prices but also on natural gas ones. Moreover, as is the case with oil, these new realities are changing the structure of natural gas markets in profound ways. One major implication of these changes is that the natural gas trade is in many instances becoming harder to politicize. As a result, while oil still claims much of the spotlight, equally dramatic changes are happening with natural gas.

Natural Gas Euphoria

The new benchmark for enthusiasm about natural gas was set in November 2011, when the IEA released a report titled *Are We Entering a Golden Age of Gas?* With as much breathless excitement as a bureaucracy can muster, the report described a gas-friendly scenario for the future that it saw as feasible, although not inevitable. In this scenario, the use of natural gas grew robustly, requiring roughly three times the gas produced by Russia (the world's second-biggest producer) in 2016 to meet additional demand. Yet, with unconventional gas of different varieties being tapped on multiple continents, the world was able to meet growing demand while keeping prices low. In the United States, as well as in China and India, natural gas replaced some coal in power generation and made further inroads into industry and transportation in this scenario. According to this vision of the future, the abundance of natural gas was one factor that helped the world advance marginally closer to its climate goals than the IEA believed would otherwise happen.

Although the IEA noted that the golden age of gas would depend on policies, technology, and market signals, the world seemed eager and poised to realize it. China embarked on an ambitious program to transform its massive gas shale reserves into natural gas output. In 2012, in its *Shale Gas Five-Year Plan*, the government announced an ambitious goal of producing significant quantities of shale gas; it aimed for the country to produce roughly the equivalent of 40 percent to 70 percent of China's total demand for natural gas in 2012 by 2020. China launched many initiatives to support this goal, including opening up the sector to some foreign investment, creating incentives for shale development, and supporting technological research related to shale.

In the United States, companies clambered over one another to seek approvals to build or revamp liquefied natural gas (LNG) terminals to export U.S. natural gas. Leaders of countries from Israel to Mozambique delighted in major offshore natural gas finds, confident that global demand would be sufficient to transform their economic and strategic circumstances. Poland also had great expectations. In 2011, Prime Minister Donald Tusk claimed, "with moderate optimism, that 2014 will be the beginning of commercial [shale gas] exploitation."

In the four years following the 2011 *Are We Entering a Golden Age of Gas?* report, the new era arrived ahead of schedule in North America. Shale gas production boomed, with U.S. natural gas production in 2015 at levels the IEA report had anticipated would not be reached before 2035. Natural gas seemed nearly ubiquitous in the United States and, despite healthy demand from revived manufacturing and power generation, prices hit their lowest mark since the turn of the century. In mid-2016, the spot price for natural gas sold in the United States was just one-seventh of what natural gas had cost at its peak price a decade earlier. Consistent with the IEA's gas-friendly scenario, natural gas had displaced coal in the United States—on a major scale. As recently as the year 2000, coal accounted for more than half of the fuel used for power generation, whereas natural gas generated less than a fifth. In 2016, largely as a result of declining natural gas prices, natural gas just overtook coal as the fuel generating the most power in the United States.

In other parts of the world, however, the golden age had yet to dawn. Europe's economic performance was lackluster and China's growth was also disappointing—both of which dulled potential demand growth for natural gas. In both places, continued government support for renewable energy also thwarted more robust use of natural gas. Additionally, a resurgence of cheap coal—because of slowing Chinese and falling U.S. demand—also tempered demand for natural gas in Europe and elsewhere.

Despite the sunny 2011 predictions of the IEA, efforts to develop shale gas outside North America also proved disappointing. In the first four years following the IEA report, for example, no countries outside North America apart from China and Argentina reached commercial production of shale gas. Expectations for future shale gas development remain, but regulatory barriers, logistical hurdles, institutional deficits, mineral rights, high costs, and other obstacles proved to be more difficult to overcome than initially expected. Just two years after rolling out its ambitious shale gas targets, the Chinese government was forced to slash them to just half of the original low-end goal. Environmental considerations have also impeded shale development in Europe, where political opposition to the development of shale gas has stymied the golden age of gas.

While hopes for more robust shale production at a global level have not yet been realized, markets became awash in natural gas for other reasons. On the supply side, other natural gas projects long in the works added new output to the market. On the demand side, as mentioned, slower economic growth and competition from coal and renewables particularly in China caused demand to be less robust than predicted. Given this situation, the world did not work off the gas overhang as IEA predicted it would by 2015 in its gas-friendly scenario. Instead, in 2016, talk of an enduring natural gas glut persisted, with the IEA and others predicting a deepening glut in the coming years as even more U.S. and Australian LNG exports come online.

Looking ahead, natural gas still is a good fuel upon which to bet. It is the only fossil fuel whose share of the energy mix is expected to rise over the coming decades according to mainstream projections. Moreover, even in scenarios where the creators present a future in which the world has become much more serious about the challenges of climate change, natural gas either maintains its share of global energy use or increases it.

There are, however, some early signs that natural gas could end up falling between two stools, at least in certain parts of the world. Either of two policy extremes could bode badly for natural gas. If the world—Asia in particular—becomes more lax about its use of coal, natural gas will be the fuel to lose out; without some policy interventions, coal remains dramatically cheaper than natural gas in most parts of the world outside the United States. On the other end of the spectrum, should the world become more aggressive in its efforts to combat climate change, policies could shift the calculus in favor of renewables, at the expense of natural gas.

While the IEA's golden age of gas has certainly not arrived globally, the changes in the world of natural gas have still been extraordinary. As with oil, there is a story about burgeoning supplies, largely attributable to tremendous technological advances. And there is the punctuation of a price collapse—no less dramatic, even if less publicized, than the oil price plunge from 2014 to 2016. And finally, as is the case of oil, there are a host of other changes—to markets, pricing arrangements, and business models—that collectively have and will continue to have a critical impact on geopolitics.

Technology Transforms

On a hazy Sunday toward the end of February 2016, four red, white, and green tugboats maneuvered the 935-foot, 100,000-ton *Asia Vision* tanker into position. The valves of the tankers had to be lined up precisely with the four loading arms of the jetty at Sabine Pass terminal, part of a one-thousand-acre facility straddling the Texas-Louisiana border. With a wrench the size of a human arm, workers secured the ship to the jetty. Natural gas, which had been cooled to −260 degrees Fahrenheit and liquefied over the course of traveling through more than a mile of steel pipe and refrigerating systems, flowed into the tanker. A few days later, at 7:39 p.m. on February 24, flying under the flag of the Bahamas, the *Asia Vision* departed Sabine Pass and began its voyage to deliver its three billion cubic feet (bcf) of natural gas to Brazil.

As routine as it might sound, this scene was historic. For Cheniere, the American company that had spent more than a decade and close to $20 billion to develop the Sabine Pass LNG facility without ever turning a profit, it was a day for celebration; the *Asia Vision* was the first vessel carrying LNG that had shipped from the lower forty-eight states since the 1960s. After decades of fretting about its burgeoning dependency on imported energy, the United States had become an exporter of natural gas.

Two drivers lay behind the new age of oil abundance explored in previous chapters. The first was technological as new extraction techniques such as fracking emerged. The second was geopolitical and came in the form of the weakening of OPEC. Together, these technological and political factors transformed the future of oil for the next decade or longer. Similarly, in the case of natural gas, there have also been two drivers of the new energy abundance—both, this time, in the field of technology.

The first is the technological advances that enabled the shale boom to emerge in North America. Although we explored these innovations in earlier chapters, it is worth once again marveling at the scale of the development in natural gas. Well into the 2000s, policymakers and strategists fretted over a growing U.S. dependency on imported natural gas, not only from Canada, but also from suppliers in Africa, South America, and even, potentially, Russia. Tapping the vast quantities of shale gas in the United States spurred a massive turnaround in America's natural gas trajectory. In 2000, the United States was producing virtually no shale gas. Just ten years later, shale production had crept up to be a quarter of the total. By 2016, shale accounted for a full 53 percent of all U.S. dry natural gas production.

Looking beyond the United States, shale gas reserves are even more ubiquitous than those of tight oil. The U.S. EIA has already assessed forty-six countries for their shale gas potential. It found that, in those countries, global shale gas resources are massive. According to those initial estimates, they roughly equal the conventional natural gas resources of the Americas, Europe, and Eurasia (including Russia). If we include other types of unconventional gas in our calculations, such as tight gas and coal bed methane, global unconventional gas resources amount to nearly four-fifths of the conventional gas resources of the whole world.

While the IEA has dialed back the timeline in which the golden age of gas will be realized, the Paris-based agency still anticipates the unconventional gas boom going global. Given the difficulties of replicating the American shale boom elsewhere, the United States will be the producer of nearly all the world's shale gas in 2025. By 2040, the United States will still be dominant, but other producers will likely have emerged apart from the United States and Canada. Australia, China, and Argentina are expected to lead the pack, but Mexico and others may not be far behind. By some assessments, by 2035, China could bring more new shale gas to market each year than any other country.

The second technological development that has recently transformed the role of natural gas in the world actually began back in the eighteenth century with a British scientist named Michael Faraday. Born into a poor family of ten children in 1791, Faraday received little formal education. However, at the age of fourteen, he became an apprentice to a French bookbinder, who permitted him to read the books on which he was working. Faraday began a voracious process of self-education and scientific experimentation, including the first successful effort to liquefy gas in 1845. In its liquefied form, gas only takes up 1/600th of the space it would otherwise, creating obvious advantages for its transportation. Yet it was not until nearly a century later that the first commercial LNG plant was built in Cleveland, Ohio, and nearly another two decades before the

Methane Pioneer, a converted World War II liberty freighter, transported a cargo of LNG from Louisiana to Great Britain.

Despite the success of this shipment and a dozen others that followed in its wake, LNG remained a niche fuel for the next forty years, with the vast majority of LNG exports feeding the energy-hungry island nation of Japan. Virtually all other natural gas was still transported through pipeline networks. After 1990, however, technological advances, such as increases in both the size of tankers and liquefaction trains, brought down the cost of LNG. In 2005, a new innovation—called a floating storage and regasification unit (FSRU)—was first deployed in the American Gulf of Mexico. Either purpose-built or converted from an LNG vessel, these units opened up possibilities for more countries to meet their gas demand through LNG. Jordan is one case in point.

The revolution sweeping through Egypt in 2011 and its aftermath had practical implications for its neighbors in more than one way. Jordan, a Middle Eastern country bereft of big energy resources, had long imported natural gas from Egypt via the 750-mile Arab Gas Pipeline snaking its way from the Egyptian delta through the Sinai Peninsula into Jordan and Syria. Instability in Egypt led to repeated attacks on this pipeline and ultimately the suspension of gas supplies to Jordan. Seeking a comparatively quick solution, Jordan negotiated for a FSRU to be docked in its Red Sea port of Aqaba, allowing the kingdom to accept LNG and address its gas shortages more quickly than had it needed to build new infrastructure.

Other countries, particularly those with smaller markets or geopolitical, political, or financial circumstances that make it difficult to secure funding for large investment projects, have also opted for FSRU units. Now, according to Leslie Palti-Guzman, head of global gas at the Rapidan Group, "you can be an LNG importer in 18 months, even if you don't have the infrastructure and you aren't creditworthy." If a country defaults on its payments, its creditors can simply reclaim the facility as collateral. These advances have made buying LNG more flexible and less capital intense to use. They require less commitment from a country that might be unwilling to invest in expensive infrastructure it hopes it will not need in future decades. In short, they have made natural gas more like oil.

Countries such as Egypt, Pakistan, and Turkey also saw these advantages and opted to build FSRUs, thereby contributing to the boost in the number of countries importing LNG—from ten to thirty-five—between 2000 and 2016. During that same time, global capacity to regasify the liquid fuel increased more than twofold and overall LNG trade more than doubled. LNG had become mainstream.

Figure 3.1: LNG Trade Volumes, 1990–2016

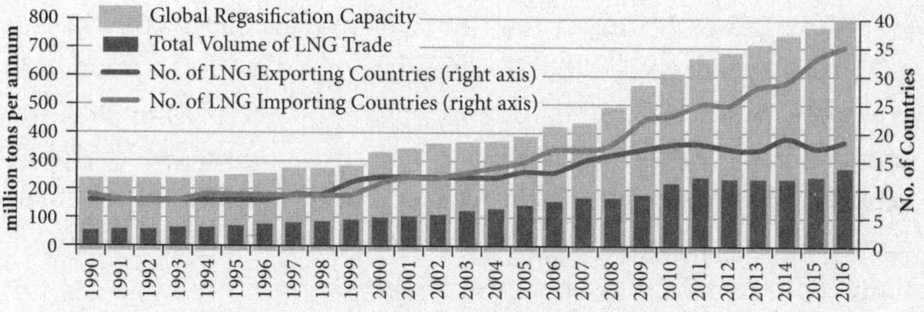

[1] Excluding Indonesia, which buys cargoes exclusively from domestic liquefaction plants.
[2] The United States is included in both totals, since it exports domestically produced LNG from Kenai LNG in Alaska and re-exports LNG from regasification terminals in the Gulf of Mexico.

Source: International Gas Union, *IGU World LNG Report, 2017,* 7.

The future for LNG looks even more robust than its recent past, particularly if the drop in its price is arrested. In the next two decades, LNG trade is poised to take on an ever-larger slice of the fast-growing pie of the overall global natural gas trade. In an absolute sense, the IEA projects the volume of LNG trade will more than double again. Remarkably, nearly half this growth in LNG supplies is expected to occur before 2020. According to BP, LNG volumes will in fact expand seven times faster than gas transported by pipe, and, by 2035, account for close to half of all globally traded gas.

The advent of U.S. LNG, made possible by shale gas, and Australian LNG, much of which is sourced by another unconventional gas called coal bed methane, are two major drivers of this expected surge. Much of the Australian LNG under development will surely make it to global markets given the huge investments already made before prices sank. There is, however, more uncertainty about the size of the expected wave of U.S. LNG.

Estimates of how much LNG America will export in the years ahead have varied, and are highly dependent on a price robust enough to make the trade commercial. Although the price of natural gas is so much lower in the United States than in most other parts of the world, the costs of liquefying the gas, transporting it long distances, and regasifying it upon arrival also need to be factored in when a potential consumer considers U.S. LNG as an option. Assuming these costs are between $5 and $8 per one thousand British thermal units (mmbtu), if the Henry Hub price of gas in the United States is around $4 per mmbtu, then prices in Asia or Europe would need to be $9 to $12 per mmbtu for the trade to be profitable. Some countries may desire to have U.S. LNG imports in their energy mix for reasons of diversification or because they see the United States as a reliable supplier. But because companies, not countries, make decisions about individual contracts, trade will occur where and when these companies see a market opportunity.

It is primarily these uncertainties that will determine just how much LNG the United States will bring to market. If *all* the applications to export U.S. LNG received by the U.S. Department of Energy for approval as of July 2016 materialized and these facilities were fully utilized, they would result in more than 58 bcf of new LNG added to global markets each day; that is nearly twice the entire global LNG trade in 2015 and five times the LNG imported by Japan, the largest LNG importer in the world, in the wake of the Fukushima tragedy in 2013. This large quantity of U.S. LNG will certainly not be commercial. Yet the volumes are set to rise from the trickle of exports in 2017 to a steady flow. According to the U.S. EIA, in 2020 the United States will be home to nearly one-fifth of global liquefaction capacity and the third largest LNG exporter in the world after Qatar and Australia.

Boon to Consumers

More superficially, there is another way in which natural gas has more closely resembled oil in recent years: the price of natural gas plunged markedly from 2014 to 2016. Thanks to American shale gas, the price of Henry Hub sank below $2 per mmbtu in the first half of 2016, down from a monthly peak of more than $13 in October 2005. European and

Asian natural gas markets also experienced striking price collapses, but for different reasons than occurred in the United States. The price to import natural gas into the European Union dropped from nearly $13 per mmbtu in April 2013 to $4 in May three years later. In Asia, where roughly 70 percent of LNG trade occurs, LNG prices were sliced in half from Valentine's Day to Thanksgiving in 2014. In Japan, the price of importing LNG fell further—from over $16 per mmbtu in 2013 to just under $6 in 2016.

Figure 3.2: Monthly Average Regional Gas Prices, January 2010–January 2017

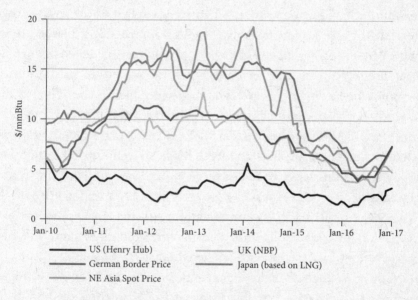

Source: International Gas Union, IGU *World LNG Report*, 2017 edition, 16.

At one level, the increase in global supplies would seem enough to explain the plummeting of these prices. Yet slower-than-expected demand—thanks in part to the financial crisis—also added to the trend. Moreoever, in Europe and Asia, where prices are partially or largely linked to oil prices, the drop in the price of oil quickly forced down the price of natural gas as well. European consumers also benefited from LNG cargo displaced by the American shale gas boom. As a result of the evaporation of U.S. demand for LNG imports from abroad, LNG exports

intended for—but no longer needed by—the United States flowed instead
first to Europe and later to Asia. Their arrival gave European utilities the
leverage to renegotiate the price they were paying companies in Norway
and Russia, in particular, for natural gas from pipelines, even under existing long-term contracts.

Yet even more was at work than these factors, with some of the additional developments poised to outlast the dip in prices. One significant
change has been made to how natural gas prices are determined. The ne-
gotiation efforts of European utilities were not solely directed at the absolute price they would pay for natural gas. They also, successfully, used
the moment to pursue a goal they had sought for a long time: jettisoning
"oil indexation." Tying the price of natural gas to that of oil was a practice
developed by Jan Willem de Pous, the Dutch minister of economic affairs
half a century earlier. The simplicity of this linkage made sense when de
Pous proposed it in 1962, given that natural gas was often substituted
for oil in the heating of homes and the generation of power. The pricing
practice, however, persisted long after this rationale disappeared, in large
part because producers such as Russia found it advantageous. It was only
when U.S. shale gas production forced more LNG unexpectedly into Europe that utilities acquired the leverage to negotiate a price formula that
placed more weight on "gas-on-gas" pricing and less on the price of oil.
From 2005 to 2015, the amount of natural gas trade in Europe that was
indexed to the price of oil declined from more than three-quarters to less
than one-half. In Asia, the shift was less dramatic, but still significant.

Critical changes to natural gas markets extend far beyond just how
natural gas is priced. Increasingly, the very distinct three regional markets of North America, Europe, and Asia are becoming more integrated.
Because of transportation, infrastructure, and other such costs, the three
markets will never completely merge into one market—like the one that
exists for oil. But what happens in one gas market is increasingly affecting what happens in the other two; the market is becoming more globalized. As piped gas accounts for a smaller percentage of overall trade, gas
is flowing more easily from one market to another, meeting demand as
needed.

In this more flexible, liquid market, more natural gas is being traded outside of long-term contracts—either on the spot market or through contracts lasting two years or less. The percentage of LNG traded in this fashion more than quintupled between 2000 and 2016; the number of countries exporting LNG on the spot market alone grew by five times in this time period. While the trend held in both Europe and Asia, it was much more pronounced in the first. Regional trading hubs are slowly beginning to emerge in Asia, allowing natural gas trade to be priced to better reflect supply and demand realities.

These trends toward greater fluidity in and between markets will accelerate in the years ahead, even if U.S. LNG only meets the most modest expectations. Yet, the transformative impact of U.S. LNG is not only a result of the increased volumes put on the global market. It also is a result of the rules and terms under which these exports will enter the market, which are much more flexible than previously traded LNG. First, the price of the overwhelming majority of U.S. LNG will not be indexed to that of oil. Instead, the base price for U.S. LNG—no matter its intended market—will be determined by Henry Hub, and therefore a product of gas supply and demand rather than oil price. Moreover, traditional global LNG contracts contain what are called "destination clauses" that restrict the ability of the buyer to resell any gas it has arranged to buy in its long-term contracts but no longer needs. This arrangement is good for producers, but less so for buyers. In contrast, U.S. LNG does not have these clauses, meaning that purchasers can resell the LNG if it makes sense to do so. Jonathan Stern, an Oxford University scholar and one of the leading experts on natural gas, predicted in 2016 that destination clauses will soon be a thing of the past. Already, consumers are resistant to signing contracts with such stipulations, signaling more changes to come in favor of even greater market flexibility.

Most of us are not likely to get excited about the prospect, still somewhat distant, of a more integrated global gas market. Consumers, however, *should* have a clear preference for such an outcome—for natural gas, in short, to become more like oil. For starters, a larger LNG market, where a significant portion of the market operates under new, more flexible terms with more players, will lead to more efficient allocation of investment resources. Moreover, increased LNG trade and more fluid LNG markets will smooth peaks and troughs in prices, helping ensure more stability, which is in the interests of both consumers and producers.

The Geopolitical Upshot

All of the innovations and developments described in this chapter add up to one big takeaway: natural gas will in many instances be a less effective political instrument than it has been in the past. In loosening the once tight relationship between many producers and consumers, the new geopolitics of natural gas favors the consumer over the producer and the market over political machinations. An expensive pipeline, undergirded by a long-term contract, where the price is linked to oil and restrictions exist on resale, no longer necessarily binds the producer and consumer together irrevocably. The consumer can now have many partners, seeking out deals in its best commercial interest and meeting its demand more effectively.

Inevitably, these new natural gas realities are changing geopolitics—the subject to which the rest of this book now turns. In the United States, the shale gas boom and the imminent shift from being a net importer to a net exporter of natural gas brings America both hard and soft power opportunites. In addition, in Europe, these energy dynamics, combined with new policies and political commitment, have created challenges for Russia and reinforced that country's desire to "turn east." China, however, no longer needs to tether itself to Russia in the interest of gaining access to sufficient natural gas resources. While Beijing will certainly seek Russian natural gas, the surfeit of potential suppliers is one important factor that has given it the upper hand in negotiations with Moscow over pipeline projects. China will want Russian piped gas, given its preference for overland routes, but the success of such projects no longer constitutes a top strategic priority for Beijing.

The American Phenomenon

FOUR

America's Unrequited Love

Don Steinberg was enjoying a rare quiet moment in his office on the sixth floor of the U.S. State Department. No stranger to conflict, Don had served as U.S. ambassador to Angola in the 1990s and later as special coordinator for Haiti and the president's special representative for the removal of land mines. Now it was November 2001, just weeks after the devastating blows of September 11, and Don was facing a different sort of challenge. As deputy head of Policy Planning, the office I had just joined, he was charged with providing assessments and foreign policy advice to Secretary of State Colin Powell.

Don was a thoughtful and approachable man, and I hoped he wouldn't mind the interruption, even if it was to ask what seemed like a much too elementary question given all else going on in the department at that moment. Like many Americans committing to public service, I was determined to be useful and to make a difference. Having spent several years at the Brookings Institution, a Washington think tank, writing books and articles on foreign policy, I had plenty of ideas about how policymakers might do things differently. But turning an idea into an action? That, I realized, was the ultimate mission and one for which I had little or no previous experience. So I turned to Don and asked, "How do we get things done here?"

As I remember, he gave me lots of great counsel in response, explaining the wiles needed to navigate the "interagency process" and other bureaucratic mysteries. But one piece of advice stuck with me: the only way to be absolutely certain a policy will be implemented is to get it mentioned in a presidential speech, or better yet, the State of the Union Address.

I took this guidance to heart, and much later—in the wake of the Iraq War—it shaped my policy pursuits. However, I was—as I am sure Don was—aware that at least in one very important domain, getting the president to commit to solving a problem in his annual speech to Congress and the nation did *not* ensure progress or even action: energy independence. For decades presidential speeches had been littered with solemn pledges and dramatic undertakings to eliminate America's dependence on foreign oil. But this goal remained elusive—until the recent energy boom in the United States. Suddenly, and unexpectedly, the prospect of achieving energy independence became real.

After decades of the pursuit of this status, one would expect the execution of the seduction to be more thoughtful. Instead, popular discourse and policy conversations are still littered with misconceptions and unrealistic or undesirable goals surrounding the concept. Moreover, many U.S. policies actually undermine the possibility of realizing the most sensible and attainable version of energy independence, which is a continental approach in conjunction with Canada and Mexico. In some circumstances, rather than aggressively pursuing the goal they claim to espouse, policymakers are thwarting America's own economic and geopolitical advantages by making North American energy independence hard to realize.

Decades of Elusiveness

Speaking as a candidate in May 2016, Donald Trump called for "complete energy independence"; only two months into his presidency, he penned an executive order titled "Promoting Energy Independence and Economic Growth." In 2010, President Barack Obama exhorted "this generation to embark on a national mission to unleash America's innovation and seize control of our own destiny" by ending dependence on oil; his predecessor, George W. Bush, urged steps to "make our dependence on Middle Eastern energy a thing of the past." Earlier presidents had been similarly preoccupied with this objective. In 1988, Vice President George H. W. Bush asserted that there is "no security for the United States in further dependence on foreign oil," while President Jimmy Carter deplored "this intolerable dependence on foreign oil" in 1979. Just four years earlier in 1975, President Gerald Ford had announced new energy programs "to achieve the independence we want by 1985." In 1973, his predecessor, Richard Nixon, had declared "Let us set as our national goal, in the spirit of Apollo, with the determination of the Manhattan Project, that by the end of this decade we will have developed the potential to meet our own energy needs without depending on any foreign energy sources."

There may be no other single issue that recent American presidents have so uniformly vowed to tackle, regardless of political party. Beginning in 1971, when U.S. oil production began to decline, every president has perceived the county's ever-growing dependence on foreign energy sources to be a major U.S. strategic vulnerability. Americans—not just their presidents—longed to be insulated from the threat of another energy embargo such as the one suffered when the Arab members of OPEC stopped exporting oil to the United States in 1973. In the 2000s, concerns arose that the United States would replicate its reliance on foreign oil with a similar dependence on natural gas.

Economists also have been quick to note how damaging energy price spikes have been to the American economy. Of the ten U.S. recessions before 2005, nine were preceded by substantial increases in oil prices. Although economists disagree on the accuracy of the models used to estimate the effect of oil prices on GDP growth, the more conservative economists suggest that the price spikes of 1979 and 2007 each cost the U.S. economy somewhat less than 1 percent of GDP—or $150 billion if a similar shock were to have happened in 2016. Insulating the country from these price spikes—whether resulting from supply disruptions, speculation, or changes in expectations regarding the future energy supply-demand balance—would remove an enormous risk to economic growth and prosperity.

The geopolitical downsides of this strategic vulnerability have been widely noted as well. Secretary of State Condoleezza Rice voiced the frustration of many when in 2006 she told Congress, "We do have to do something about the energy problem. I can tell you that nothing has really taken me aback more, as Secretary of State, than the way that the politics of energy is . . . 'warping' diplomacy around the world. It has given extraordinary power to some states that are using that power in not very good ways for the international system, states that would otherwise have very little power." That same year, *New York Times* columnist Tom Friedman captured the discomfort of many when he lamented how the rising price of oil enabled authoritarian leaders to get the world "to look the other way at genocide, or ignore an Iranian leader who says from one side of his mouth that the Holocaust is a myth and from the other that Iran would never dream of developing nuclear weapons, or to indulge a buffoon like Chávez, who uses Venezuela's oil riches to try to sway democratic elections in Latin America and promote an economic populism that will eventually lead his country into a ditch."

Despite such concerns, American progress in paring back its dependence on foreign oil was fleeting in the thirty-five or so years after the Arab oil embargo in 1973. The Energy Policy and Conservation Act of 1975 introduced fuel economy standards and energy conservation programs. It also sought to increase domestic coal production and gave the government the authority to order power plants to burn coal instead of natural gas or oil—the very opposite of what is happening today! Tax incentives for the augmented production of other fossil fuels—and renewable energies as well—were increased substantially. In line with other developed economies, the U.S. government established the Strategic Petroleum Reserve to reduce vulnerability to supply shocks.

These measures paid off temporarily. U.S. net imports of crude oil from 1977 to 1985 fell nearly in half and American dependence on foreign oil dipped to about a quarter of total petroleum consumption. But such policies could not keep up with economic growth and declining U.S. oil production. Earlier gains were overwhelmed over the course of the next decade. By 2007, American imports of oil were a staggering 12 mnb/d—significantly more than the entire daily output of Saudi Arabia and nearly three times what they had been in 1985. Nearly two-thirds of America's daily oil consumption of oil was met by imports from foreign countries. The notion of U.S. energy independence had become more fantasy than reality.

By this time, the obsession with America's energy exposure went well beyond the corridors of the White House. At the end of 2006, more than a dozen senior American business executives and retired military officers banded together to put pressure on the U.S. government to do more to tackle the country's vulnerability. Co-led by FedEx chairman Frederick W. Smith and former Marine Corps commandant P. X. Kelley, this group formed the Energy Security Leadership Council and launched an intense effort to get their message out through television and print ad campaigns as well as lobbying Congress and the White House. Around the same time, as U.S. and Iraqi casualties in Iraq were skyrocketing, political consultant James Carville conducted a survey on the national security priorities of voters. "Reducing dependence on foreign oil" was ranked the number one concern by 42 percent of those polled; "combating terrorism" and "the war in Iraq" clocked in at a distant second and third. Carville, a Democratic strategist, urged his party to convince Americans to associate "energy security" with Democrats in the way the public associated "tax cuts" with Republicans. "This is not something to add to the stew—this is the stock," he told candidates looking to win the upcoming midterm elections.

Serendipity

Enter serendipity. The word *serendipity* was coined by the historian and politician Horace Walpole in 1754 after he read a Persian fairy tale in which three princes of Serendip "were always making discoveries, by accidents and sagacity, of things they were not in quest of"—thus the meaning: a fortunate happenstance or a pleasant surprise. While most of the world was fixated on eking out a renewable energy revolution that, according to President George W. Bush, would move the United States "beyond a petroleum-based economy," the boom in unconventional oil and gas was brewing—and with it, an unexpected shot at deliverance from the strategic vulnerability created by America's dependence on foreign energy.

As noted in earlier chapters, U.S. crude oil production nearly doubled between 2009 and 2015 alone, clawing its way back to 1970 highs of nearly 10 mnb/d. Natural gas production increased by 40 percent over the decade up to 2015. The United States became the largest energy producer in the world, outpacing Russia in natural gas in 2007 and Saudi Arabia in total oil production in 2013. The world's only remaining superpower had once more become an energy super producer—a status it had relinquished more than forty years earlier.

Given a further leg-up by declining U.S. demand for oil for many of the years in question, these production increases reduced American net oil imports at an astonishing rate, from nearly two-thirds of consumption in 2007 to only one-quarter in 2016.

The specter of a growing, never-ending dependence on foreign natural gas was extinguished at the same time. Demand for natural gas continued to grow, but burgeoning domestic production all but squelched imports of LNG by 2012. Companies such as Cheniere Energy, which had gambled big in 2003 to build facilities on the U.S. Gulf Coast to *import* expected waves of LNG, took a second multibillion-dollar wager. Banking that this reversal of energy fortunes would turn the United States into an exporter of natural gas, they convinced investors to support their efforts to convert these facilities from importing LNG to exporting it. In the policy world, the debate seemed to shift seamlessly from handwringing over imports to concerns over exports, be it the lifting of the U.S. ban on the export of crude oil or the hoops American companies needed to jump through to export natural gas.

Americans today can be confident their energy position has changed materially for the coming decades. As discussed in Chapter Two, this is true even taking into account the uncertainties about the exact path of future American production and despite the fact that U.S. demand for oil has grown again after years of looking as if it would soon head in the opposite direction. (Witness the gusto with which Americans have been purchasing SUVs and light-duty trucks in response to cheaper gasoline.) As can be seen in the graphic below, regardless of what exact production path materializes, Americans can still expect their dependence on imported oil to be lower—in many cases dramatically so—in the coming decades than it was in the mid-2000s. There is even less uncertainty about future natural gas production, with nearly all companies and agencies forecasting continuous growth through 2035.

Figure 4.1: Projected U.S. Crude Oil Production Under Different Scenarios, 2005–2040 (million barrels per day)

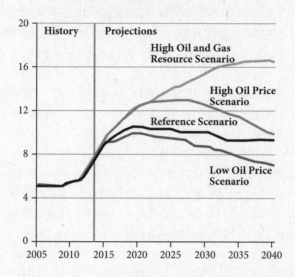

Source: Energy Information Administration, *Annual Energy Outlook 2015*, www.eia.gov/todayinenergy/detail.cfm?id=20892.

Figure 4.2: Projected U.S. Net Petroleum Product Imports Under Different Scenarios (percent of U.S. petroleum product supplied)

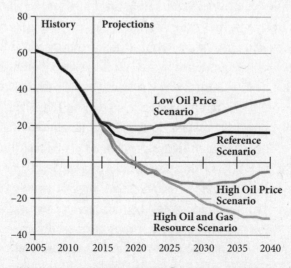

Source: Energy Information Administration, *Annual Energy Outlook 2015*, www.eia.gov/todayinenergy/detail.cfm?id=20892.

The scale and swiftness with which the United States has moved from being an energy supplicant to an energy super producer is stunning; in modern history, it is rare for the strategic position of a single country to change so dramatically in such a short period. So rapid has this transition been that many of America's policymakers and opinion shapers have been left in the lurch, still fixated on the vulnerabilities that used to come with large dependence on other countries for energy supplies.

Not a Bargain

In 2005, Andre Hoth, a computer geek and amateur musician from California, won a song contest sponsored by the nonprofit Americans for Energy Independence. Hoth's entry, "A Crude Energy," beat out contestants from six other states who had vied to pen a "rallying song" that would capture the importance of ending U.S. reliance on foreign sources of energy. As the winner, Hoth got to headline a CD of the top ten songs, including titles such as "Feedin' Them Oil Hogs" and "EI Now" (a shortened reference to energy independence).

At the time of this contest, energy independence seemed a long way off. Like an unrequited love, its unattainable nature prevented Americans from really evaluating whether its acquisition would be worth the pursuit. But a world in which the United States could meet all its energy needs seemed like an unequivocal good, certainly something worth singing about.

The new energy abundance requires the United States finally to scrutinize the energy independence for which it has long pined, and to do so with a head cleared of illusions. What would energy independence really look like? Two very different ideas are commonplace. The first defines energy independence as a situation in which the United States meets all of its own energy needs, without any reliance on international trade or global energy markets. In this scenario, America would be completely "self-sufficient" or "autarkic." The second form of possible energy independence is one in which the United States is not separated from global markets, but simply produces more energy than it consumes.

In the first type, an autarkic America would realize many of the benefits of those who have dreamed about energy independence for decades. Completely set apart from any international trade (except perhaps to export excess energy), the United States could be inoculated from potential global supply disruptions, threatened embargoes, or the price spikes that have so often precipitated economic calamity or at least frustrated sound economic planning.

Alas, America is not yet close to reaching this level of self-sufficiency. It is true that virtually all the coal, nuclear, and renewable energy consumed in the United States is produced or generated indigenously. It is also true that, with relatively little effort, this could be the case for natural gas as well. But oil self-sufficiency is far from inevitable. While U.S. oil imports have diminished dramatically, the United States remains a considerable distance from producing all the oil it needs on its own. Positioning America so it could sever its dependence on any form of petroleum or oil imports in the near term would take dramatic government interventions to staunch imports, increase production, reduce consumption, and transport oil long distances domestically from producing areas to consuming ones.

Two historical episodes give us a good sense of what efforts to achieve oil self-sufficiency might look like—and how costly they might be. The first dates back to when Dwight Eisenhower was president of the United States. Like his successors of recent years, Ike was concerned about the national security implications of America's growing dependence on foreign oil. In the face of cheap oil from the Middle East, Venezuela, and North Africa, imports had nearly doubled during the 1950s, reaching close to one-fifth of U.S. consumption by 1957. Ultimately, Eisenhower was swayed by the arguments of small American oil companies—and legislators from oil-producing states, such as Senator Lyndon Johnson and House Speaker Sam Rayburn, both of Texas—that cheap foreign oil was undercutting the domestic U.S. oil industry, which could have not only economic, but also national security, implications. After failed attempts to induce voluntary reductions in the use of oil, on March 10, 1959, Eisenhower issued a presidential proclamation limiting the imports of foreign oil to roughly 10 percent of domestic consumption in order to preserve "a vigorous, healthy petroleum industry."

Clarence B. Randall, a well-known conservative economist and then Eisenhower's chairman of the Commission on Foreign Economic Policy, argued against the new policy. He stressed that it would push up domestic consumer prices while simultaneously accelerating the depletion of domestic resources. Subsequent years validated his predictions that this policy would be costly. A task force initiated by President Nixon before he lifted the import quota in 1973 calculated the costs of Eisenhower's policy in 1969 to have run to the tune of $5 billion ($33 billion in 2017 dollars).

To be sure, the quota and its accompaniments did also contribute to the more rapid development of American resources; crude oil production increased by a third over the span of this policy, before peaking and then beginning its descent in 1971. Yet these gains contributed to a geopolitical calamity that was already brewing. Predictably, limitations on the import of oil by the United States created a glut in international oil markets, driving down global prices. Feeling the pain, one of the Seven Sisters—the handful of international oil companies that effectively controlled the global market at the time—made the unilateral decision to cut the price it would pay Middle Eastern producers for every barrel they produced. This action, in turn, led Iran, Iraq, Saudi Arabia, and Venezuela to gather in Baghdad in September 1960 and establish the Organization of the Petroleum Exporting Countries or OPEC. While their stated objective at the time was to restore the price of oil to the level it had been before the cut, their goals morphed over time and their collective power grew until it culminated in the 1973 embargo.

That sounds like ancient history today, but we do not need to go back so far in time to see the high costs of government interventions aimed at isolating American energy from international markets. We have an excellent case study from recent years: the ban on the export of crude oil from the lower forty-eight states of the United States that was in place until December 2015. This ban was a relic of times past, when real and perceived supply shortages in the 1970s threatened America's sense of well-being. In 1975, when Congress instituted the ban, the notion of sending precious and limited American crude abroad seemed antithetical to U.S. interests. There seemed to be no practical reason for any U.S. administration to lift the ban over the next forty years; as imports doubled, then tripled, after Eisenhower's restriction on imports was removed, exporting U.S. crude seemed inconceivable. So the export ban persisted, virtually forgotten, from its inception in 1975 until the 2010s.

Burgeoning U.S. tight oil production led to a reexamination of this policy, and Congress agreed to remove it in 2015. However, the extensive public debate that occurred before Congress made its move generated a thorough examination of the significant costs associated with separating the United States from global oil markets. These costs brought together bedfellows as unlikely as Republican senator and Tea Party stalwart Ted Cruz and President Obama's former economic advisor Larry Summers. They—and many analysts and institutions in between—stressed the costs to the U.S. economy of continuing the ban. If left to the market, they anticipated that some of the crude oil produced from U.S. tight oil fields would more naturally be shipped abroad than be refined at home, given differences in the quality of crude and refining capacities. Instead, because there were no alternative markets for their oil, the delta between the price U.S. producers could get for their crude and a common global price was as much as $15 per barrel. Producers, therefore, had less incentive to invest. American consumers ended up paying more at the pump because gasoline prices correspond to global prices, which would have been lower if U.S. crude were flowing to international markets and thereby increasing global supply. John Hess, CEO of Hess Corporation, and others fervently made the case that the export ban—by threatening jobs and investment in the oil industry—created risks for the larger U.S. economy. Scholars and institutes from across the political spectrum— from the liberal-leaning Brookings Institution to the conservative Heritage Foundation—advocated lifting the ban, citing the significant gains to be reaped by American consumers in the form of lower energy prices, more jobs, higher wages, and increased economic growth.

If becoming energy self-sufficient or autarkic is too costly and inefficient, then perhaps it is preferable to strive for the second form of energy independence often discussed: a situation in which the United States simply produces more energy than it consumes. The United States may well still import and export some oil, but the key in this scenario is that the country would be a "net exporter" of energy. Certainly, becoming a net energy exporter is a more feasible and achievable goal. If one standardizes all energy sources and counts them simply in terms of their energy content (in British thermal units, or btus), America could become a net energy exporter in less than a decade according to the U.S. EIA, depending on the assumptions one makes about prices, economic growth, and resource development. Demand for nuclear power, coal, and renewable energy would continue to be met largely by domestic sources, as they are today. The United States would likely export significant volumes of natural gas, which would cancel out the continued import of crude oil.

In addition to being a net energy exporter, the United States could also be a "net oil exporter." To claim this status, the United States would need to export more oil than it imports. But unlike in the autarkic scenario, the United States would be both an importer and exporter of oil, with the market and existing infrastructure and refineries to a large extent determining volumes and directions. Various companies and agencies differ as to whether the United States is likely to become a net exporter of oil, depending on their views about the resource base, coming efficiency measures, and other factors. The more "optimistic" scenarios do not see the United States reaching this milestone until the 2030s. Others see America as coming close, but not clinching the net exporter status.

If markets alone cannot deliver net oil exporter status, the U.S. government could certainly help. Allowing the market to determine what is exported and imported would reduce the inefficiencies associated with the autarkic option, so the policies needed to achieve this net exporter status most likely would not entail the potentially large costs and inefficiencies associated with the pursuit of self-sufficiency.

A reasonable debate might ensue about whether such policies are worth their costs—if in fact net exporter status were a meaningful measure of energy security. However, despite the raw appeal of exporting more than one imports, net exporter status would not protect the United States from either the supply shocks or the price spikes that make energy independence attractive in the first place. The United States, as both an exporter and importer, would still be tied to the global market and to the global price set for oil. As a result, any disturbance to that market would still reverberate throughout the U.S. economy. If Iranian production, say, were drastically curtailed for any reason, the global price would rise, boosting the price Americans pay to fill their cars with gasoline—even though the United States does not import a drop of Iranian oil. Although there may be other reasons to pursue the government policies mentioned above, the goal of becoming a net exporter of energy is not one of them.

Alas, the appeal of energy independence does not survive scrutiny in the harsh light of day. While the benefits of energy self-sufficiency may be considerable, the prescriptions for getting there are more toxic than the affliction of energy dependence itself. In contrast, being a net energy or oil exporter may be less costly to achieve, but its value is much more limited than it might at first seem. In the words of a March 14, 1909, *New York Times* editorial on a totally different subject, "Cheapness without usefulness is not a bargain."

Fortunately for those fixated on energy independence, there is a third variety that comes in the continental form. Heralding "North American energy independence" has become more fashionable as the downsides and limited benefits of a more narrow U.S. energy independence have been exposed. This broader concept envisions the United States, Canada, and Mexico collectively meeting all their energy needs from resources on their shared continent. It would still require the United States to be dependent on other countries, but most Americans would be comfortable that neither Canada nor Mexico is likely to undergo change so revolutionary that it could disrupt production. Nor is either country likely to use energy to blackmail the United States. Moreover, the three economies are already so closely integrated that for any one country to use energy to extort or inflict costs on another would be in some measure an act of suicide. Unlike some other forms of U.S. energy independence, North American energy independence could well be within reach. Its prospects depend not only on U.S. energy fortunes, but also those in Mexico and Canada—two countries that have also seen many changes in recent years.

A Historic Set of Reforms

Enrique Peña Nieto, a young politician with movie star charisma and a record as governor of Mexico's most populous and politically important state, became Mexico's fifty-seventh president in 2012. At the time, he faced multiple challenges, any number of which could easily tank his otherwise promising six years in office. High on his agenda was reforming Mexico's energy sector. In its heyday as recently as 2004, Mexico was the fifth largest oil producer in the world, producing nearly 3.5 million barrels of oil per day—roughly the same amount Iran was producing before intensive international sanctions were put in place against it in 2012. Cantarell, a shallow offshore field named for the fisherman who in 1976 reported an oil slick off the Yucatán Peninsula that led to the field's discovery, was once the second largest producing oil field in the world after Ghawar in Saudi Arabia.

For decades, oil had been the undisputed powerhouse of the Mexican economy. Yet, President Peña Nieto inherited a limping energy sector. The production of Cantarell had begun a precipitous decline in 2005. Despite Mexico's large oil reserves and the $70 billion the country plowed into investment in exploration and production between 2008 and 2012, technical shortfalls and incompetence kept Mexico from bringing enough new production online to keep its daily production numbers afloat. Petróleos Mexicanos, or PEMEX, the country's national oil company, was amassing major losses, and was mired in corruption and inefficiencies so dramatic that in 2012 it produced just twenty-five barrels of oil for every person it employed; by contrast Exxon produced fifty-five per person, while Norway's oil company, Statoil, churned out nearly eighty. Over the course of just five short years, Mexico's daily production of crude oil and lease condensates dropped by a quarter. A year before President Peña Nieto took office, a study by the Baker Institute at Rice University warned that Mexico could become a net oil importer within a decade. Galloping domestic use of oil and natural gas translated into waning oil exports and, in the case of natural gas, growing imports despite the country's own resource wealth. Notwithstanding its significant natural gas deposits, Mexico began to import expensive LNG from Nigeria, Qatar, and Egypt in 2006.

Efforts to reform Mexico's energy sector had been tried before despite the enormous associated political challenges. Mexico was the first country in the world, in 1938, to nationalize its oil industry. State control over oil resources remains so central to the identity of the country that Mexico celebrates the act of nationalization each year on March 18 with a civic holiday called Aniversario de la Expropriación Petrolera. Felipe Calderón, Peña Nieto's predecessor, struggled unsuccessfully against history and tradition to produce meaningful reform.

Peña Nieto was dealt a different and more promising hand. He was able to bring Mexico's largest political parties together to support wide-ranging political and economic reforms through what was known as Pacto por México. And the boom in unconventional energy sources in the United States gave the country more reasons to push hard toward reform. According to initial surveys, Mexico could also claim huge deposits of both unconventional oil and gas. Reform now made sense not only to avoid the threat of becoming a net importer of energy, but also as a way of capturing a new opportunity, one with a model ready for emulation, across the border both on land and in the Gulf of Mexico. If Mexico wanted to attract the foreign investment it needed to shift its energy fortunes and capitalize on its newly discovered unconventional wealth, it needed to offer international oil companies terms competitive with those they could find just to the north.

Reform was essential and ambitious reform was the only kind worth doing. Peña Nieto and his historic political coalition captured this moment. They introduced reforms to the oil, gas, and electricity sectors in the form of constitutional amendments, including language allowing for foreign investment in Mexico's resources. Instead of the vague language many skeptics anticipated, the amended constitution was now explicit in welcoming a range of contract models, including the types that international oil companies have generally found most attractive. The reforms ended the long-time state monopoly PEMEX had maintained as the sole custodian of Mexico's oil and gas reserves as well as its related infrastructure and induced meaningful competition into the sector for the first time. The electricity sector was also reformed, making way for greater private participation there and for the easing of electricity prices, which were on average 73 percent higher in 2012 than those in the United States. In August 2014, Mexico's Chamber of Deputies passed the necessary supporting legislation in a marathon seventy-three-hour session, the longest in the history of the body and the inspiration for any number of bizarre protests. Neither the discovery of a snake, which had been inadvertently smuggled into the chamber in a display of flowers for a "mock" funeral for PEMEX, nor the spectacle of opposition legislator Antonio García Conejo stripping down to his black underwear briefs in protest, was able to derail the proceedings. The implementing legislation that followed was passed with astonishing speed.

The realization of North American energy independence is to some degree dependent on the success of these reforms going forward. Their passage significantly enhanced prospects for Mexico eventually reversing its production declines and recapturing its position as a significant oil—and possibly gas—producer. Despite the misfortune of coinciding with the global collapse in oil prices, Mexico's first bid rounds have been reasonably successful in attracting the largest and most experienced oil companies to invest in Mexico. ExxonMobil, Statoil, Chevron, BHP Billinton, and the Chinese National Offshore Oil Corporation have all committed to exploring and developing Mexico's resources. While some of the most optimistic expectations about how quickly the reforms can transform Mexico's energy sector have been dampened, analysts and investors still anticipate a significant turnaround in the coming years.

Companies and agencies significantly revised their previous expectations for Mexican oil production since the reforms passed. The U.S. EIA suggested that Mexico could stabilize its oil production by the end of the decade and, in the decades that follow, increase it by a third to reach 2008 levels by 2040. Mexican natural gas production is also expected to increase significantly, although Mexico's immediate focus has been more on importing cheaper U.S. natural gas, rather than on developing its own. As is always the case, the magnitude of these improvements depends to some extent on global prices. Yet, even with such uncertainties, Mexico is poised to stem the downward trajectory of its energy industry and recapture at least some of its earlier glories. Mexico's success will mean American success, particularly for those who see real value in securing North American energy independence.

A Lemonade Stand with One Customer?

Canada's energy fortunes also play into whether a North American energy independence will be achieved. While Mexico's energy reforms captured headlines, Canada's energy fortunes have also been in flux—bolstered and buffeted by the unconventional boom. For Canada, "unconventional resources" refer less to those extracted by fracking and more to the vast oil sands—or bitumen—exploited through very different extraction processes. (Remember that the term "unconventional" does not refer to the specific types of molecules in the resource, but the method by which they are extracted.)

Canada's unconventional boom well preceded that of the United States. Between 1980 and 2014, Canada's oil output more than doubled, with virtually all the new production coming from the oil sands and a significant share coming online since 2000. Calgary, the capital of Alberta—the province where virtually all the oil sands are located—boomed. Its population grew by more than a third just between 2000 and 2014. Dating advice columns urged single Canadians to "go west" to Alberta, the only place in the country where men outnumbered women under the age of sixty. Demand for housing skyrocketed, forcing up real estate prices by 50 percent in 2006 alone. Calgary's annual Stampede festival continued to grow in scale and extravagance, with a record 1.4 million people attending the ten days of rodeos, parades, concerts, and chuck wagon racing in 2012.

The proven ability to extract the resources locked in its oil sands not only boosted Canada's overall output, but also catapulted the country into the top three holders of oil reserves in the world. In 2003, almost without fanfare, the annual rankings of global oil reserves listed Canadian reserves at 180 billion barrels—up from only 5 billion just a year earlier. A Canadian trade group had convinced a ranking agency to include Canada's oil sands in its reserves number for the first time, and, literally overnight, Canada went from ranking twenty-second in the world for its oil reserves to being third behind Saudi Arabia and Venezuela. Several years later, after much lobbying from the industry, the American Securities and Exchange Commission also revised its definition of reserves of oil and gas to include unconventional resources.

Yet Canada's unconventional boom has had its own particular vulnerabilities. As long as the price of oil stayed over $80 or $90 a barrel, Canadian oil sands production looked set for robust growth, which would more than make up for the continued decline in Canada's conventional oil production. But the oil price plunge beginning in 2014, fed by the supply glut brought on by U.S. tight oil production, put Alberta—and its resource-dependent economy—on an austerity diet. Given the major investments and long time horizons required to develop oil sands, Canadian production will likely continue to climb for the medium term even in the face of low prices, but investments in future production have been curtailed as investors see less of a rationale for developing relatively expensive resources with energy prices not expected to return to earlier highs. As of June 2016, despite a downward revision, the Canadian Association of Petroleum Producers was still expecting positive growth in Canadian oil production out to 2030, but at lower levels than anticipated.

America's shale gas bonanza has, at least in the short and medium terms, also meant more hardship than opportunity for Canadian energy. As of 2017, the United States remained virtually the only foreign customer of Canadian oil and gas. Nearly three-quarters of Canada's oil production and almost half of its natural gas production flow to its southern neighbor; the rest is consumed domestically. As a result, Canada's only export market for its natural gas rapidly dried up as shale gas began to flow in the United States and America began to meet more and more of its own needs. Having no other outlet for its gas, and a small domestic economy unable to absorb the excess, companies in Canada were forced to rein in production, which dropped by a fifth from 2005 to 2013. Although projections for the future suggest burgeoning Canadian natural gas production and export, realizing these visions will require capturing new markets and building new infrastructure to deliver the gas that the United States no longer craves. As of 2016, Canada had no capacity to export LNG beyond trucking small quantities of it across the border to New England, despite multiple active plans and proposals for the construction of pipelines that would make LNG exports possible.

Despite these uncertainties, North American energy independence among the United States, Canada, and Mexico is feasible, even without costly policy interventions targeted to this goal. North America as a whole is on track to become a net exporter of energy—including oil and gas—by 2020. This would be not just at an aggregate level in terms of btus, but in each type of energy source. Indeed, the United States, Canada, and Mexico had the ability to collectively meet all their natural gas needs in 2015 and could at least theoretically collectively meet their demands for petroleum and other liquids by 2020. Mexico's growing need to look beyond its borders for natural gas will be more than met by increased U.S. production. Both Canada and Mexico produce more petroleum and other liquids than they consume, with the excess more than meeting the future remaining thirst of the United States for imported oil beyond its own resources.

Enter the Politicians

American president after president has declared the need to pursue energy independence. Yet, one might be surprised that recent leaders have frequently adopted or advocated for policies that work against the most feasible and attractive version of this concept—North American energy independence. Rather than prioritizing the ability of the continent to collectively meet all of its energy needs, other objectives have gained hold in Washington that seem to override the prospect of attaining North American energy self-sufficiency anytime soon.

Figure 4.3: North American Net Imports of Petroleum and Other Liquids (million barrels per day)

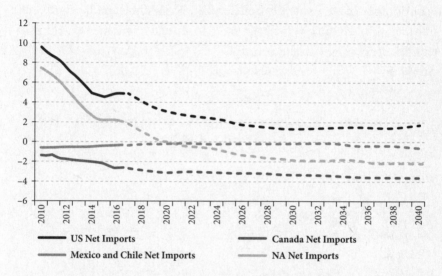

Note: The U.S. EIA reports Mexico and Chile together; Chile's numbers are, however, stable and small; they do not distract greatly from the bigger picture above. Solid lines are actual, whereas dotted lines are projections.

Source: Derived from U.S. Energy Information Administration, *Annual Energy Outlook 2017*.

One recent indication of this dynamic was the rejection by President Obama of a permit application to build Keystone XL—a pipeline intended to deliver 730,000 barrels of Albertan oil sands to U.S. refineries on the coast of the Gulf of Mexico, either to be consumed in the United States or exported. After a laborious, multiyear process of evaluation, President Obama announced the decision to reject proposals for this pipeline in part based on the need for more robust action to combat climate change. The argument was *not* that Keystone XL would significantly add to climate change problems; multiple State Department assessments of the pipeline proposal found that the new infrastructure was unlikely to alter global emissions because a U.S. rejection of it would not prevent the development of the oil sands. Instead, President Obama's statement rejecting the pipeline focused on how it would damage America's ability to lead on the issue of climate change. This did not have to be the case. President Obama might have taken the approach that Canadian Prime Minister Justin Trudeau pursued a couple of years later; he could have approved the pipeline and coupled it with steps to support renewable energy or other climate friendly policies. Yet, President Obama's words were unqualified: "America is now a global leader when it comes to taking serious action to fight climate change. And frankly, approving this project would have undercut that global leadership."

Figure 4.4: North American Net Imports of Natural Gas (trillion cubic feet per year)

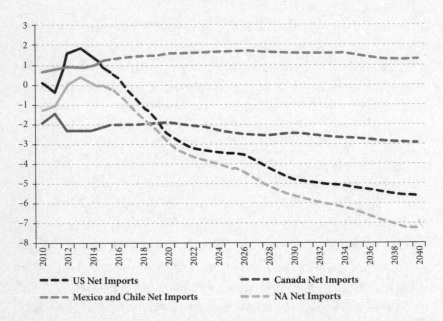

Note: The U.S. EIA reports Mexico and Chile together; Chile's numbers are, however, stable and small; they do not distract greatly from the bigger picture above. Solid lines are actual, whereas dotted lines are projections.

Source: Derived from U.S. Energy Information Administration, *Annual Energy Outlook 2017*.

Soon after taking office, President Trump reversed this decision and revived prospects for the Keystone XL Pipeline. It is, however, harder to undo the lesson that Canadians took from the multiyear episode before its permit was first rejected—and the setback to North American energy independence that followed. Decisions and delays made in Washington convinced the Canadian government that it cannot be energy secure as long as the United States remains Canada's only market for its energy exports. Previous to the experience with Keystone XL and America's own unconventional boom, "Canadians just took it for granted that the U.S. could take all the oil and gas that Canada could ever produce," said Preston Manning, once a political ally of former Canadian prime minister Stephen Harper.

But after a tense 2011 phone call in which President Obama informed Harper of another delay in Keystone's approval, and a subsequent heated exchange between the two men at a picnic table on the margins of the Asia-Pacific Economic Cooperation summit that same week, Harper reportedly ordered his cabinet to begin the exploration of alternative export routes so that Canada could sell its energy to non-U.S. customers. Neither Harper nor his cabinet members were shy about their plans to court Asian consumers for their oil. As former U.S. ambassador to Canada Gordon Giffin explained to me, "There was a patriotic response that Canadians were not going to let Americans decide whether or not they can move their oil."

Bringing Canadian oil and gas to "tidewater" for export beyond the continent has proven more difficult to achieve than anticipated, particularly in the low-energy-price environment. Since that cabinet meeting in 2011, Canada has run into its own domestic difficulties in moving its resources east and west for export. Numerous pipeline proposals have faced challenges from environmental groups and First Nation communities. The low oil price environment has made it harder to accommodate the demands of these groups while still keeping the projects commercial.

Yet, even with these obstacles—and even with the reversal on Keystone XL—Canadian efforts to develop the capacity to export their energy to other markets continue and will eventually succeed. Bringing Alberta's oil and gas to the Canadian tidewater would free the country's energy exports from the whims of U.S. markets and American politics.

At that point, reaching North American energy self-sufficiency would still not be impossible, but would be harder to achieve given that Canada would be sending its oil exports to many countries, not just the United States. Yes, Canada might in the future have sufficient excess production capacity to meet both U.S. demands and a portion of Asian ones, but it is unlikely to forget the lesson of Keystone quickly. Diversifying its energy markets, rather than doubling down on American ones, will remain a political priority. Meanwhile, without the full participation and cooperation of Canada, North American energy self-sufficiency would become harder to attain.

Similarly, for an administration so vocally committed to achieving energy independence, many of the Trump Administration's early positions could actually work in the opposite direction, by undermining the prospects for North American energy independence. In particular, many of the policies articulated toward Mexico could jeopardize the progress that country is making in reviving its oil sector—thereby making it harder for the continent to meet all its own energy needs.

Most worryingly, the populism and nationalism of the Trump era has spurred a reaction in Mexico's own domestic politics, empowering political candidates opposed to the historical energy reforms. Andrés Manuel López Obrador—known by the moniker AMLO—is Mexico's own combative populist. He is a lifetime politician who first came to national prominence by organizing protests against environmental damage caused by PEMEX and exposing electoral fraud of the long-ruling Mexican party, the Institutional Revolutionary Party or PRI. He lost the presidential elections narrowly in 2006 and again in 2012. But the rhetoric of Donald Trump as candidate and president has bolstered López Obrador's popularity as Mexicans search for a figure who can be a forceful leader for Mexico in potentially adversarial times. As of early 2017, Mexican polls showed AMLO as the individual most favored to win the presidential elections of 2018. If López Obrador takes the head office, Mexico's energy reforms could stall or at least lose momentum. Not only has López Obrador long been an opponent of these reforms, but he has made clear his intentions to undo them if given the opportunity. In 2013, shortly before the reforms were finalized, López Obrador sent a letter to then–ExxonMobil CEO Rex Tillerson—as well as nine other CEOs of international oil companies—telling him that investing in Mexico "would be like buying goods without a receipt, something crooked, tantamount to piracy." As the IEA assessed in a 2016 study of Mexico's energy sector, the Mexican economy stands to gain more than $1 trillion as a result of the reforms; in contrast, the "no reform case" explored by the IEA creates a major drag on the Mexican economy as a whole, which would have many implications for the United States, including significantly less demand for its natural gas exports to Mexico.

Even should López Obrador's presidential ambitions once again be thwarted, the nationalism of President Trump's "America First" approach has made many in Mexico—as was the case in Canada—think twice about complete dependence on the United States for critical natural gas imports. Lourdes Melgar, a soft spoken but razor-sharp former Mexican deputy secretary of energy, told me in 2017 that, "People used to think I was crazy when I warned against complete reliance on the United States for the natural gas that is a critical fuel for generating Mexico's electricity. But, today, many share my view that Mexico must have other options than American gas, at least as an insurance policy." With almost two-thirds of all U.S. natural gas exports flowing to Mexico in 2015, and Mexico's demand for natural gas climbing at a steep rate, the United States can only lose from a growing national sentiment that Mexico should minimize dependence on its northern neighbor.

Finally, the fate of the North American Free Trade Agreement (NAFTA) has major implications for American energy companies working in Mexico and—most relevant to the question of North American energy independence—Mexico's ability to reach new production levels. Throughout the 2016 U.S. presidential campaign and the early months of the Trump administration, President Trump spoke passionately about the need to renegotiate NAFTA, calling it "the single worst trade deal ever approved in this country" in the first presidential debate; in other exchanges, he made clear his willingness to withdraw from the trade pact if necessary. Abandoning NAFTA could crimp Mexican energy development, particularly by American companies, in at least two ways. First, it would remove some of the efficiencies that make it so easy for U.S. companies to operate in Mexico. Thanks to streamlining of customs and approvals under NAFTA, companies can move equipment and people

across the border almost seamlessly. A senior industry official based in
Mexico explained this to me vividly. "If I need a highly specialized piece
of equipment from the United States, I can have it anywhere in Mexico in
36 hours," he said. "If I were seeking to use that equipment in Colombia,
it would take more like 36 days, and in Argentina, closer to three and a
half months."

Second, and of even greater concern, are the investment protections
under chapter eleven of NAFTA that American companies could lose.
The mechanisms established in this chapter of the trade pact are intended
to ensure that all parties have access to an impartial tribunal if there is
a dispute over an investment; investors claiming a host government has
violated the terms of their investments have recourse to one of several
international arbitration mechanisms. While some opponents of NAFTA
have objected to these provisions as a violation of sovereignty, they have
effectively provided companies investing in Mexico an alternative to addressing disputes under outdated and less friendly local laws. If NAFTA
is scrapped, companies would lose this important layer of protection,
again, making it less attractive to invest and operate in Mexico.

Now that some form of energy independence is at least feasible, the
American conversation about it needs to move beyond aspirational language. A close examination of the concept—and the steps likely needed
to achieve some variations of it—should dampen American fascination
with the idea. Energy independence is not the unadulterated good that
many Americans suppose. However, one form—North American energy
independence—is possible without unreasonable and costly contortions.
Yet, even there, its pursuit involves tradeoffs across other policy priorities. Politicians and policymakers who move beyond general exhorta-
tions of energy independence to understand its pros and cons will be
best positioned to make these tradeoffs. Some will match their rhetorical
love of energy independence with action, while others will not. Either
way, an analytical understanding of the concept needs to be part of the
calculation.

Another approach altogether may make more sense. Rather than
focusing on the narrow drive for energy independence, policymakers,

pundits, and voters could open their eyes to the real benefits of the energy boom and see the ways in which the new energy abundance is already bolstering American strategic and economic positions. With some notable exceptions, the new energy landscape has altered the international landscape to the advantage of the United States; this reality will become apparent in the third section of this book. But the new energy abundance has also, for the most part, augmented American sources of strength. The next two chapters reveal how this energy environment is reinforcing the foundations of American power, fortifying its hard power and buttressing its soft power.

FIVE

Hard Power Accelerator

W hen President Ronald Reagan entered office in 1981, the United States was already in a downward economic lurch. The real median family income had begun to decline in 1979, dropping more than 7 percent by 1982. Meanwhile, the poverty rate was creeping upward and would increase by a third between 1978 and 1983. Interest rates were soaring; the prime rate had peaked in 1980 at 21.5 percent. The stock market continued its anemic performance. Then, months after Reagan stepped into the Oval Office, the situation got worse. A severe recession began. Unemployment peaked at over 10 percent at the same time that inflation reached double digits.

Resuscitating the U.S. economy was Reagan's first priority. The new president saw a healthy economy as critical in two ways. The first was self-evident: a stronger economy would ease the hardships facing Americans at home. The second reason was external. Without a strong economy, the United States could not tackle looming challenges overseas. As Reagan recalled in his memoir, *An American Life*, "In 1981, no problem the country faced was more serious than the economic crisis—not even the need to modernize our armed forces—because without a recovery, we couldn't afford to do the things necessary to make the country strong again or make a serious effort to reduce the dangers of nuclear war. Nor could America regain confidence in itself and stand tall once again. Nothing was possible unless we made the economy sound again."

Reagan's words reflected the experience of great world powers over time: economic vitality is the absolute cornerstone of hard power. In the modern era, a nation cannot maintain its status as a great military power without a robust economic base. Great Britain, for example, rode to global prominence in the eighteenth and nineteenth centuries on the back of the Industrial Revolution. But by the mid-twentieth century, Great Britain's financial and economic weakness led directly to a declining role on the global stage—a connection that became painfully apparent in the 1956 Suez Crisis. The Soviet Union's economic woes in the 1980s were at the root of its demise and ultimate dissolution. Bernard Brodie, a military theorist developing the framework for nuclear deterrence in 1959, summed it up when he wrote "strategy wears a dollar sign."

A strong economy also underpins the willingness of the American people to expend treasure to address global problems. Times of economic weakness are frequently accompanied by calls to rein in American international involvement. The link between Americans' economic pain in the 1930s and the growing isolationism manifest in the Neutrality Acts of 1935, 1936, and 1937 is perhaps the most extreme, but not the only, example of this connection.

In recent years, a similar dynamic has emerged. Battling the worst recession since the Great Depression, American unemployment doubled from the end of 2007 until late 2009. On average, household consumer spending—a traditional source of U.S. economic growth—dropped by nearly 8 percent from 2007 to 2010. The situation was so dire that in October 2008, billionaire financier Warren Buffett described the country as facing "an economic Pearl Harbor." The climate in Washington in the early days of the recession was one of near panic. Hank Paulson, secretary of the treasury at the time of the Lehman Brothers bankruptcy, describes in his memoir how the anxieties of the time brought him to the verge of physical breakdown. The entire focus was on staving off a full economic collapse, well before attention could turn to resuscitating growth.

In the midst of this collapse, Americans elected a president who many thought would steer the country away from international military involvements to focus on domestic problems. In 2011, with the economic recovery still moving slowly, President Obama's call "to focus on nation building here at home" again underscored the urge to turn away from international involvement during times of economic distress. In a 2013 public opinion poll, close to half of Americans responding felt that the United States did too much to solve global problems; half of those expressing that opinion cited economic woes as the primary reason for their view.

An American Renaissance

Amidst the economic gloom and doom, one sector seemed to be thriving: the oil and gas industry. In fact, the infusions of new technologies were leading to what many called the "American energy renaissance." The oil and gas sector was boosting growth, employment, tax revenues, and domestic investment—all at a time when most economic indicators were still pointing downward. According to one study by the consultancy IHS, unconventional oil and gas production added almost 1 percent to GDP each year from 2008 to 2013, making it responsible for approximately 40 percent of all GDP growth during that period. A 2015 Harvard Business School/Boston Consulting Group report used a more inclusive methodology and calculated that oil and gas produced by fracking contributed $430 billion—or just about 2.5 percent of GDP—to the U.S. economy in 2014 alone. This amount translates into roughly $1,400 for each American in a single calendar year and is equal to more than half the entire stimulus package passed in 2009 to fuel investment in infrastructure, education, renewable energy, and health over the course of the following decade.

The burgeoning energy sector also meant significant increases in jobs at a time when economists were discussing a "jobless recovery" from recession. Estimates of the impact on job creation vary, depending on the benchmark used. Moody's Analytics, a consultancy focusing on global financial markets, calculated that more than a quarter of a million jobs were *directly* created by oil- and gas-related industries between 2006 and early 2015, with most stemming from the shale gas and tight oil sectors. Each of these directly created jobs was estimated to have spurred another

3.4 related jobs, making a total of over one million *new* jobs attributable to the boom. These new jobs were roughly equivalent to half the number of American manufacturing jobs lost from December 2007 to June 2009, the official length of the recession, according to the National Bureau of Economic Research. In contrast, the 2015 Harvard Business School/Boston Consulting Group report looked at jobs *supported* (rather than directly created) by the unconventional boom, and found this number to be 2.7 million in 2014.

This growth in GDP and in jobs includes gains from growing U.S. competitiveness spurred by the energy boom. In April 2017, ExxonMobil and Saudi Basic Industries Corp announced plans to invest $10 billion to create a massive plastics and petrochemical plant in Corpus Christi, Texas. The two companies describe the project as "a unique opportunity created by the abundance of low cost U.S. natural gas" and note that it will create thousands of temporary construction jobs and six hundred permanent ones. This announcement follows on the heels of other such investments. Big River Steel in Osceola, Arkansas, is another big project made possible by the energy boom. The new $1 billion steel mill broke ground in September 2014, close to the prolific Fayetteville shale formation. In early 2016, the mill began hiring 425 people, with salaries averaging $75,000 a year—more than twice the per capita income of Arkansas.

In other instances, low cost natural gas has helped motivate the return of factories that had been transferred overseas. For instance, in 2008, the U.S. company Dow Chemical shuttered American plants and moved production to the Middle East in order to be closer to cheap energy inputs. Several years later, Dow spent $4 billion to build new factories in Texas and to reopen old ones it had sealed shut in Louisiana. Such stories are indicative of a larger trend noted by President Obama when, in his January 2014 State of the Union address, he referenced nearly $100 billion of investments in "new factories that use natural gas." While some of this investment was domestic, much of it flowed in from abroad.

Of course, low energy prices do not provide a leg-up over international competitors in every manufacturing sector. The advantages are limited to industries where energy inputs are a substantial percentage of overall costs. But these sectors, which include chemicals, metals, and paper, constituted not quite a third of all U.S. manufacturing and 3.5 percent of the U.S. economy in 2016—which is greater than the size of the U.S. automotive industry.

Other economic benefits also accrued on account of the unconventional boom. Increased oil and gas production raised government revenues in 2014 alone by $111 billion, with roughly half going to the federal government and the other half to state and local coffers. For context, this increase in government revenues is slightly more than a fifth of the whole federal defense budget for the same year, or somewhat more than a tenth of the defense budget if one just considers the boost in federal revenues from the boom.

While North Dakota, Texas, Ohio, and other states with shale gas and tight oil reserves boomed visibly during the worst years of the recession, consumers nationwide enjoyed the benefits of low energy prices. The Brookings Institution found that, collectively, residential, commercial, industrial, and electric power customers saved a whopping $74 billion in 2013 thanks to the shale gas boom. Homeowners around the country saw residential gas prices drop by a fifth from 2006 to 2014. The average customer of Public Service Electric & Gas Company, New Jersey's largest utility company, saved even more. In part by purchasing its supply from the nearby Marcellus shale formation, the company was able to report in 2015 that it had cut residential gas bills virtually in half since 2009. One firm estimated that in 2014 the average American household pocketed between $425 and $725 a year thanks to lower energy costs attributable to the boom in shale gas production.

Such numbers are eye-popping, but they still underestimate the contribution of the boom to the U.S. economy. The benefits—to employment, revenues, and investment—need to be appreciated in the context of an otherwise fragile economic recovery. Imagine what the economy coming out of the recession might have looked like in the absence of the boom. Energy prices would have been significantly higher. As a result, the recession would have been deeper, and recovery from it likely longer and even more painful. In 2013, political unrest in the Middle East knocked unusual amounts of oil out of production in the wake of the Arab revolutions. Had American innovators and entrepreneurs *not* previously figured out how to extract tight oil at a reasonable cost, oil prices would have shot up. By one estimate, oil prices might have risen as high as $150 in 2013—rather than the $109 they actually were. Similarly, another study suggested that, had the surge in shale gas not materialized, U.S. natural gas prices would have risen to $7.07 per mmbtu in 2013, rather than resting at half that price.

Higher energy prices like these would have hit not just American but global growth hard. The International Monetary Fund (IMF) estimates that a 10 percent increase in oil prices can trim 0.2 percent to 0.3 percent off global growth. When the world was growing at 3.4 percent, or when growth in the Euro area was negative as was the case in 2013, such haircuts can be particularly damaging. In this hypothetical world without tight oil production, the U.S. trade deficit would have been bloated by rising costs of ever more expensive oil imports. One calculation suggests that the tight oil boom alone slashed the U.S. oil import bill from $360 billion to $215 billion in 2014. Estimates of the savings incurred from virtually phasing out natural gas imports are less precise, but energy guru Daniel Yergin places them in the realm of $100 billion a year.

The low prices of the oil and gas downturn beginning in late 2014 did throw a damp towel over this economic exuberance. Fortunately, the slowdown didn't start until 2015, when the overall American economy had regained much of its economic composure. As lower prices for oil took away the incentive to produce, towns across North Dakota, Texas, and other states began to shrink as quickly as they had sprouted up several years earlier. But the boom had already helped pull the U.S. economy through the worst of the recession.

While the slowdown in the U.S. oil industry was pronounced, it did not decimate the economic gains of earlier years. In fact, some economists looking years and even decades beyond the slowdown are quite sanguine about the unconventional energy boom, its contributions to the American economy and, therefore, America's hard power. Even in the context of somewhat lower oil prices, they still anticipate significant future economic gains from the unconventional boom. The Harvard Business School/Boston Consulting Group report published in June 2015, as oil prices hit six-year lows, projected that overall economic gains of the unconventional boom—apart from the benefit of lower prices but including potential oil and gas exports—could reach $590 billion annually by 2030. U.S. government revenues from the boom should continue to rise, possibly reaching more than $150 billion a year by 2030.

More Than Wine for Cloth

The energy boom provides the opportunity for even greater economic benefits should the United States and Mexico take advantage of it to grow more closely together, rather than succumb to pressure to build physical and other barriers. Already the two economies are much more closely integrated than most Americans might appreciate. Looking only at the volume of bilateral trade, in 2015, the United States and Mexico traded more than $1 million of goods and services every minute. In 2016, Mexico was the first or second largest export market for 28 of the 50 states; for the same number of states, it was the first, second, or third largest source of imports. But, even more important than the size of bilateral trade is the unique nature of the manufacturing relationship between the two countries. Rather than simply trading in final products, the United States and Mexico build goods together, utilizing complex supply chains that crisscross the border. Take this example from the Mexico Institute at the Woodrow Wilson Center. Mexican oil might be shipped to the United States where it would be refined and transformed into plastic in Louisiana. It could then be sent to the American Midwest, where that plastic could be used to create parts for the dashboard of a car. Those components could return to a factory in Mexico and be incorporated into the assembly of a car, which might then be sold to an American consumer north of the border. Such an example clarifies how 40 percent of the value of U.S. imports from Mexico actually originate in the United States—meaning that greater Mexican manufacturing exports also boost U.S. exports, jobs, and the American economy. Because of this almost unique trading relationship, an increase in productivity on one side of the border actually benefits the other side as well.

Given these complex realities, it is far too simplistic to just focus on the bilateral trade deficit that the United States has with Mexico as a litmus test of the health of the relationship. There is a risk that U.S. policymakers will do just that—either by withdrawing from NAFTA or seeking "remedies" that could jeopardize the nearly 5 million American jobs reliant on U.S.-Mexico trade. But should American policymakers assess the U.S.-Mexico trade relationship on broader grounds, they will see that there are even more possibilities for mutual benefit.

In particular, the energy sector offers more opportunities to integrate these two economies further and, hence, to enhance the competitiveness of both countries. This is in part because when NAFTA was negotiated in the early 1990s, Mexico insisted that energy be left out of the accord given the high political sensitivities among its own public. In addition, the recent energy boom creates possibilities for new synergies. Together, the United States and Mexico could foster a "manufacturing renaissance"— a much broader, more sustained association that would have even greater economic impact. To this marriage, the United States could contribute cheap and abundant energy, particularly natural gas. Mexico could bring to the table some of the lowest labor costs in the world. By some assessments, Mexican manufacturing labor costs are not only on par with those of China, but are lower than most other countries in the world apart from India, Indonesia, and the Philippines.

Progress has already been made in this direction. More than twenty natural gas pipelines already crisscross the border, moving molecules south from the United States. Since NAFTA was signed in 1993, U.S. natural gas exports to Mexico have risen dramatically from almost zero. The volume of exports nearly doubled between 2009 and 2013, and then again from 2014 to 2016. Unless politicians take steps that cause investors to change plans, companies intend to spend more than $14 billion to expand Mexico's pipeline gas connections; this will help Mexico to nearly double its capacity to import U.S. natural gas over the next three years.

Further expanding access to the cheapest natural gas in the world, just across the border, will be a huge boon to the Mexican manufacturing sector as well as be good for the United States. In 2014, Mexican industries paid 80 percent more for electricity than U.S. ones did. As a result, high-energy-intensive industries in Mexico have done poorly when compared to low-energy-intensive ones. Companies of all varieties suffer shortages and spikes in electricity prices, hurting productivity and dragging down GDP. Princeton scholar Jorge Alvarez and IMF economist Fabián Valencia measured gains that would accrue to Mexico if, as they deemed plausible, Mexican electricity prices eventually converged with U.S. levels, on account of the Mexican energy reforms; increased imports of cheap U.S. natural gas would also contribute to such a price convergence dynamic. These scholars envision that this cheaper electricity would drive increases in the Mexican manufacturing sector by as much as 14 percent and expansion in GDP by as much as 2.2 percent. Again, given the synergies between the U.S. and Mexican economies, this boost for Mexico would not only mean more U.S. exports of natural gas, but it would also create wider gains for the American economy.

Oil as a Strategic Interest

On February 15, 2003, a giant puppet depicting President George W. Bush holding buckets of oil bobbed above crowds of freezing protesters marching through Manhattan. More than 100,000 people pressed against barriers outside the United Nations, all there to protest the looming Iraq War. This gathering, boisterous as it was, was relatively small compared to others occurring simultaneously around the world. In cities including London, Damascus, and Sydney, an estimated six to ten million protesters participated in hundreds of antiwar rallies. The Rome protest reportedly involved approximately three million objectors, winning it a place in the 2004 *Guinness Book of World Records* as the largest antiwar rally in history. Speaking to London's largest demonstration ever, then-mayor Ken Livingstone said, "This war is solely about oil. George Bush has never given a damn about human rights." From Switzerland to Taiwan, people took to the streets to chant "No Blood for Oil." A group of scientists stationed in Antarctica even protested at the edge of the ice on the Ross Sea. Academics writing about the date years later called it "the largest protest event in human history."

Although their efforts to prevent a U.S.-led invasion of Iraq failed, at least some of the millions of protest participants must now hope that the new energy abundance will discourage future "oil wars." Those who believe the United States invaded Iraq to control oil likely anticipate Washington will refrain from such wars now that its thirst for oil can be nearly sated by domestic or continental sources. These people, however, will be disappointed, once again.

To understand why, it is important to distinguish between *commercial* interests in oil and *strategic* interests in oil. Commercial interests might best be understood as seeking to physically control the production or sale of oil, while strategic ones pertain more to ensuring that such resources are not controlled by one's adversaries. The Axis powers of World War II did perceive commercial gains or physical control over oil resources to be the key to their energy security. They aimed to expand the areas over which they had outright control and from where vital resources could be directly extracted. Hilter's generals once presented him with a cake decorated with a Baku oil rig, reflecting his earlier comment that "unless we get Baku's oil, the war is lost." Japan was equally fixated on the oil-producing Dutch East Indies—and was more successful in gaining control of them than Hitler was in getting control of Baku.

In contrast, when it comes to the resources of other countries, the United States has generally taken a different approach, prioritizing *access* to the energy produced, rather than direct control of those resources. U.S. military power has rarely, if ever, been used to advance commercial gains by securing the physical control of oil. Both Iraq wars actually support this premise, rather than refute it. Had the United States sought actual possession of oil in the 1991 Gulf War, it could have refused to relinquish control over some of the most oil-soaked territory in the world that it had acquired in southern Iraq while pushing Iraqi forces out of Kuwait. Or it might have restored sovereignty to Kuwait only with provisions giving American companies preferential treatment in investment in or operation of that country's massive oil fields. Instead, the United States did neither. Foreign investment in the development of Kuwaiti oil fields is virtually prohibited even to this day.

In the wake of the second Iraq War, the American-led coalition in charge of governing Iraq from 2003 to 2004 might have demanded favorable terms for U.S. companies looking to invest in post-Saddam Iraq. Serving in Baghdad throughout the occupation, I worked directly with the Iraqi Governing Council—an Iraqi body intended to serve as the counterpart to the coalition until sovereignty was restored. Although the coalition proposed and prepared many draft ordinances for the council to debate, sign, and promulgate, we intentionally steered clear of any directives relating to the disposition of Iraqi oil. Rather, we determined that the fate of Iraq's oil could only be decided, later, by an elected

Iraqi government. Even more compelling, had the United States really sought commercial benefit for its oil companies above all else, it would not have needed to go to war. In the early 2000s, shortly before the invasion, Saddam Hussein was happily trading sweetheart oil deals with Russia, France, and China in exchange for political support for the lifting of U.N. sanctions on Iraq. The United States—the main holdout on the sanctions—could have easily struck such a lucrative deal, thereby getting control of Iraqi oil without sending a single soldier to that country.

While commercial gains surrounding oil have not been a major motivator of U.S. military action, strategic ones have. When Iraq invaded Kuwait in 1990, Baghdad gained control over one-fifth of the world's proved oil reserves. A large part of the rationale for subsequently going to war against Saddam Hussein was to reverse the leverage that a ruthless, unpredictable dictator had by holding such a huge percentage of the world's oil. There was also the additional risk that Iraq would continue its march into neighboring Saudi Arabia. If it had, such an invasion would have given Iraq's regime sway over more than 40 percent of the world's oil reserves at the time. In the 2003 war, oil figured less prominently at a strategic level, although policymakers were concerned about Saddam's ability to translate his country's oil wealth into the pursuit of weapons of mass destruction and other nefarious activities.

The United States will continue to have—and be motivated by—such strategic interests related to oil, even if its own import needs have changed dramatically. Who controls the world's oil reserves will continue to be a legitimate U.S. concern. American tight oil production is not so great as to fundamentally change what is considered strategic amounts of oil. In 2015, U.S. tight oil accounted for less than 1 percent of the world's total proved oil reserves—the oil that can be commercially produced with today's technology. (The distribution of oil *production*, which often varies significantly from reserves, tells a similar story; in 2016, U.S. crude production from tight oil was 6 percent of global crude oil production.) Perhaps in the future, substantial unconventional oil production from China, Argentina, Algeria, Mexico, Canada, and Russia—in combination with U.S. production—could lessen the strategic significance of oil concentrated in any one part of the world. However, for the foreseeable future, due to its vast reserves, low costs of production, and collective output, the Middle East will remain the strategic heart of global oil. In fact, as we will explore in Chapter Eleven on the Middle East, a persistent low-price environment could actually lead to an even greater global reliance on oil from that part of the world.

Reducing Military Constraints

Rather than sparing the U.S. military from huge military missions, the new energy abundance could unburden it in lesser and more subtle ways. For instance, in a surprising fashion, new energy realities could create the circumstances for greater burden sharing of efforts intended to stabilize the Middle East. In the United States, America's new energy prowess may create political pressures for it to do less in that part of the world (despite continued U.S. interests there), making the United States a less reliable guarantor of regional stability. At the same time, China will become more dependent on Middle Eastern oil and, therefore, further invested in the region's stability. The combination of these two factors could lead to greater U.S.-Sino cooperation in the region, as explored in Chapter Ten.

More immediately, the new energy abundance diminishes the cost and reduces the difficulty of procuring the energy that, in the words of General David Petraeus, "is the lifeblood of our warfighting capabilities." The Pentagon is the largest single consumer of oil in the world; as of 2013, its operations and the maintenance of its facilities consume more energy on a daily basis than all but forty-four countries in the world. The Pentagon's oil intake constitutes more than 90 percent of all oil used by the federal U.S. government. In 2007, when American forces were in both Iraq and Afghanistan, the Pentagon gulped down a third more fuel than Israel consumed, almost twice as much as Ireland, and twenty times more than Iceland. In 2013, the Pentagon consumed virtually the same amount of oil on a daily basis as Nigeria—a country of 160 million people with the second largest economy in Africa at the time.

The Pentagon's military operations are particularly oil intensive. Some of this is attributable to inefficient, if tried-and-true, equipment. For instance, the Arleigh Burke class destroyer, a backbone of the navy's fleet, travels at 30 knots (roughly 35 miles an hour) and burns about 1,000 gallons of fuel an hour. The changed nature of warfare has also upped the

Pentagon's dependence on oil. During World War II, on average, the military consumed about 1.7 gallons of fuel a day per soldier. In Iraq, where heavy equipment was transported to remote areas and efforts were made to cool both soldiers and gear, sixteen times that amount was used per soldier in the field.

In the late 2000s, the Pentagon changed how it calculates the cost of fuel. Whereas previously it had only considered market charges, it now assesses what is known as "the fully burdened cost of fuel." This more expansive assessment takes into account not only open market prices, but the full costs of transporting and securing the fuel. In some of the most remote and rugged terrain in the world, the cost of getting one gallon of fuel to soldiers "deep within a battlespace" could cost hundreds of dollars, according to the Defense Science Board.

Given its voracious appetite for fuel, the price of oil matters to Pentagon planners. In the fiscal year of 2013, the Pentagon consumed almost 103 million barrels of petroleum products, at a total cost of more than $15 billion. Two years later, the costs had gone down significantly to less than $9 billion, partially as a result of lower consumption but more due to the drop in price. The $6 billion saved equaled more than 1 percent of U.S. defense spending.

Over the medium or long run, a consistently lower energy bill could help the Pentagon put more resources into productive endeavors, such as research and development, the readiness of the forces, or modernizing equipment. As former Deputy Secretary of Defense Bill Lynn noted upon the release of the Department of Defense's first *Operational Energy Strategy* in 2011, "A dollar spent on increased energy costs is a dollar not spent on other warfighting priorities." Already, lower fuel costs have translated into small gains, such as increased training time. As Senator Jack Reed commented in 2015 when he was the ranking member of the Senate Armed Services Committee, "One of the benefits of the lower gasoline prices, for as long as they last, is they [military planners] can be buying fuel so they can be allowing pilots to fly more hours."

The real boon to U.S. military operations, however, lies in the future. It will come from energy-efficient technologies that greatly reduce the demand for oil, or be realized when some other form of energy is found or devised that can replace oil-derived fuels on the battlefield. On June 7,

2011, General David Petraeus, then commander of forces in Afghanistan, wrote a memo to all American forces underscoring how the voracious appetite of the military for fuel creates real risks and opportunity costs for soldiers. In calling on his commanders to "take ownership" of the energy demands of their units, he stressed that "a force that makes better use of fuel will have increased agility, improved resilience against disruption, and more capacity for engaging Afghan partners."

Some military studies were less diplomatic in their calls for conservation, explicitly linking casualties with operational fuel demands. With nearly half of all convoys in Iraq and Afghanistan devoted to hauling fuel, and a significant percentage of coalition casualties occurring on convoy duty, the math is stark. One study by the army concluded that there was one casualty for every twenty-eight fuel supply convoys in Afghanistan and one for every thirty-eight such convoys in Iraq. Given the number of fuel convoys in the two combat theaters, such calculations suggest that there were 170 American casualties attributable to securing fuel convoys in 2007 alone. Numbers in the same report indicate there was one U.S. casualty for every 55,702 gallons of fuel consumed in Afghanistan. A second study concluded that a 1 percent decrease in fuel consumption could keep 6,442 troops off convoy duty. Such statistics put in context the plea of then–Lt. General James Mattis to the Pentagon, after he commanded troops in Iraq, to "unleash us from the tether of fuel."

Honing the Coercive Tool of Sanctions

Early one morning in January of 2006, Stuart Levey was having breakfast in Bahrain. A lawyer by training, Levey had just recently been appointed to the newly created role of undersecretary for terrorism and financial intelligence at the U.S. Treasury Department. Scanning the local newspaper, his eyes gravitated toward an article about a large Swiss bank that had cut off business with Tehran. Levey clipped the article and tucked it into his papers.

For much of his career, Levey had helped the U.S. government construct new legal obstacles to prevent countries and companies from doing business with regimes America viewed as nefarious. Iran was at the top of the list, given its perceived relentless efforts to pursue a nuclear weapon outside its commitments to the Treaty on the Non-Proliferation of Nuclear Weapons. The article from breakfast got Levey thinking. Perhaps his focus on legal mechanisms was too strict? Maybe there was greater scope to bring in the private sector and to build on its natural aversion to conducting transactions with questionable entities? This seed of an idea eventually blossomed into a multiyear effort spanning both the Bush and Obama administrations to convince the private sector to eschew doing business with Iran. Levey's objective was to limit and ultimately terminate Iran's ability to do business outside its borders. "That could spark the right internal debate in Iran," Levey eagerly anticipated. With the backing of the highest echelons of government, Levey traveled to more than eighty countries in the first couple of years of his efforts. By 2008, he had successfully convinced scores of banks to curtail their business with Iran.

Levey was an important engine in the American—and subsequently international—effort to exert severe economic pressure on Iran in order to bring it to the negotiating table to discuss its nuclear pursuits. Central to Levey's initiatives—and to the broader quest to limit Iran's economic links with the world—was the passage of sanctions. U.S. and international sanctions on trade, investment, and financial transactions created enormous pressures on the Iranian economy, which shrank by 10 percent from March 2012 to March 2014. Iran's oil exports fell 60 percent from 2011, slicing Iran's revenues by more than two-thirds in 2013 and even more in 2014, as the falling oil price further diminished export earnings and exacerbated Iranian economic hardship. Tehran's financial difficulties were compounded by the fact that, due to American financial sanctions, most of the payments for Iran's dwindling oil sales could not be made in hard currency. Instead, China settled its accounts with Iran with goods, such as car parts. Iran effectively received Indian wheat, pharmaceuticals, rice, sugar, soybeans, and other products in exchange for much reduced volumes of oil. The rial, Iran's currency, lost half of its value in the two years between January 2012 and January 2014, forcing inflation up to well over 50 percent.

In July 2015, after an interim agreement in 2013 and years of negotiations, China, France, Russia, the United Kingdom, the United States, and Germany (a group known as the P5+1) and Iran agreed to a deal to limit Iran's nuclear program in exchange for the lifting of sanctions. This outcome is attributable to many factors, but the essential role of sanctions and economic pressure is undisputed. The economic hardship suffered by average Iranians permeated many walks of daily life, shaped the country's domestic politics, and created the conditions for negotiations. Hassan Rouhani, a relative moderate, ran in the 2013 presidential elections on the themes of "hope and prudence." He promised both to restore Iran's standing in the international community and to free the country from economic strictures. A landslide vote in his favor not only brought him to power, but also helped provide him with a mandate to pursue a deal surrounding the nuclear program.

Before sanctions proved so critical to securing a deal with Iran, many policymakers, businesspeople, and academics were quick to dismiss these tools as little more than "chicken soup diplomacy"—a tool to be used to make one feel better, without a real expectation of it having any effect. What enabled sanctions to go from a home remedy to the foreign policy tool clinching what many consider to be President Obama's greatest foreign policy achievement?

The new energy abundance lies at the heart of the answer to this question. While not the only factor to consider, the energy environment was critical in securing the necessary sanctions. Efforts by previous U.S. administrations to pressure Iran rarely were successful in securing the cooperation of others to impose sanctions. During the tenure of President William J. Clinton, the United States largely failed to convince other countries that the benefits of multilateral sanctions outweighed the potential costs. Efforts to corral others to impose sanctions in the 1990s had, in fact, led to diplomatic rows between the United States and its European allies, who resented U.S. efforts to use "secondary sanctions"—the imposition of sanctions on third parties, usually countries or companies doing business in the targeted country—where diplomatic entreaties had failed. Getting multilateral cooperation to limit Iran's oil sales in the mid-2000s was a Herculean feat. With global oil demand growth continuously outstripping that of global supply, prices edged upward, adding to the sense of resource scarcity and impending price spikes. Efforts to marginalize Iran, at the time the third largest *exporter* of oil, from global markets seemed nothing short of economic suicide. Some analysts even suggested that doing so would be counterproductive; decreasing supplies in a tight oil market would push up prices so dramatically that Iran might actually earn as much or more from selling a smaller volume of oil! As the perception of the threat posed by Iran grew, the administration of George W. Bush helped secure three sets of U.N. sanctions between 2006 and 2008, although none of them touched Iran's oil.

As Tom Donilon, former national security advisor to President Obama, remembers, the conversations with customers of Iranian oil were still difficult in 2010 and 2011 even after unease about Iran's nuclear pursuits had grown markedly. The late 2009 exposure of Fordo, a covert uranium enrichment facility near the ancient holy Iranian city of Qom, had largely ended earlier debates among Western countries about the intentions behind Iran's nuclear pursuits. Despite Iran's insistence that its program was purely peaceful, the country's actions strongly suggested otherwise. But oil markets were still tight, and the global oil price was still high. To make matters worse, the United States was in the midst of a slow and very fragile economic recovery, while the European Union was tipping back into recession. While Donilon and his national security team debated options for putting the squeeze on Iran's economic jugular, their economics counterparts in the administration continually reminded them of the high stakes for America's economy.

The boom in tight oil helped change this equation. Burgeoning U.S. tight oil production was critical in convincing both domestic and international skeptics that sanctions on Iran could be imposed and incrementally tightened—all without sparking an economically debilitating price spike. When U.S. officials such as Stuart Levey traveled the world, seeking to convince countries to curb their purchases of Iranian oil, they had two strong arguments to make. In Beijing, New Delhi, and Tokyo, these officials pointed to repeated public statements by Saudi oil minister Ali al-Naimi and other Saudi officials that Saudi Arabia was willing and able to up its production to cover any shortfall that could result from increased pressure on Iran. U.S. officials were also armed with statistics about U.S. tight oil production, whose growth had exceeded even the most optimistic projections each year since 2008. They could argue with confidence that the combined oil resources of the United States and Saudi Arabia could more than compensate for any Iranian oil that came off the market.

They even had an insurance policy. A provision in U.S. legislation gave President Obama the ability to ease up on sanctions if the global oil market seemed inadequately supplied. Sufficiently assured that greater pressure on Iran need not create an economic calamity, China, India, Japan, and South Korea all reined in their imports of Iranian oil over time. Europe instituted its own ban on Iranian oil imports in 2012. And, thanks to Saudi cooperation and continued American energy prowess, President Obama never had to rely on the legislative provision to curtail sanctions as a result of tightening oil markets.

While the Iranian example is the most dramatic and consequential, it is not the only case in which the unconventional boom has helped the United States gain international cooperation for sanctions. In the wake of Russia's annexation of Crimea, for example, the United States and its European allies sought to increase economic pressure on Moscow, both as a means of deterring further Russian incursions and of cajoling Russia to engage in talks to limit violence in eastern Ukraine. In order to maximize the impact of these sanctions, Washington sought Tokyo's cooperation, given the close financial ties between Japan and Russia.

Tokyo was initially unenthusiastic about enacting such measures. At the time, Russia was the fourth largest supplier of Japan's natural gas and fifth largest supplier of crude oil. The amount of natural gas Russia provided Japan had risen by 40 percent in the wake of the Fukushima disaster when, overnight, Tokyo sought to replace nuclear-generated electricity with that powered by natural gas. Multiple joint investments in Russian LNG projects such as Sakhalin-1 and Vladivostok in the Far East were intended to help ensure that Japan had a diversified supply of natural gas. The Japanese worried that, over time, sanctions could undermine Russia's ability to export energy, as the country would struggle to secure the financing required to build needed facilities. Prime Minister Abe also prided himself on a deep personal relationship with President Putin. The two had met five times between when Abe was sworn in during the final days of 2012 and the Sochi Olympics in February 2014.

Nevertheless, despite Tokyo's wariness, the Japanese government agreed to join the United States in imposing financial sanctions and an embargo on arms sales to Russia in 2014. The ongoing dispute between Japan and Russia over the Kuril Islands, a 1,300-kilometer volcanic island chain north of Japan, certainly factored into Tokyo's desire to see a clear message sent to the Kremlin about annexation of a neighbor's territory. But, also important was the expectation that the United States would soon be a significant exporter of natural gas. Knowledge of America's burgeoning energy revolution helped Japanese policymakers get comfortable with the idea of sanctions, given their confidence that a secure source of U.S. LNG would soon be flowing.

Sanctions have long been a popular tool of foreign policy, as policymakers often turn to them to express outrage or disapproval over an issue important enough to warrant action, but not egregious enough to justify military force. Popularity, however, has not always been synonymous with effectiveness. In their book, *Economic Sanctions Reconsidered*, economists Gary Hufbauer, Jeff Schott, Kim Elliott, and Barbara Oegg claim that sanctions work only 34 percent of the time. University of Chicago professor Robert Pape is tougher on the tool, claiming that sanctions are almost entirely ineffective—only 5 percent of sanctions have the desired effect. One reason for the relatively poor performance of sanctions is that they are frequently unilateral. In a globalized world, even an economic power like the United States does not have sufficient heft to isolate a country from the global economy on its own. It must have other countries and players, such as the U.N. or regional bodies, join it in its efforts. By assuaging concerns that third parties have about the deleterious effects of constraining energy supplies—either through restrictions on exports, investment, or technology—the new energy abundance has handed American (and other) policymakers a tool that has greater importance and effectiveness than it did in the last decade. While its rejuvenated use may pertain strictly to energy matters, a look at the roster of U.S. sanctions imposed since the end of the Cold War reveals clearly that the most robust sanctions efforts have been on energy producers.

Hard power will remain the cornerstone of American influence in global affairs. Although the U.S. GDP constitutes less of the global economy than it did decades ago, the American economy continues to be a vital source of strength for its own population and for its ability to endure hardship in the pursuit of other foreign policy and national security goals.

In providing the U.S. economy a vital boost during the slow economic recovery in the years following 2009, the energy boom played a critical role in restoring American economic health. It will continue to be an engine of growth, although not necessarily the outsized one it was in the period immediately following the Great Recession. Similarly, the energy boom has many positive implications for American use of military force. As long as the global economy is powered on oil, the United States will have strategic interests in parts of the world with huge oil endowments. But given the energy-intensive nature of combat, the new energy abundance does provide some meaningful relief to those engaged in or preparing for such missions. Finally, the new energy environment has resuscitated an important nonmilitary coercive tool in America's tool kit: economic sanctions. While the experience of Iran will not necessarily be replicable in other circumstances, the "end of peak oil" will continue to lessen international opposition to using these tools to achieve strategic ends.

SIX

Soft Powering Up

Writing in 1990, Joe Nye, a distinguished professor of international affairs at Harvard University, coined the term "soft power." He defined the concept, simply, as "the ability to get what you want through attraction rather than coercion or payments." According to Nye, the soft power of a country depends largely on three dimensions: the culture of the country, its political values, and how it generally conducts itself in the world and at home. To the extent that these three factors are attractive to others and seen to be consistent with its own actions, a country can expect to wield considerable influence abroad. This soft power comes in addition to, and often complements, more traditional notions of hard power: the military, the economy, and coercive elements of policy. The two, taken together, are what have made the United States such a meaningful force in global geopolitics. Just as the new energy abundance has given American hard power a boost, as will be explored in this chapter, it provided a leg up to U.S. soft power.

America has always relied heavily on soft power to influence others. In the decades since the end of World War II, the United States used its soft power to help build international institutions—such as the United Nations, the World Bank, the International Monetary Fund, and the World Trade Organization—that have provided the scaffolding for the global system. During the Cold War, the attractiveness of American ideas, values, and policies helped motivate those living under Soviet rule to oppose it and enabled the United States to sustain alliances in the face of decades-long pressure from Moscow. At other times in U.S. history, violations of American ideals have reverberated globally, diminishing U.S. standing and influence worldwide. Revelations of the humiliating ways in which U.S. prison guards were treating Iraqi inmates at the Abu Ghraib prison in 2004, for example, had repercussions well beyond Iraq's borders, suggesting to some that America was no longer the champion of human dignity. And, lately, America's intense partisanship at home has led many abroad—from Europe to Asia—to question the attractiveness of the U.S. political model.

The United States unquestionably still claims vast stores of soft power, from hosting nearly one million foreign students each year to its continued dominance of the global film industry to American-invented devices such as the iPhone and iPad. If anything, as military force becomes more difficult and less desirable to employ in response to global problems, the need for soft power to advance U.S. interests is only growing. In this context, the considerable boost that the unconventional energy boom has given to American soft power is particularly welcome. In a wide variety of ways, the attractiveness of American culture, values, and policies has been augmented by the boom.

An Antidote to American Decline

The U.S. energy boom has become a powerful antidote to the notion of systemic American decline. For much of the 2000s, it was fashionable to characterize the United States as "despondent, hopeless and pessimistic" in the words of the German newspaper *Der Spiegel*. The financial crisis of 2008 and the doubt it cast on the U.S. economic model, the dysfunctional politics of Washington, and more than a decade of exhausting combat in Iraq and Afghanistan cumulatively fed almost gleeful pronouncements by U.S. adversaries of the end of the American era. In 2009, China's government-run *People's Daily* declared, "at least one thing is certain: the U.S. strength is declining at a speed so fantastic that it is far beyond anticipation." In 2011, while speaking at a summer youth camp, Russian president Vladimir Putin described Americans to his

young listeners as "living beyond their means and shifting . . . their problems to the world economy. . . . They are living like parasites. They are
leeching on the world economy." Such narratives encroach on the ability
of America to inspire others to emulate its system or adopt its values.

Out of this negativity arose an irrepressible phoenix, the boom in
unconventional energy. The very nature of the boom underscored the
vitality of the U.S. private sector, the ability of the country to innovate,
and the resilience of the American economy—elements that have long
made America attractive in the eyes of those abroad. While the U.S. government did contribute to stimulating the boom, small companies and
relentless entrepreneurs seeking to capitalize on high energy prices par-
layed government support into extraordinary technological advances.
Appreciation for America's unique institutional environment and capital markets grew further as other countries sought to replicate America's unconventional energy experience. Gregory Zuckerman, author of
The Frackers, a book about the individuals who pioneered the boom,
said, "When I talk to foreigners, they're even more impressed than many
Americans by this renaissance. They understand that it only could have
happened in America."

Salvaging the Liberal International Order

In his far-reaching book *World Order*, Henry Kissinger describes the period between 1948 and the turn of the century as "a brief moment in
human history when one could speak of an incipient global world order
composed of an amalgam of American idealism and traditional concepts
of balance of power." That moment was a product of enormous effort on
the part of the United States and its Western allies to create and sustain
the set of norms, institutions, and frameworks undergirding an international order based on the values of participatory governance, respect for
state sovereignty, and liberal economic interaction. Much of the world
did well under this system. Democratic governance, by no means universal, is treasured by those who have it and desired by many who do
not. While nonstate actors have risen in importance, states, by and large,
have remained the main protagonists on the international scene. The liberal economic order has been even more successful, driving a period of

unprecedented economic prosperity in which more than a billion people have been lifted out of extreme poverty since 1950. Wars have not ceased, but as pointed out by Harvard professor Steven Pinker, organized violent conflicts of all kinds have diminished in the past quarter century.

This liberal international order, however, is under strain from both of the two tendencies that eventually challenge every world order in the view of scholars and statesmen. First is self-doubt; one could make the case that "those charged with maintaining the system"—primarily the United States—now question the values serving as the foundation of the order. The 2016 U.S presidential election underscored how many Americans resent what they view as the excesses of globalization and explicitly or implicitly question the applicability of their political arrangements to other systems. Moreover, the positions articulated by President Trump suggest that he is the first U.S. president since World War II who sees the costs and responsibilities of underwriting the global order as greater than the benefits that accrue to the United States. The second challenge—also clear in today's world—is when the international order struggles to accommodate significant changes in relations between major powers.

China's relationship to the current world order is particularly problematic. While perhaps no single country has benefited more from the economic structures in place over the last seventy years, China rejects the political constructs that often accompany them. Moreover, China had little say in the development of the rules of this order, given its preoccupation with its own civil war after World War II—the time of "the creation" in the words of Secretary of State Dean Acheson. Now, as a global power, Beijing is skeptical about both the legitimacy of these rules and their applicability to China and much of the world. At a minimum, China desires a greater say in how the system works, precipitating the need for the international order to adapt to new realities, or to crater under their pressure.

Scholars and policymakers feverishly debate whether it is possible for the United States and its allies to maintain the current system under this strain. Doing so is very much in the interest of the United States. America has been the country most invested in this international order, and the spread of its advantages to others. But the order has also been the vehicle through which America has been able to influence and shape the world to its benefit. Its stewardship of the system—with its international norms and behaviors—is the ultimate soft power vehicle. The new energy abundance can help the United States with this challenge of adapting the order while still preserving it in several ways.

First, the new energy abundance reinforces one of the mainstays of the international order: well-functioning markets. As described in Chapter Three, the new energy landscape is transforming—and will continue to transform—the nature of natural gas markets from being segregated, rigid, politicized, and in many ways inefficient to being more integrated, fluid, and better able to allocate resources in a predictable manner.

In addition, the energy boom has increased the confidence of critical players, most notably China, in the market. When energy was perceived as scarce, there was a trend of growing reliance on nonmarket measures to secure energy needs. Given these doubts about the sufficiency of global energy supplies, China therefore pursued a wide range of deals— from equity investments to secure ownership of energy resources in other continents to bilateral arrangements in which oil was sold outside the market—to ensure access to this most strategic of commodities. The new energy abundance has changed China's approach. It is now more comfortable relying upon the markets, which are the core of the liberal economic order, to meet its energy needs.

The new energy abundance also diminishes certain challenges to the international order that have persisted since the turn of the century. For instance, Beijing's strategy of acquiring oil and gas resources in Africa and Latin America—intentionally or not—led China to support its own gallery of rogues; it cultivated relationships with governments acting decidedly outside the norms of the international order. In protecting regimes from Venezuela to Zimbabwe, China created space for governments that flouted international norms to exist and prosper. Now, as discussed in Chapter Ten, in the face of the new energy abundance, China may see less need to pursue resources in this manner and at these costs, potentially removing or minimizing a persistent irritant to the international order. More obviously, an era of low prices will weaken regimes for which oil and gas revenues are a lifeblood; from Russia to Venezuela to Iran, many of these countries are the most frequent challengers to the international order.

Finally, the new energy abundance offers the United States and others more breathing room to conduct the reform of global energy governance structures that is required to meet new realities. In February 1974, with the world still reeling from the Arab oil embargo, Secretary of State Henry Kissinger convened an energy conference in Washington, D.C. The goal was to establish some sort of institutional framework to galvanize cooperation among energy-*consuming* countries. A new institution—the International Energy Agency—did emerge over the course of the year following the conference. The new Paris-based body was constructed under the auspices of the Organization for Economic Cooperation and Development (OECD), an organization established in 1961 to promote markets and economic development among politically like-minded countries. By its charter, all members of the OECD—and therefore the IEA—must be led by democratic governments with a commitment to market economies. For many years after the establishment of the IEA, this subset of countries coincided nicely with the large energy- and oil-consuming countries of the world.

In the past fifteen years, however, non-OECD countries have developed growing and—in the case of China—massive interests in global energy governance. In 2012, China began to clamor for new global energy governance institutions. It was not simply pushing back on the dominant role of the IEA, but also lamenting the large number of noninclusive, fragmented, and uncoordinated international institutions addressing energy governance. Most recently, China has focused on the G20—an international gathering of the governments of twenty of the largest economies in the world—as a potential vehicle for establishing greater international cooperation on energy issues. President Xi Jinping has reportedly said privately, and bluntly, that if China is to be part of the international energy order, it must have influence over how it is run.

The West and its institutions were slow to respond to China's explicit calls for new international structures to manage global energy issues. However, a development in late 2014 lit a proverbial fire under those who had been nodding passively at China's exhortation for new institutions or the reform of old ones. On October 24, 2014, China and representatives from twenty-one other countries signed a charter for the establishment of the Asian Infrastructure Investment Bank (AIIB). The Obama administration and others in the United States saw this move as a direct challenge by China to the Bretton Woods system that established the World Bank and the International Monetary Fund to help with the development of infrastructure and other projects. While some believe that China would have propelled the AIIB forward under any circumstances, others saw the move as a direct result of the sluggish pace at which the Bretton Woods institutions were willing to grant China greater influence in them, given that its economy has grown more than twentyfold since their establishment in 1944.

U.S. and other officials were suddenly nervous that if Beijing's persistent entreaties for reform of global energy governance were not addressed, China would launch its own parallel set of energy institutions, directly challenging yet another element of the existing international order. The combination of these anxieties, and recognition of the need to create more inclusive, coordinated structures, stoked a significant effort on the part of the IEA to bring China and other large energy consumers into the fold. Upon taking the helm of the IEA in September 2015, Executive Director Fatih Birol made Beijing the destination of his first official trip. There, he explained that strengthening ties between the IEA and emerging powers would be a key objective of his tenure. "China," he told an audience at the Chinese Academy of Sciences, "is at the very top of this list." In the months that followed, the IEA opened a joint energy center in Beijing. China and other Asian countries also activated an "association status," granting them access to a wide variety of data, training, and discussions at the IEA.

Will such steps be sufficient to keep China—and potentially others— in the tent of existing international energy institutions and from creating competing institutions? That remains to be seen, but there is no question that agreeing to principles and terms of reference for new or existing structures will be less contentious in an environment of abundance than in one of scarcity and competition. Moreover, many of the steps that might be required of aspiring members or associates—such as the obligation of creating a strategic oil stockpile equal to ninety days of the country's consumption—are more easily done when prices are relatively low, rather than sky-high.

Restoring U.S. Leadership on Climate Change

The fifteen ambassadors from the European Union regularly met for lunch in Washington, D.C. Often, they would invite a member of the U.S. administration to this private gathering for an exchange of ideas about pressing foreign policy issues. In March 2001, with the Bush administration still settling into office, the EU ambassadors considered it a stroke of good fortune to have National Security Advisor Condoleezza Rice join them at the Swedish ambassador's residence. They were nervous about some of the signals the new administration had sent on the issue of climate change and wanted to register how important ratification of the Kyoto Protocol—the international climate treaty negotiated in 1997—was to their governments. Rice didn't mince her words, "Kyoto is not acceptable to the administration or Congress . . . Kyoto is dead."

If the reaction at the luncheon was, as reported, subdued, it bore no resemblance to the international furor that followed the administration's public disavowal of Kyoto two weeks later. Leaders from around the world met the Bush administration's decision not to send the treaty to the Senate for ratification with sharp words. The French minister for the environment, Dominique Voynet, called the decision "completely provocative and irresponsible," and warned the United States against "continuing the work of sabotage" if other countries sought to pursue Kyoto without the United States. Romano Prodi, the former prime minister of Italy and president of the European Commission, the executive body of the European Union, at the time, chided, "If one wants to be a world leader, one must know how to look after the entire earth and not only American industry."

In her 2011 memoir, Rice expresses regret over how the administration handled Kyoto, calling it a "self-inflicted wound." The combination of the failure to offer an alternative to the treaty and the abrupt manner in which it was handled dealt a blow to American soft power. The rejection of Kyoto by the United States ran counter to the idea that America would be at the forefront of solving global problems and left the impression that America cared more about maintaining its "lifestyle" than dealing with the potentially existential threats climate change could pose to other nations. Even though the Bush administration had some understandable objections to the protocol, its positions were dismissed as narrowly self-serving. Its concerns that the burdens of cutting emissions fell exclusively on industrialized countries were perceived as the United States using the inaction of others to justify its own reluctance.

By 2015, the Obama administration had largely reversed the American position on climate change and reestablished the United States as a global leader on the issue—thereby regaining much of the soft power lost in the Kyoto episode. Speaking in Alaska in August 2015, the president stated, "I've come here today, as the leader of the world's largest economy and its second-largest emitter, to say that the United States recognizes our role in creating this problem [climate change] and we embrace our responsibility to help solve it." In addition to markedly different rhetoric, the Obama administration placed the United States at the forefront of the 2015 United Nations Climate Change Conference held in Paris. It not only submitted to the United Nations a plan detailing how the United States would meet its goals of reducing greenhouse gas emissions 26 to 28 percent below 2005 levels by 2025, but it also took an active role in getting others to make comparable commitments. In a classic use of soft power, the administration sought to leverage its willingness to embrace ambitious goals to get other countries to do the same. Melanie Nakagawa, a State Department official in charge of assisting other countries as they make the transition to low-carbon economies, highlighted "a real pull from other countries for U.S. assistance to help them get on a glide path to meeting commitments in the wake of Paris." Environmental Protection Agency (EPA) administrator Gina McCarthy said much the same when she commented that "if we don't take strong action, we know we're not going to get global action." McCarthy pointed to India and Brazil, alongside China, as countries inspired by U.S. pledges to make more ambitious commitments of their own.

Perhaps ironically, none of this would have been imaginable in the absence of the unconventional energy boom. The surfeit in shale gas production in particular lubricated a resumption of U.S. leadership on climate change in two specific ways. First, it gave the United States credibility as a serious mover on emissions reduction. This was especially important in the context of negotiating the November 2014 U.S.-China Joint Announcement on Climate Change—a bilateral accord committing both the United States and China to specific actions to address carbon emissions and a springboard to the agreement forged in Paris the following year. In urging China—and eventually others—to commit to reducing its own emissions, the United States could point to the significant progress it had made in bringing American emissions down to their lowest absolute level in twenty years. As discussed in greater detail in the next chapter, this decline was largely due to the advent of cheap natural gas and the switch away from coal that it inspired. In private, administration officials talked of how, in the wake of the bilateral agreement with China, several large countries approached the White House in the hopes of securing a similar, high-profile deal.

Without the shale gas boom, achieving the emissions cuts that the United States did would have required significant government intervention and involved high costs—all in a period of tepid economic recovery and high partisan rancor. In short, a hard scenario to imagine. Most likely, there would have been little in the way of emissions reductions.

In addition, the shale boom also facilitated renewed U.S. leadership on climate change by providing the Obama administration with a more palatable approach for meeting future U.S. pledges to reduce carbon emissions. The ability of the United States to meet the goals announced in the 2014 U.S.-China climate accord and later submitted to the United Nations in advance of the 2015 Paris Conference relied heavily on new EPA regulations to limit the carbon dioxide emissions of power plants. Even in a low-cost energy environment, such efforts to achieve further reductions in carbon emissions are politically controversial and will carry costs. Had they required utilities to shift to high-cost natural gas (rather than low-cost natural gas), or more expensive solar or nuclear, they would have galvanized even stronger domestic opposition.

On June 1, 2017, President Trump announced he would withdraw the United States from the Paris Agreement. He lamented the costs of meeting U.S. targets and appealed to the sense of some that the rest of the world was taking advantage of American naïveté. His administration could—and should—have taken a different approach. Instead of forsaking the agreement and the soft power that accrued to America as one of its vanguards, the Trump administration could have called upon American innovators to find another—perhaps less costly and more efficient—way to meet U.S. commitments. Nevertheless, Americans in support of the Paris accord can still take heart. Not only are sub-state actors galvanized to meet American goals, but continued advances in natural gas will help frustrate coal's fortunes in the U.S. economy and open opportunities to meet commitments at lower costs than would exist in a world of energy scarcity, rather than abundance.

Promoting American Ideas about Markets, Transparency, and the Rule of Law

A lot was riding on Obama's first official trip to China. Across Washington, U.S. officials huddled in various agencies, brainstorming about what "deliverables" could flow from his visit to Beijing in 2009. The likely agenda was broad and expected to include everything from global financial stability to the Iranian and North Korean nuclear programs. But lurking behind these more tangible issues was the larger question of what kind of role China would assume in the international system and to what extent the United States and China could work together to address global issues.

David Goldwyn had recently assumed a new post as special envoy and coordinator for international energy affairs, reporting to Secretary of State Hillary Clinton. He and his team were concerned about China's energy trajectory and the many ways in which it might cause the United States and China to clash in the years ahead. By all accounts, Chinese dependency on Middle Eastern oil stood to grow substantially in the coming years, and Beijing was nurturing relations with countries in the region. Just as the United States was seeking the help of other countries to curtail Iranian oil sales, China was signing significant energy deals with Tehran. Moreover, China's burgeoning coal use posed an environmental hazard not only for its own people, but for the global climate agenda. And China's mercantilist approach to energy, best exemplified by its equity investments in pariah regimes in Africa, undermined American and international efforts to resolve certain conflicts and promote transparency and good governance.

Given the circumstances, President Obama's visit to China seemed unlikely to open a constructive conversation on such delicate issues. What's more, in late 2009, China seemed to have the upper hand. The United States was just beginning to crawl out of the Great Recession, which had reached bottom earlier that summer. In contrast, China had skipped over a short, sharp economic slowdown the previous year and was again growing rambunctiously, in contrast to the anemic U.S. economy. The Chinese real estate and stock markets were booming, and China was far and away the best performing major economy, driving much of global growth.

Goldwyn and his team had a clever idea. For all its economic difficulties, the United States was enjoying a massive boom in shale gas production. Other countries, including China, were no doubt wondering if this gas surge could be replicated within their own borders. As part of the deliverables for President Obama's November 2009 trip, Goldwyn crafted a proposal the Chinese could not refuse: a resource assessment of China's own shale deposits and a subsequent workshop to provide the Chinese with information about how to develop and manage whatever wealth was discovered. The Chinese seized the offer, and the launch of the U.S.-China Shale Gas Resources Initiative was announced as part of the joint U.S.-China statement following the trip. Its execution provided not only an important node of bilateral energy cooperation, but also created the opportunity for extensive conversations—the first of their kind—between U.S. officials and their Chinese counterparts about oil markets, the role of the market in procuring energy, pricing mechanisms for natural gas, and other matters. "The only reason the Chinese had any interest in having these conversations with us was because we were committed to helping them find a way to safely develop their own shale resources," Goldwyn told me. "The initiative was an enormous door opener to even bigger and more strategic discussions."

China, it turns out, was not the only country interested in benefiting from American shale gas expertise. The visit of India's prime minister Manmohan Singh to Washington followed quickly on the heels of President Obama's 2009 trip to Beijing—and a similar agreement was made between the two governments. America's offer to help India assess its own shale potential helped pave the way for conversations about India's heavy reliance on coal, prospects for the Iran-Pakistan-India pipeline to which Washington objected, and a panoply of needed reforms such as paring back energy subsidies.

Pretty soon, Goldwyn's colleagues in other bureaus at the State

Department were clamoring for similar overtures to countries around the globe. Could they offer Poland assistance with its regulatory structure? Could they help Morocco and Jordan, two energy-poor countries in the Middle East, determine if they had any shale resources? What about Ukraine, Romania, and Bulgaria, whose dependence on Russian gas was a strategic vulnerability? In August 2010, Goldwyn launched the State Department's Global Shale Gas Initiative. With a mere $5,000 to spend, Goldwyn and his team attracted representatives from nearly two dozen countries—from Pakistan to South Africa—who listened intently for two days to U.S. officials describe every element of shale gas development; they discussed everything from assessing the geology to executing the permitting to managing water resources. The U.S. Commerce Department invited company representatives to present their own experiences in one session only; State Department sensitivity to criticism that the effort was more about business promotion than technical assistance led Goldwyn and his team to ask the participants from companies such as Devon, Chesapeake, and Halliburton to come only for their panel and then leave.

Today, that innovative initiative—now known as the Unconventional Gas Technical Engagement Program—has provided various forms of government-to-government assistance to dozens of countries from Hungary to Morocco to Jordan. Some received help assessing their resources, while others benefited from briefings, workshops, or aid in drafting regulations.

If evaluated solely on the amount of energy produced, the program cannot be considered a raving success. Initial high hopes, energized governments, and American encouragement boded well at the beginning. Yet the efforts of Poland, Ukraine, and other countries in Eastern Europe subsequently fell woefully short due to a combination of complicated geology, restrictive property rights that discourage landowners from allowing exploration, and environmental concerns. In other places, shale gas efforts are in the early stages; with a few exceptions, commercial production is far from imminent.

The program, however, has been an unequivocal success in terms of enhancing America's soft power. For starters, it gave the United States a seat at the table in what would otherwise be domestic-only conversations around the world related to energy security and the environment. It has enabled U.S. diplomats and area experts to engage foreign officials on these topics, not as lecturing outsiders, but as those bringing value added to the discussion. "These conversations provided the entry point for us to discuss bigger issues of how countries run their economies, how they treat investment, and the problem of corruption," according to Goldwyn. "We couldn't effectively address these issues from a theoretical point of view. We needed to have some skin in the game, to have the potential to be truly helpful."

In a less direct way, the unconventional revolution also helped the United States gain traction with countries on a much broader set of issues related to energy markets, transparency, corruption, and the rule of law. Through the Energy Governance and Capacity Initiative—another brainchild of Goldwyn's—the State Department sought out countries that, according to the U.S. Geological Survey, had real potential to produce significant volumes of *conventional* oil or gas. It offered U.S. advice on how a country that was not yet a producer might build a framework for responsible production of these resources. Guyana, Liberia, Sierra Leone, Uganda, Papua New Guinea, and Timor-Leste were just some of the countries that seized the opportunity. Vanishing natural gas imports and dwindling oil ones made the United States a more comfortable partner for countries in these conversations. According to Goldwyn, "the less energy the United States needed to import from abroad, the less our conversations were colored by host country perceptions that we were helping them merely as a way of getting their oil."

Unfortunately, despite their generation of soft power, such programs have not been developed to their full potential. Part of the reason is the growing realization that the American model of developing unconventional resources is not easily transferable. Another constraint on these programs came from the U.S. side, due to the Obama administration's sensitivity to accusations that such programs were simply vehicles to promote American businesses and—even more controversially—that the United States was promoting practices abroad whose environmental safety had not yet been confirmed at home. While the Trump administration is less likely to be swayed by such concerns, maintaining and augmenting such programs to improve energy governance over time will require more—rather than less—resources devoted to civilian agencies in the U.S. government.

The Soft Power of Spare Capacity

Along the Gulf coastlines of Texas and Louisiana lie sixty-two massive underground salt caverns. They were made by a process of injecting freshwater into wells drilled thousands of feet into the caverns, then pumping out the saltwater that resulted. One cavern alone is easily large enough to accommodate Chicago's Willis Tower, the second tallest building in North America. These caverns house America's massive Strategic Petroleum Reserve (SPR), a pool with a capacity of more than 700 million barrels of crude oil that is held by the U.S. government to address sudden supply disruptions. The amount of oil held in them is not arbitrary; as a member of the IEA, the United States is required to hold at least ninety days of net imports in reserve.

Many view the new energy abundance as making these gigantic caverns obsolete. Just as America's rising import dependence during earlier decades spurred efforts to expand the size of the reserves, the long-term decline in import dependence could be an impetus to shrink them. Certainly, this has been the logic of some who have seen America's energy prowess as an opportunity to sell off existing oil reserves in order to finance other expenditures.

A more creative mind might have a different future in store for the SPR. Rather than using its contents to generate windfall revenue, these strategic reserves might bolster American soft power by allowing the United States to exercise a pale Saudi-like influence over global oil markets in times of crisis. Chapter Two explored how tight oil production itself is not really a replacement for OPEC or Saudi Arabia's traditional role of actively balancing the global oil markets by calibrating its production levels. But a radical rethink of how the United States uses its SPR might allow it to come a lot closer.

The bar has been set intentionally high for the release of SPR stockpiles into the market. The legislation President Gerald Ford signed in 1975 to establish the SPR laid out specific and stringent circumstances under which the United States could tap into the reserve. Initially, the president was authorized to tap into the SPR only in the event of a "severe energy supply interruption." In the wake of the 1989 *Exxon Valdez* spill, Congress moved to give expanded authority to the nation's chief executive to allow sales from the SPR under circumstances where a shortage had not yet occurred, but was anticipated, and where "action taken . . . would assist directly and significantly in preventing or reducing the adverse impact of such shortage." Subsequent revisions further loosened the criteria but still in a very conservative fashion.

In fact, the SPR has only been used on three occasions in coordination with other IEA members for the specific objective of addressing an actual or possible shortage. In January 1991, within hours of the onset of the bombing of Iraq during the first Gulf War, President George H. W. Bush announced the sale of some crude from the SPR. In 2005, following Hurricane Katrina, President George W. Bush did the same. And, six years later, in the face of international military action again Libya in 2011, President Barack Obama authorized the sale of crude oil from the SPR.

Advocates of maintaining such a high bar for tapping into the SPR have strong arguments in their favor. They are concerned that politicians might be tempted to use the SPR for short-term political purposes. Defenders of the status quo also contend that using the SPR to tame rising prices—not just to respond to changes in supply induced by earthquakes, revolutions, or short-term events—will only delay an inevitable spike in prices once the SPR is exhausted or after releases from the reserves stop.

These arguments deserve deeper examination in light of the new energy environment. In a world of tight oil, the benefits of using the SPR to smooth peaks and troughs—rather than just respond to actual or anticipated crises—are at least worthy of more serious consideration. As U.S. policymakers become more confident in and knowledgeable about the supply response of tight oil to price changes, they could potentially calibrate releases from the SPR in response to rising prices—with a declared commitment that such interventions will not be without end. A sharp increase in oil prices—whatever its source—will have the effect of inducing higher tight oil production. The use of the strategic petroleum reserve to shave off the spike in the price, without eliminating it altogether, could soften the blow of the price rise for the months (not years) required for the tight oil supply response to come online. Concerns about political manipulation could be addressed through the establishment of an independent body—perhaps along the lines of the Board of Governors of the Federal Reserve System—that would be charged with deciding when to release stocks.

There are complications to this proposal. Most obviously, in shaving off a price spike, an SPR release would remove at least some of the incentive would-be shale producers had to invest in more production. But there are also potentially significant benefits. Together, more responsive U.S. tight oil production and the more liberal use of the SPR could prove to be at least a partial substitute for the loss of Saudi spare capacity, if the kingdom ultimately concludes that it will leave responsibility for balancing supply and demand to others. Should the United States find a working mechanism to calm global oil markets, it would be yet another way for the country to project soft power. Yes, good arguments exist against becoming an activist on this front. But policymakers should weigh these downsides against the upside of moderating the serious economic toll that oil price spikes can take on the U.S. and other economies, and the reality that the old way of dealing with them—leaning on Saudi Arabia and OPEC—may be less viable than at any time in the past forty years.

In 2007, Brazil discovered enormous oil reserves lying far beneath the ocean floor under salt formations more than a mile thick. President Luiz Inácio Lula da Silva announced the discovery from his presidential palace in the capital of Brasília, declaring that it "proves that God is Brazilian." Few American political figures have been so bold or brash to declare that the energy boom is an indication of God's favor. Yet unquestionably, the new energy abundance has endowed the United States with much more than just large physical quantities of oil and gas. It has, in addition, brought significant strategic advantages in bolstering many dimensions of America's soft power. Whereas the Trump administration has openly described itself as being most interested in "hard power," the unconventional boom grants it the ability to augment both. In an era where many see America's soft power in decline and where international respect is generated not only due to military strength, but also to leadership on global issues, the new energy prowess of the United States provides a welcome opportunity to bolster the foundations of American strength. It will, however, require political leadership that both views the new energy abundance in an expansive way and parlays these opportunities into tangible outcomes.

SEVEN

Energy Abundance, Climate, and the Environment

At the turn of the century, environmentalists and "Big Oil" appeared to be moving toward one another. With the world anticipating a global climate change agreement, companies like Shell and BP made large-scale investments in renewables. Shell was running the world's largest solar power–producing plant in Germany in 2004. BP had not only invested billions of dollars into renewable energies and low-carbon technologies, it also redesigned its logo to resemble the sun and even declared in 2000 that its name stood not for British Petroleum but "better people, better products, big picture, beyond petroleum." Big Oil was no doubt motivated to explore cleaner energy sources in part by "the end of easy oil," whereas the greens saw a future reliant on fossil fuels as being environmentally unsustainable. Environmentalists and Big Oil did not necessarily agree on the problem, but they seemed in consensus on the prescription: the need to develop alternatives to fossil fuel energy.

Alas, this moment of unusual alignment was short-lived. The majors largely sold off their renewable energy investments several years later, consolidating efforts behind finding and producing fossil fuels. Some explained, quite simply, that—at the time—the renewables business was not profitable. Faced with a global recession, industrialized countries shied away from policies that would have slapped a price on carbon. Instead of fretting over the need to seek out more difficult-to-extract oil in more difficult-to-operate countries, the oil and gas industry was suddenly able to access huge quantities of unconventional oil at commercial prices in countries with low political risk. The consensus to develop alternative fuels had the economic rug pulled out from under it. Soon after, in the United States, the national security rug followed, as unconventional oil eased the fears of Americans formerly worried about increasing reliance on foreign fuels.

No assessment of the interaction between energy and geopolitics today would be complete without consideration of the relationship between the new energy abundance and efforts to tackle one of the greatest transnational challenges of the times: climate change. Although great scope for debate remains, a serious effort to disentangle rhetoric from reality at least allows us to draw some preliminary answers to some pressing questions. First, fears that the new energy abundance will blow a hole in efforts to address carbon emissions—at least for now—seem exaggerated; if anything, the evidence suggests that there have already been modest benefits to the climate from more shale gas production. Second, with additional support and direction from policymakers worldwide, the new energy abundance could be an important part of the global effort to address climate change. Finally, taking the need to address climate change seriously does not need to come at the expense of encouraging the oil and—particularly—gas boom in the United States and elsewhere; in fact, too much disregard for climate and the environment could actually have a boomerang effect on America's newfound oil and gas prowess.

The New Energy Abundance and Carbon Emissions

Can two entirely reasonable people make completely contrary arguments about the relationship between the oil and gas boom and carbon emissions and climate change? Yes. In fact, it happens all the time. The first might present a compelling case for why this new energy abundance harms efforts to rein in carbon emissions—focusing, for example, on how more abundant oil and gas leads to increased consumption of these fossil fuels and, therefore, greater emissions. She might argue further that unconventional oil is associated with higher levels of carbon emissions, and she could also highlight how lower prices for oil and gas reduce both the political will and the economic rationale for investing in renewable and alternative energies like nuclear. The second person might make the opposite case, heralding the virtues of the new energy boom in helping the world come to terms with climate change. She would be more likely to focus on how in the United States—where the energy revolution has been the most pervasive—absolute levels of carbon emissions have declined since fracking became a household word.

With one exception, both interlocutors would be correct to some degree. The one exception relates to the claim that the carbon intensity of unconventional oil is greater than conventional resources; the facts suggest this argument is more wrong than right. As described in the beginning of the book, unconventional oil is an umbrella term that pulls together a number of resources whose commonality is that they all require unusual (or unconventional) methods to extract them from the earth. These resources differ considerably, including in their carbon footprints.

One form of unconventional oil—oil sands—*is* associated with higher emissions, about 17 percent greater than most other conventional crudes. These higher emissions are largely due to the greater energy required both to extract and to process this oil. It is, however, useful to keep in mind that oil sands—while accounting for close to two-thirds of Canada's oil production and for about one-third of all types of unconventional oil—only constitute about one-fortieth of overall global oil production. Extra-heavy oil and oil shale—two other unconventional oils—also are associated with higher emissions, but they constitute an even smaller amount of global oil production.

By contrast, the production of tight oil—which accounted for more than half of all types of global unconventional oil production and half of U.S. crude oil production in 2017—involves significantly lower emissions than those associated with oil sands. The reasons are twofold. First, tight oil is extracted through fracking, a far less energy-intensive process than that used for oil sands, which are produced either by what is essentially mining or by heating the viscous resources to high temperatures while it is still in the ground. Second, tight oil in the United States (the only place besides Argentina and Canada where it is being produced commercially as of

2017) is what is known as "light oil," meaning that it needs minimal refining before it can be used as gasoline or other end products by consumers.

In fact, emissions associated with tight oil are similar to and, in many cases, lower than those associated with conventional production. Some research outfits have calculated that tight oil actually has even lower carbon emissions than the lighter conventionally produced crudes from Nigeria and elsewhere; it is these imports to the United States that tight oil has tended to displace. Stanford professor Adam Brandt, although focusing on the frequent flaring of natural gas produced alongside oil, makes essentially the same point when he says, "if flaring were controlled, the Bakken [tight oil] crude would have lower emissions than conventional crude."

The carbon intensity of unconventional oil aside, both debaters would have some support for their positions. Common sense, for example, suggests that if the price of oil were lower, people would use more of it. This point is also supported by two contrasting scenarios of the future created by the EIA in 2017. One scenario, the "High Oil and Gas Resource and Technology Case," imagines that the United States produces dramatically more oil overall, and more tight oil in particular, in both 2020 and 2050. This higher production leads to lower prices than would otherwise exist and *does* result in greater usage of oil across the decades examined. The implications for carbon emissions are not surprising. In both the reference (or "most likely") scenario and the high-resource scenario, emissions from the energy sector are lower in an absolute sense in 2020 and 2050 than they were in 2011, due to greater efficiency and the use of more climate-friendly fuels. But the high-resource case involves 3 percent greater carbon emissions in 2050 than the reference one due to greater consumption. The data on the extent to which the low oil prices of 2014–2016 have spurred consumption growth are both limited and mixed; while the world as a whole saw faster demand growth in 2015 and 2016 than in the two years before the price crash, in the United States, one year saw robust demand and the other much more modest demand.

Another likely impact is that the downward pressure on the price of oil will make it more difficult for alternative fuels to be competitive in transportation. The Paris-based IEA concluded in a 2013 study that until technology advances, few other fuels will be cost competitive so long as oil remains below $90 a barrel. Of the twenty possible alternative fuels examined, only natural gas and coal-to-liquid technologies had lower production costs than oil when it is at $60. The climate effects of each are very different. While moving natural gas into transportation could lead to a diminished global carbon footprint, a move toward coal-to-liquids (without capturing carbon) would take the environment in the opposite direction.

For these reasons, some people will conclude that addressing climate change is easier in a high oil price environment with the threat of scarcity, rather than one of abundance and lower prices. Nevertheless, it is worth remembering that high oil prices also brought with them strong arguments to pursue other technologies that were *less* carbon friendly than oil. Throughout the mid-2000s, when the price of oil was high and climbing, there were multiple efforts in the U.S. Congress to introduce greater support for technologies that could convert coal-to-liquids despite the fact that diesel and jet fuel made from coal emit twice as many carbon emissions as oil. One such effort was the Coal to Liquid Fuel Energy Act of 2007—cosponsored by fourteen U.S. senators including Senator Barack Obama—which would have provided government loan guarantees to projects advancing these technologies. Enthusiasm for coal-to-liquids was not limited to the United States, but included South Africa and China; Beijing reportedly sought to ramp up its efforts to turn coal into liquid fuel by twenty-fold between 2010 and 2020. Generally viewed as only being commercial if the price of oil is around or above $60, coal-to-liquids efforts may have taken off had the new energy abundance not squelched talk of such pursuits.

We have more facts to draw upon when assessing the impact of the boom in shale gas on carbon emissions. In the United States, the trends are clear. The advent of shale gas enabled the United States to bring down its emissions to their lowest absolute level in twenty years. Between 2005 and 2015, U.S. CO_2 emissions related to the energy sector declined by 12 percent. In 2015, U.S. per capita CO_2 emissions were as low as they have been at any point since the early 1960s. As made clear by the International Panel on Climate Change—an organization considered by many anti-fracking activists to be the gold standard—fracking was "an important reason for a reduction of GHG [greenhouse gas] emissions in the United States." The IEA agreed, declaring in 2013 that "The decline in energy-related CO_2 emissions in the United States in recent years has been one of the bright spots in the global picture. One of the key reasons has been the increased availability of natural gas, linked to the shale-gas revolution."

Why was this the case? The boom in cheap, plentiful American natural gas changed the economics of power plants. Due to the shale gas bonanza, the Henry Hub price of natural gas in the United States dropped from roughly $15 per mmbtu in 2005 to approximately a fifth of that price ten years later. Utilities did what made simple economic sense: they switched from fueling their power plants with coal to firing them up with natural gas, which has only half as many emissions associated with its use.

As a result, coal's historical dominance in the United States took a beating. After rising consistently since the 1960s, the use of coal began to decline in 2007—and natural gas gobbled up its share. Over the following decade, natural gas usage grew from about a quarter of the overall energy mix to a third, while that of coal decreased from about a quarter to less than one sixth. The shift was particularly dramatic in the generation of electricity, where coal's supremacy was most marked. In 2005, half the electricity in the United States was generated by coal and only a fifth by natural gas. A decade later, natural gas was virtually tied with coal as the largest source of electricity. The number of American coal mines declined by more than four hundred in the five years before 2015. The industry lost 98 percent of its value from its peak in 2008 until early 2016. Indicative of future expectations for coal, shares in Peabody Energy, the world's largest private coal company, dropped from a high of $81 in 2008 to just over $2 eight years later. According to David Victor, a professor and energy expert at the University of California, San Diego, the impact of the switch from coal to natural gas on U.S. emissions each year was equal to "about twice the *total* effect of the Kyoto Protocol on carbon emissions in the rest of the world, including the European Union." This is no small feat. But the key point to keep in mind is that without the shale gas boom and the subsequent drop in natural gas prices, the market simply could not have driven on its own a shift from coal to gas and, with it, such a dramatic decline in emissions.

Figure 7.1: Annual Share of Total U.S. Electricity Generation by Source, 1950–2016 (percent of total)

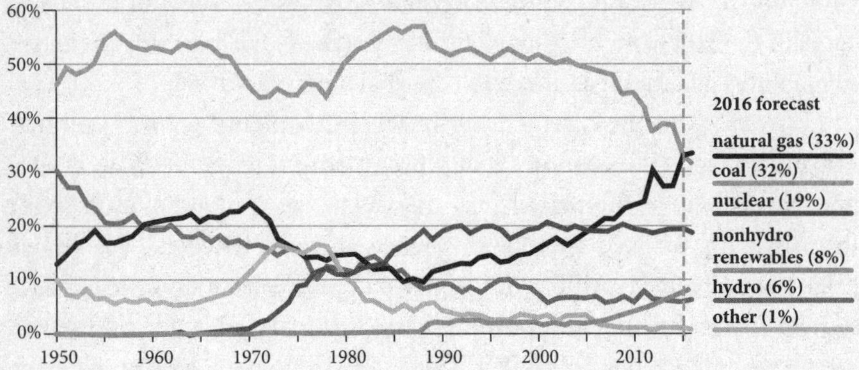

Source: Energy Information Administration, "Natural gas expected to surpass coal in mix of fuel used for U.S. power generation in 2016," March 16, 2016.

Looking beyond the experience of the United States in the past decade, the extent to which the unconventional boom ends up hindering or helping efforts to address climate change depends largely on how two key dynamics play out. The first, the so-called substitution effect, measures the extent to which cheaper oil and gas replace lower carbon energy sources. The second, the "consumption effect," gauges the extent to which lower-priced energy leads to greater use of more carbon-intensive fuels.

Given the relative newness of the unconventional boom, there is not yet enough actual data for us to conclude definitively whether the consumption effect or the substitution effect will have a bigger impact on overall emissions. But we do have some basis for insight—at least for the question of unconventional gas, if not unconventional oil.

A number of studies have sought to give greater clarity to the vexing question of whether the boom in unconventional gas is a blessing or curse for climate change. After a series of complex computations and modeling efforts, Richard Newell and Daniel Raimi, then both of Duke University, concluded that increased shale gas supply has a positive effect on climate by lowering greenhouse gas emissions in the U.S. economy as a whole. But their study suggests this decrease is extremely small (1.4 percent). (This amount is, however, negligible when compared to the reductions required to avert climate change.) Lower emissions that result from less coal usage are largely—although not completely—offset by the displacement of renewable and nuclear energy and greater overall energy consumption. Interestingly, depending on which figures are used to calculate just how much methane is released into the atmosphere during natural gas production—still a subject of great debate—these results could move in either direction.

The IEA sought to make a comparable assessment at the international level. In a 2011 report discussed earlier in this book, *Are We Entering a Golden Age of Gas?*, it compared expected global emissions in 2035 with emissions that would emerge in a scenario in which there was even more extensive global gas development. Natural gas displaces coal worldwide, with nearly half of the coal displaced coming from China. But, similar to the other studies focused on the United States, the net benefit in terms of overall emissions reductions is small, because of a decrease in the deployment of renewable and nuclear energy and an overall rise in energy consumption.

The key takeaway is that the boom in shale gas can in fact be an ally in the battle to combat climate change. But several steps are required in order to ensure that natural gas plays this positive role. Talking with me on the margins of a large energy conference in Houston, Richard Newell, now the president of Resources for the Future, underscored what was needed for natural gas to "be part of the climate solution, not part of the problem." Perhaps most important, it is essential that natural gas substitute for coal. While this occurred seamlessly in the United States, the market does not necessarily ensure this will always be the case. With abundant natural gas, policies such as a carbon tax or a cap and trade system allowing companies or governments to trade allowances in carbon emissions can be cheaper to realize and can help ensure it is coal—not renewables or other alternative energies—that lose out from a surge in natural gas. Ultimately, technologies such as carbon, capture, and storage (ccs) will be essential if natural gas and other fossil fuels are to remain part of the global energy mix over the longer term; most scenarios depicting a global energy system where the threat of catastrophic climate change has been averted envision a "zero emissions" power sector.

In addition, policy measures can be important in seeing that investment in renewable energy does not flag in the face of cheaper fossil fuels, especially natural gas, which is the obvious substitute for alternative energies. Thus far, there is not strong evidence to support fears that low fossil fuel prices will come at the expense of continued investment in renewables and other alternative energies. While global investment in renewables hit all-time highs in 2015 as oil and gas prices were scrapping lows, the numbers—at first blush—for 2016 were less encouraging; investment in non-hydro renewable energy fell by almost a quarter from the record high of the previous year. A closer look, however, suggests that the cause of this drop is not necessarily cheaper oil and gas, but the sharply declining costs in the renewable energy industry. A report by the United National Environment Program that examined these trends described 2016 as a year of "more for less"; even though overall investment in renewables was down in an absolute sense, the world added a record amount of renewable energy capacity that year. These robust numbers are at least in part a consequence of government policies supporting such investment. For instance, in the United States, a deal forged in Congress at the end of 2015 extended tax credits for renewable energy investment out an additional five years, helping reduce uncertainty associated with these investments.

Figure 7.2: Global New Investment in Renewable Energy by Asset Class, 2004–2016 (billions of dollars)

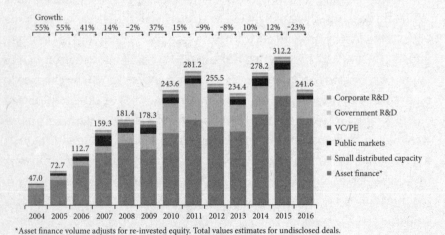

*Asset finance volume adjusts for re-invested equity. Total values estimates for undisclosed deals.

Source: UN Environment Program and Bloomberg New Energy Finance, *Global Trends in Renewable Energy Investment 2017: Key Findings*, 12.

In Tension but Not Necessarily in Full Scale Opposition

At the headquarters of the Environmental Protection Agency on the afternoon of March 28, 2017, President Donald Trump was flanked by coal miners. With an air of excitement in the room, he signed an executive order titled *Promoting Energy Independence and Economic Growth*. The order focused on repealing regulations that, in the words of the president, "threaten(s) our miners, energy workers, and companies" and constitute a "crushing attack on American industry." The Clean Power Plan—President Obama's signature effort to regulate greenhouse gases from power plants—was the most prominent, but only one of the regulations the order sought to reverse.

In sharp contrast to the mood at the turn of the century, when Big Oil and environmentalists found common cause, there is an emerging political strain in the United States that perceives a zero-sum game between nurturing the unconventional boom in American oil and gas and protecting the environment and combating climate change. There are some places where the two are in obvious tension—such as with the question over whether federal lands should be open to oil and gas development. But, in general, an approach that sees the two as completely in opposition—and overwhelmingly prioritizes energy development over climate and the environment—is not only misguided, but carries real risks to U.S. interests generally and even to the unconventional boom specifically.

One of the greatest benefits of the rapid expansion of American oil and gas production has been in the realm of geopolitics. As discussed in Chapter Six, the shale boom and the consequent decrease in U.S. carbon emissions was essential to America's ability to re-exert leadership at the global level on climate change. The fact that President Obama was able to approach China with a proven track record of decreasing emissions helped lubricate a U.S.–China climate agreement in November 2014; this bilateral accord was a springboard to the agreement forged at the Paris meeting of the United Nations Framework Convention on Climate Change thirteen months later. In the months that followed, 175 of the 193 UN member states signed the agreement and pledged specific steps to lower their own carbon emissions. Not only did this turn of events provide China and America a rare platform for cooperation, but it generated significant soft power for the United States. In playing this catalytic role on an issue of importance to so many countries in both the developed and developing world, the United States affirmed its interest and its ability to lead globally. These geopolitical gains are at risk since President Trump decided to formally withdraw the United States from the convention. Even if technology and civic activism enable America to decrease its emissions outside the climate agreement, the United States will have still lost the soft power it gained as its champion.

More tangibly, backtracking from the Paris agreement might—perhaps surprisingly—actually have a dampening effect on America's unconventional boom. Although other countries have pledged to abide by the agreement even in the absence of U.S. leadership, concerns remain that other large emitters such as China, India, and Brazil will walk back their efforts to rein in carbon emissions. Should they do so, the world would consume more coal and almost certainly less natural gas; as discussed earlier, future global demand for natural gas in part depends on the seriousness with which the world approaches the question of climate change. This shift toward a less climate friendly approach could dampen the expected surge of U.S. LNG, the export of which is completely consistent with the Trump Administration efforts to bolster oil and gas production and the number of jobs associated with it. Howard Rogers, a scholar at the Oxford Institute for Energy Studies, demonstrates that should demand for natural gas in Asia be less than anticipated in the five years out from 2017, the market would balance in one of three ways, including the curtailment of U.S. LNG exports.

Those seeking aggressive deregulation in the interest of promoting the oil and gas boom should also keep some other risks in mind. While some regulations of the industry can and should be removed, the industry also needs to maintain a "social license" to operate. This license is not just a formal law or regulation, but is the trust and confidence of the communities in which the companies produce energy. In Europe, companies have essentially lost, or never gained, the social license for fracking and, as a result, European shale gas development is at a virtual standstill. Even in the United States, proposals to severely restrict or ban outright fracking—or the development of infrastructure to transport oil and gas—are on the rise. Thus far, successful statewide bans on fracking have largely been limited to areas where little fracking has yet occurred or where resources are not believed to exist in commercial quantities. Maryland, for example, prohibits fracking although developers have shown little interest in the state. By contrast, New York sits atop part of the prolific Marcellus shale formation, but has never allowed the practice. Yet over the course of 2015 and 2016, a raft of proposals surfaced across the country that, if implemented, would place significant curbs on fracking. A large increase in fracking prohibitions—particularly in resource-rich states or on the national scale—would severely undercut the assumption of a new era of U.S. energy abundance.

The industry could lose its social license to operate in a variety of ways. The first I was reminded of on an October morning in 2016. My handheld Geiger counter was beeping frantically, alerting me to the fact that I was receiving far more than ten times the normal levels of radiation. This was not surprising to me, given that I was standing two hundred yards in front of the metal and concrete "sarcophagus" covering the Chernobyl nuclear reactor in northern Ukraine. More than thirty years after its meltdown, the reactor was still emitting radiation, in part through the cracks and fissures that had emerged in the sarcophagus that was built hastily in 1986—without screws, bolts, or other means of keeping it together. A line of workers formed outside the gate of the reactor, waiting to enter the grounds where they were employed to maintain the existing structure or to build the new one that would be placed over the original just weeks after my visit. Alexey, the Ukrainian guide who accompanied me during my visit, told me that there were fears the aging sarcophagus would collapse, releasing tons of radioactive dust before it could be replaced. But he reassured me the odds of this happening while we were visiting were "extremely remote."

That prospect should have terrified me, but I was thinking of something else entirely: how an environmental disaster can derail the prospects of a new and promising energy source. The Soviets had planned to build twelve nuclear reactors at Chernobyl. Less than two miles away an entire town, Pripyat, had been constructed in 1970, explicitly for those who worked at the nuclear plant. The Soviets expected the town to boom in subsequent years. Instead, driving through it now, we saw eerie streets overgrown with vegetation, and schools, apartment buildings, and civic centers hurriedly evacuated and never reinhabited. The Ferris wheel in Pripyat's amusement park—opened the day before the town was evacuated—still stands almost intact, in poignant tribute to a future abandoned.

Pripyat and Chernobyl—ghost town and ghost reactor—remind us powerfully never to take for granted the growth and uninterrupted expansion of a promising new energy source or technology. John Deutch, a former director of the CIA and emeritus professor at the Massachusetts Institute of Technology, does not see any possibility of a physical "disaster" comparable to Chernobyl or Fukushima. Yet, he did express concern to me that "if industry and government regulators do not pay attention to the major environmental impacts of unconventional oil and gas production, it is possible that public opinion will turn against the practice—just as inattention to safety, proliferation, and waste management has turned the public against nuclear power."

Fracking has in many ways already tried the patience and risk tolerance of some American communities. Just ask the residents of Sparks, Oklahoma, whose Saturday night on November 5, 2011, was interrupted by an earthquake—later attributed to the disposal of waste water from fracking—that registered 5.6 on the Richter scale. Or solicit the views of Hugh Fitzsimons, a bison rancher in Dimmit County, Texas, who attributes the two-thirds drop in his well water that occurred from 2009 to 2012 to the large volumes of water used for fracking operations in neighboring counties. Or you may wish to talk to Mike Lozinski, an air traffic controller in Denver, who complains that the noise from local fracking operations has disturbed his sleep so fundamentally that he is sometimes no longer capable of doing his job safely. Or even chat with the residents of Pavillion, Wyoming, who in 2010 were told by the federal government not to shower without proper ventilation while it investigated concerns that fracking had caused serious water contamination.

Even as the benefits of the energy boom to the United States have mushroomed, public discomfort with fracking has increased. Many question whether any amount of regulation can make the practice safe. Others, such as "Keep It in the Ground" environmental groups, seem less interested in this question, focusing instead on stopping the production of fossil fuels regardless of the alternatives. They have grown in strength and prominence, successfully influencing policymaking and public opinion at both the national and subnational levels.

This activism will increase with the drive to deregulate the energy sector. Under the Obama Administration, many environmental activists felt that the federal government generally was looking out for their interests; they believed they had an advocate for their views at the highest levels of government. The reversal of regulations on the energy industry—while welcomed by the oil, gas, and coal industries—has had the opposite effect on environmental communities. No longer confident that the federal government is protecting and advancing their interests, environmental and other such groups are likely to be much more active at a local level, seeking not only to hold companies accountable to higher levels of conduct, but also to disrupt the production of fossil fuels as a goal in and of itself; depriving companies of their social license to operate is the objective of many of these groups. In 2016 and into 2017, Americans were captivated by news stories of the members of a Sioux Native American tribe and thousands of others who joined them in enduring sub-zero temperatures, tear gas, and other affronts in order to protest the construction of the Dakota Access Pipeline near the Standing Rock Indian Reservation. This dramatic standoff between law enforcement, the courts, and indigenous people could be the harbinger of future confrontations across the country.

A Goldilocks Solution?

Fortunately, the tradeoff between advancing America's oil and gas abundance and protecting the environment and addressing climate change is not as stark as either some in the Trump Administration or in the environmental community would have you believe. There is certainly room to remove some regulations on oil and gas production and to support the construction of oil and gas infrastructure that will help increase production and benefit the economy more generally—without assuming that these steps need to come at the expense of the environment. In other instances, new technologies and practices may provide the answer to legitimate environmental concerns surrounding fracking. For instance, several years ago, operators insisted that fresh water was essential for fracking, but today many companies are experimenting with recycled or brackish water; Halliburton claims it can create workable fracking fluid from nearly any quality of water. General Electric is testing new technologies that allow for the reuse of water after on-site treatment similar to a desalination process. GasFrac—a small Canadian company—asserts that it has devised a fracking gel that not only is "greener," but also eliminates the need for water altogether—and therefore its reinjection into disposal sites. Companies such as Houston-based Apache have focused on substituting natural gas for the traditional diesel used for drilling and pumping, making their overall operations cleaner and cheaper. Others have started to use infrared cameras to identify leaks at fracking sites, and experiments using drones to monitor methane leakage may ultimately prove effective in doing the same.

Yet even with advances in technology, certain regulations will be important to maintain. Advocates for oil and gas industries should think of these regulations as dual-purpose: they both protect the environment and help companies maintain the social license to operate. Happily, such regulations need not be "job killers." In fact, a number of initiatives have identified a set of best practices and measures that can help ensure the development of tight oil and shale gas in a responsible, sustainable way. In 2011, the U.S. secretary of energy tasked a subcommittee of his advisory board to explore the question of whether shale could be safely developed. John Deutch chaired this effort and led a process that incorporated feedback from the public, industry, regulators, and other experts. The report's bottom line was that while fracking has environmental drawbacks, many can be mitigated to preserve its benefits. Other organizations such as the IEA in its *Golden Rules* report or the Center for Responsible Shale Development have come to similar conclusions and advanced comparable recommendations. These reports point the way toward a constructive approach that will make fracking more sustainable and, therefore, part of the American energy landscape for the future.

Concerns that the implementation of these best-practice measures will undermine the economic competitiveness of the boom are also understandable, but not well founded. The work of Michael Porter of

Harvard Business School demonstrates that embracing such best practices would increase the costs of developing these resources by only a nominal amount. In a 2016 presentation, he demonstrated that complying with the performance standards suggested by the Center for Sustainable Shale Development in the development of an average well in the Marcellus would only cost 1–2 percent of the lifetime revenues of the well. This is less than the average daily price change in the Henry Hub spot price.

———

Americans will continue to differ on the question of what is the appropriate amount of regulation and even whether the state or the federal government is the desirable entity to legislate and enforce regulations. This book is not a plea for a specific form or level of regulation, but it is an entreaty to policymakers, business people, activists, and citizens not to view nurturing the energy boom and protecting the environment and addressing climate change as pursuits that are in total opposition to one another. Yes, there are places where the two will be in obvious tension. But in many other domains, they have complementary objectives. A drive to keep oil and gas in the ground or to obstruct any and all pipelines could backfire, especially if the alternatives are coal and the transport of oil through riskier means like railways. Similarly, oil and gas development with little consideration for the environment or climate change carries costs and risks—from diminishing America's soft power to potentially curbing LNG exports to risking the loss of the social contract companies require to operate. Moreover, whereas some deregulation creates more space for economic activity, other regulations have in the past spurred incredible innovations and helped keep U.S. businesses competitive abroad. In a world increasingly cognizant of the need to meet its energy needs while protecting its environment, America's global companies will not only be focused on domestic requirements, but will also keep one eye on the need to meet standards in their foreign markets.

What America needs now is not the zeal of either energy extreme, but the moderation of a Goldilocks solution. Such an approach is not as far-fetched as American political discourse might make one think. It is not only in the realm of academia or think tanks where such compromises can be envisioned. Several U.S. companies, including ExxonMobil and General Motors, have voiced support for a carbon tax. Some environmentalist groups—such as the Environmental Defense Fund—have dedicated themselves to working with industry to find mutually agreeable solutions. For those interested in finding a Goldilocks solution, the building blocks are in place.

The International Environment

EIGHT

Europe

Catching a Break

Dressed in a three-piece suit, striped bow tie, and rimless glasses, Estonian president Toomas Ilves brought to mind a mild-mannered academic when he spoke in Brussels on March 21, 2014, just hours after Vladimir Putin signed a law completing Russia's annexation of Crimea. But Ilves's words were far from measured as he belittled Europe's initial response to Russia's latest provocation. Calling European sanctions on not quite two dozen individuals "piddly" and "a slap on the wrist," Ilves blamed Europe's significant economic ties with Russia for being an obstacle to stronger action. "The things we hated most about the 20th century" were occurring again, he warned. "Europe is sitting here and watching it happen and saying there are 21 people who can't visit us anymore. I don't think that's really an appropriate response," Ilves scoffed. Rising to the bait, Italy's new foreign minister, Federica Mogherini, shot back, "So, let's bomb Russia? What is the solution then?" she asked. "Italy can do without Russian energy, but can Ukraine?"

Russia's flagrant actions in Ukraine forced Europe to face many uncomfortable realities. Foremost among them was the fact that norms and agreements that had governed international behavior in Europe since World War II—including the sanctity of borders—had crumpled. At the same time, as Mogherini's retort suggested, Europe had to face a separate hard truth about energy. European politicians, analysts, and businessmen had long taken comfort in the idea that Europe's import of vast quantities of Russian energy created a mutual dependence. The enormous loss of revenues Russia would suffer from curtailing or ending exports to Europe were, many Europeans assumed, enough of a consequence to make any large-scale disruption inconceivable. This view prevailed among many—particularly those in Europe's business community—even after two episodes in 2006 and 2009 when Russia cut natural gas exports to Ukraine, creating havoc not just there but further downstream in parts of Europe. Many argued that these crises were provoked by Ukraine's failure to pay its bills—and therefore did not invalidate the theory of energy trade being protected by a sort of mutually assured destruction.

Crimea punched a hole in this thinking. By boldly annexing Crimea over the unequivocal objections of Europe and the United States, Putin demonstrated that some political objectives were much more important to him than any economic costs associated with reaching them might be. As Western sanctions increased beyond Europe's initial tepid measures and the costs to Russia's economy mounted, the lesson came into even sharper focus. Europe could not be complacent about Russia's willingness to use its natural gas exports as a political tool. A letter penned by Putin to eighteen European countries three weeks after the annexation of Crimea made this even more clear. In a regretful tone, Putin explained how Russia would be forced to take the "extreme measure" of cutting off gas to Ukraine—with consequent implications for the rest of Europe—if Kiev did not pay for the gas in advance. "The fact that our European partners have unilaterally withdrawn from the concerted efforts to resolve the Ukrainian crisis, and even from holding consultations with the Russian side, leaves Russia no alternative," he wrote.

Europeans were also reminded of a second harsh reality. The global nature of the oil market meant that potential Russian disruptions of the sale of its *oil* to Europe could be managed relatively easily. But Europe was in a much tougher position when it came to natural gas. Russian gas accounted for a third of all the gas Europe consumed; seven European countries, including Bulgaria, the Baltics, and Hungary, looked to Russia for at least four-fifths of their gas. But even more important, many countries in Europe had few, or no easy, alternatives to Russian gas should the Kremlin put the brakes on sales. The nature of their infrastructure tied them to Russia, in some cases exclusively.

This situation was, of course, not news to Europe, but its policies did not fully reflect these geopolitical realities. In fact, European energy policy had traditionally not viewed energy security as its most important objective. Instead, for years, the energy focus of the European Commission had been on using its regulatory power to build upon and improve the smooth functioning of markets. In the aftermath of the 2006 and 2009 crises, Europe had begun to shift its approach away from using its influence primarily to promote markets to also addressing geopolitical vulnerabilities. But other priorities seemed more pressing, be it the drive to decarbonize Europe's economy, or the need to boost economic growth in the wake of the global recession and the European stagnation that followed.

There had been some positive developments. Lithuania had built more gas storage. Poland, Estonia, Greece, Croatia, and others focused on constructing terminals to import LNG. New reverse-flow pipeline capabilities allowed Ukraine to import at least a small portion of the natural gas it consumed from Germany via Poland and Hungary. In December 2013, the final investment decision was made to build a pipeline to connect the Shah Deniz II gas field in Azerbaijan with Southern Europe. But, despite these steps, large-scale, Europe-wide endeavors had flagged.

For instance, the Nabucco Pipeline—envisioned to bring Caspian gas to Europe—succumbed at the end of 2013 to a slow death of a thousand cuts, some inflicted by European energy companies that found the project simply too expensive in light of cost overruns and national decisions to subsidize renewable energy. In the trio of priorities among environment, economic competitiveness, and energy security, Europe—at least as a whole—often seemed to rank energy security last. In the words of Richard Morningstar, a former American official who had devoted much of his long career to helping Europe diversify its sources of energy throughout the 1990s and 2000s, "Often, the United States seemed to care more about the diversity of Europe's energy supplies than Europe did."

By 2014, when Putin swept into Crimea, Europe's position was exactly the scenario Ronald Reagan had feared when he picked his first major fight with Europe less than a year after he took office in 1981. Germany, Italy, and other Western European countries were struggling under the high price of oil in the wake of the Iranian Revolution that had occurred two years earlier and were eager to transition their economies to natural gas where possible. They saw a 3,300-mile gas pipeline from Soviet Siberia to Western Europe as the answer to their dilemma—and agreed to finance and help construct it. Reagan dismissed the ongoing détente with the Soviet Union in which many Europeans were still invested and decried the pipeline effort, not only because of the subsidies but also due to the leverage that such dependency would give the Soviets over Europe. Even if the Soviets never actually terminated the flow of gas, their ability to do so would provide its own quiet form of influence.

Much has happened in the three short years since Russia's annexation of Crimea compelled the European Commission to "securitize" its energy policy and make it more geopolitical in nature. As discussed below, Europe remains, and will continue to be, a major consumer of Russian gas. But Europe is unquestionably more energy secure and can be confident that the age of Russian natural gas dominance is coming to an end. A variety of factors has come together to deliver this result.

First, Europe's regulatory efforts have, over time, constrained Gazprom—Russia's biggest business, the world's largest natural gas company, and a majority-owned entity by Moscow—and fundamentally altered how it interacts with Europe. European initiatives to promote competitive markets have successfully leveraged the power of the Single European Market to force Russia to abide by certain rules in the sale of its gas to Europe. For instance, the 2009 "Third Energy Package" of regulations for natural gas and electricity markets in the European Union gradually forced Gazprom to adhere to new laws that prevent it (as well as others) from owning both the pipeline infrastructure and the natural gas that flows through it. Even more telling is the proposed settlement of a long-standing antitrust suit in which the EU prosecuted Gazprom for anticompetitive practices, such as charging unfair prices and inhibiting cross-border gas trade. Under the proposed settlement still under discussion as of mid-2017, Gazprom would commit to specific measures for a duration of eight years—in order to avoid potentially billions of dollars in fines; Gazprom would have to allow countries to resell gas and to bring the high prices paid to Gazprom by Central and Eastern European countries more in line with those paid by its other customers.

Second, policies to support infrastructure improvements have had similarly transformative effects on Russia's ability to use gas as a weapon. In particular, Europe has plowed its resources into building interconnectors between countries reliant on Russia for their natural gas. As a result, were Russia to halt natural gas exports to one country today, the target nation would be able to meet its needs by turning to others—perhaps to supply Russian gas, but at least to do so through a less direct route. Natural gas has traditionally flowed from east to west, but the ability to make a quick adjustment and send gas in the opposite direction now exists.

Nowhere has the impact of such interconnectors been so dramatic as in Ukraine. I met Andriy Kobolyer on a sunny Kiev morning in October 2016. I did not know much about him, except that he had taken over as CEO of Naftogaz, Ukraine's state energy company, a month after protests dislodged Ukrainian president Viktor Yanukovych from power and a week after Kobolyer's predecessor was arrested for corruption. At thirty-six, Kobolyer found himself in charge of over 175,000 employees and inherited a company with more than $5 billion of unpaid bills, most of them to Russia, a neighbor waging a covert war against Ukraine. Yet, two and a half years later, Kobolyer was able to share some good news about what he had been able to achieve. For years, Ukraine depended on direct purchases of natural gas from Russia for an overwhelming proportion of its imports—a relationship that proved problematic for both sides. But, thanks to gas market reforms, a slowing economy, and shifting gas market realities, Ukraine imported exactly zero natural gas directly from Russia in 2016. The growing use of reverse pipelines was the most important factor in delivering this outcome. Sure, Ukraine still consumed some Russian natural gas, but it purchased all that gas through other countries such as Poland and Slovenia. Moreover, Kobolyer had high hopes that Naftogaz would further benefit from the surfeit of global LNG. He was seeking a pipeline from Poland dedicated to gas sourced from LNG. "Gas is so abundant now," he explained. "If our suppliers do not provide adequate amounts of gas, others will immediately step in to fill the gap. Gazprom has lost its ability to dictate price and volume."

Finally, the new energy abundance has also played a significant role in loosening the Russian noose around European energy consumers. As Kobolyer's words suggest and the rest of this chapter explores, the global advent of shale gas and the remaking of gas markets has created a wide array of new opportunities for Europe to become more impervious to Russian efforts to use energy for political ends. Yet, as has been the case in other parts of the world, these breaks are not of the variety most commonly anticipated—and the liberation from Russia not necessarily of the nature most crave.

No Easy Fix

At first blush, Europe's own shale gas would seem to be the obvious answer to the continent's energy security woes. The U.S. EIA estimated in 2013 that Europe, including Ukraine but not Russia, holds 598 trillion cubic feet (tcf) of shale gas, or 8.3 percent of the global shale gas reserves. A 2013 study by Germany's Federal Institute for Geosciences and Natural Resources suggested that Europe's shale gas resources were nearly three times the size of its entire reserves of conventional natural gas. If all European countries with unconventional gas were to overcome political and other obstacles and develop their resources according to a certain set of best practices, the Paris-based IEA estimates that Europe could produce 0.4 tcf a year of unconventional (mostly shale) gas by 2020 and almost 2.8 tcf a year by 2035. This latter number is impressive, equaling not quite half the amount of gas Europe imported from Russia in 2012.

Unfortunately, Europe will not realize such production, barring a dramatic and wholly unexpected shift in attitude. Despite great expectations at the turn of the decade, Europe has been unable to unleash its own shale potential. A range of factors has hampered Europe's ability to replicate the U.S. experience. Geology, for starters, has proven more complicated and less promising than originally hoped. Initial estimates of Europe's reserves have been scaled back significantly since the heady days of the first EIA report in 2011.

One might expect Poland to be first out of the blocks in the commercial production of shale. Given its history, its heavy reliance on Russia, and the belief that its technically recoverable shale reserves were the largest in Europe, Poland salivated at the prospect of becoming self-sufficient in natural gas. International investors flocked to the country in 2011 to scoop up concessions; one even reportedly paid for some on a credit card. The government awarded more than ninety licenses, and those receiving them spent billions drilling dozens of test wells. Yet, in March 2012, the Polish Geological Institute assessed Poland's shale gas basins to hold only a fraction of what the EIA had estimated a year earlier—albeit using a very different methodology. One executive from a major U.S. oil company summed it up tersely: "The rocks aren't there." Far from turning Poland into a "second Norway"—a reference to Norway's status as an energy powerhouse—as predicted by Foreign Minister Radoslaw Sikorski in 2011, there was no commercial-scale shale gas production in Poland as of 2017.

Poor geology, though, is not the only reason major international companies—ExxonMobil, Chevron, ConocoPhillips, Shell, Eni, and others—have scaled back or terminated their European efforts to find and produce unconventional energy. Fiscal and regulatory frameworks have been tough to navigate, and have consistently undermined the incentive to operate. Where laws and regulations do not limit fracking, other factors—such as dense populations, property rights that deter exploration, and lack of available drilling rigs and infrastructure—often do. In Austria, simply complying with all regulations makes shale gas uncommercial to develop.

Moreover, Europe's ambivalence toward natural gas as a fuel has stymied shale gas development. As described by Oxford scholar Jonathan Stern, unlike in other parts of the world, advocates of natural gas in Europe have been unable to make a strong case for the fuel. Doing so has been especially difficult in light of high gas prices from 2011 to 2014, government support for renewables, a failure in European carbon pricing and cheap coal, and concern about security of supply. The resulting coolness toward natural gas has further raised the bar for European shale gas development in the face of growing environmental concerns. As a result, the trend is distinctly in the direction of growing restrictions on fracking and less shale development rather than more.

As of 2017, each country in Europe had its own laws governing shale development; the European Commission does not have a mandate to unilaterally set binding community-wide policies related to shale. France, Belgium, the Netherlands, Luxembourg, Bulgaria, Ireland, and the Czech Republic have banned fracking. Germany has moved to the left on the issue, with its coalition government and parliament agreeing to a ban on fracking except where state governments give express permission for exploratory drilling.

Poland and the U.K. have been the most encouraging of shale gas exploration, although the approaches taken by different parts of the U.K. have varied widely. Restrictions exist on fracking in Northern Ireland, Wales, and Scotland, despite the fact that Scotland is believed to have sufficient shale gas resources to meet Britain's natural gas needs for three decades. London has been the most ardent voice against the imposition of uniform regulations on shale development by the European Commission. The U.K.'s departure from the European Union may set the stage for more aggressive exploration and production in at least parts of the island. But elsewhere in Europe, without Britain railing against regulatory restraints, bureaucrats in Brussels may have better luck in securing the authority to set continent-wide restrictions on the practice.

Figure 8.1: Status of Fracking in Europe as of July 2015

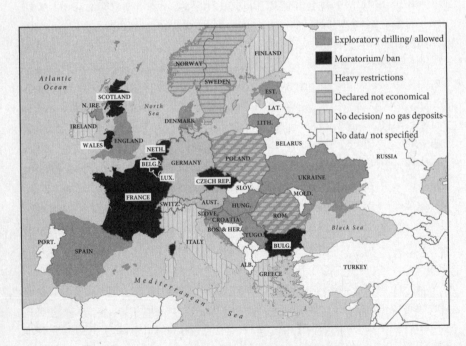

Source: Brigitte Osterath, "Whatever happened with Europe's Fracking Boom?" Deutsche Welle, July 20, 2015, www.dw.com/en/what-ever-happened-with-europes-fracking-boom/a-18589660.

This environmental activism is in line with European sensibilities and Europe's status as a leader on climate change and the environment. But some see a more nefarious hand in Europe's definitive move away from the production of shale gas. Former Denmark prime minister Anders Fogh Rasmussen harbors such suspicions—and he said so publicly in 2014 while he was secretary general of NATO. Rasmussen sees the fervor with which political parties and other environmental groups have sought to discredit fracking in Eastern Europe as a function of aggressive, quiet funding by Russia. Long before there was talk of Russia meddling in the politics of the United States, France, and Germany, Rasmussen told an audience in London that he has "met allies who can report that Russia, as part of their sophisticated information and disinformation operations, engaged actively with so-called nongovernmental organizations—environmental organizations working against shale gas—to maintain European dependence on imported Russian gas."

Given Europe's current trajectory, no rational actor would expect Europe to soon reach the levels of shale gas production the IEA deemed possible back in the early 2010s. Even if Europe did ramp up its shale production, this output would be more likely to fill in for declining European *conventional* gas production and mitigate any further dependence on Russia, rather than substitute for large volumes of Russian imports. In fact, even in the most optimistic shale gas scenario, where unconventional gas production grows from almost zero to nearly half of Europe's indigenous production in 2035, net imports are still expected to go up. In this way, the new energy abundance falls short of its potential to transform geopolitics in Europe. Looking for the real impact of the energy boom on European political dynamics requires us to look elsewhere—perhaps to U.S. LNG?

A New Game in Town

A resident of the Baltic Sea port of Klaipėda might have once boasted to visitors about the forty-eight-bell carillon hanging in the city tower that is the largest musical instrument in the entire country. But on October 27, 2014, Klaipėda received a much larger national treasure: the country's first LNG storage vessel, aptly named the *Independence*. Led by Klaipėda's mayor, hundreds of Lithuanians turned out to welcome the arrival of the floating storage regasification unit in a gathering punctuated by cheers and cannon salutes. It was not just a local event. The prime ministers of Lithuania and Latvia attended, as did members of the U.S. Senate. Norway, Estonia, Finland, Sweden, and other countries also sent representatives. Lithuanian president Dalia Grybauskaitė spoke emotionally, acknowledging that until that moment, her country had been entirely dependent on imports from Russian government–owned Gazprom. She declared, "The liquefied natural gas terminal is an important strategic project and a great victory of our state. It is not only energy independence, but also political freedom. From now on, nobody will dictate [to] us the price for gas or buy our political will." U.S. Secretary of State John Kerry sent a congratulatory letter, which was read aloud by a U.S. diplomat.

Secretary Kerry's letter did not specifically mention the possibility of U.S. LNG flowing to Lithuania. But, Kerry's statement that "The United States looks forward to continuing our joint efforts with Lithuania . . . to further strengthen energy security in Europe" could be interpreted as both an oblique reference to future American exports and an indication that the United States was in the European natural gas market to stay. Whatever was actually intended, Secretary Kerry's letter spoke both to the potential for and the limitations on U.S. involvement in Europe's ongoing energy struggles with Russia and Gazprom.

Certainly, hopes on both sides of the Atlantic had been high. Just seven months earlier, in the wake of Russian soldiers flowing into Ukraine in March 2014, Speaker of the U.S. House of Representatives John Boehner wrote an opinion piece in the *Wall Street Journal* in which he optimistically declared "the ability to turn the tables and put the Russian leader in check lies right beneath our feet, in the form of vast supplies of natural energy." Four foreign ambassadors—from Hungary, Poland, Slovakia, and the Czech Republic—seemed to agree; they sent a letter to the U.S. Congress the day after Boehner wrote, requesting that the United States take steps to expedite the sale of American natural gas to Central and Eastern Europe. In fact, the match between Europe's desire to lessen its dependence on Russia and America's newfound shale gas seemed so perfect that it fueled deep suspicion among more than a few Russians. When I traveled to Russia just a week after the formal annexation of Crimea, many Muscovites shared with me their view that the United States had provoked the Ukraine crisis for the express purpose of frightening Europe and creating a market for American natural gas.

However desirable the goal may be to Europeans and Americans, those expecting U.S. exports to bump Russia out of Europe's natural gas market will be disappointed. The size of U.S. LNG flowing to Europe will depend far less on Washington's wishes than on energy markets. Even under the most favorable conditions, U.S. LNG will not displace Russia as a major supplier of natural gas to Europe.

Most analysts scoffed when then House Speaker Boehner called for U.S. LNG to provide an escape from Russia to Europe. They rightly pointed out that American and European companies, not governments, would make decisions about whether to buy U.S. LNG or to sell it to a European market. They stressed that the potential for profit, not geo-political factors, would be the most important factor. Such trade will only materialize if the difference between the price of gas in America and the price of gas in Europe is large enough so that—even after paying additional costs for liquefaction, transport, and regasification—firms can make a profit on the trade.

This remains true. When Boehner penned his op-ed, expectations of significant LNG trade between Europe and the United States were modest given what was perceived to be a small window for arbitrage between the price of gas in America and that in Europe. Yet, in the years since, some companies have in fact assessed that it is commercial to send U.S. LNG to Europe; the continued very low price of Henry Hub gas still allows a profit to be made. If anything, the opportunity seems more attractive to companies in 2017 than it did in 2014, as much lower LNG prices in Asia mean that the opportunity costs of exporting to Europe (rather than further east) are significantly less than they were a few years ago. Indeed, many of the first shipments of U.S. LNG did flow to Europe. By February 2017, Italy, Malta, Portugal, Spain, and Turkey had all received some amount of U.S. LNG.

This modest flow of U.S. LNG exports to Europe, however, does not presage a flood. A report from Columbia University anticipated that, based on expected future U.S. LNG exports, American natural gas will claim just shy of 12 percent of European gas supplies by 2020 or 2025. In a much more robust scenario, where total U.S. LNG exports are double this amount, American gas claims 19 percent of Europe's market. As shown in the figure below, such inroads will come at the expense of Russian exports, to some extent. But one should also note that a doubling of U.S. LNG exports does not automatically translate into a penetration of Europe's markets that is twice as big. U.S. LNG is likely only to displace natural gas from suppliers whose gas may be more expensive to produce, and that is almost certainly not Russia. Moreover, Gazprom is not a static player and it could well drop the price of its piped gas to defend its market share if it thought it could squeeze out other competitors.

Some experts are even more optimistic, anticipating U.S. LNG will do even better in Europe. In 2016, one consultancy projected that the United States would be sending more than half of its LNG to Europe by 2020. Releasing another report at the same time, a German bank estimated that the United States could send as much natural gas to Europe as Russia does within a decade. While it is difficult to know what exact scenario will play out, we can be confident of two things. First, U.S. LNG exports will continue to help diversify European gas supplies—and diversity is a key element of national security and political independence. Second, under any scenario, Europe will remain a large consumer of Russian natural gas.

Figure 8.2: Impact of U.S. LNG on European Gas Supplies, 2020–2025

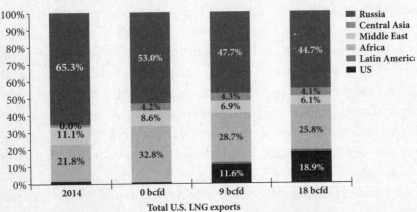

Source: Jason Bordoff and Trevor Houser, "American Gas to the Rescue? The Impact of US LNG Exports on European Security and Russian Foreign Policy," Columbia University, Center on Global Energy Policy, September 2014, 29, http://energypolicy.columbia.edu/sites /default/files/energy/CGEP_American%20Gas%20to%20the%20Rescue%3F.pdf.

Exceeding Original Hopes in Unexpected Ways

The positive impact of the new energy abundance on Europe's energy situation is not so much in the volume of U.S. LNG that flows, or in the European shale gas that is produced, but rather in the transformative effect that the boom has had on reshaping natural gas markets. What matters much more than the size of Russia's market share in Europe is the extent to which the continent—as a whole and as individual countries—is resilient in the face of a Russian gas shutoff and is therefore less vulnerable to the political leverage Russia sometimes seeks to exert through its gas trade.

At the time of Putin's ominous April 2014 letter, Europe had few options to quickly or easily obtain alternative gas supplies if the Russian leader had followed through on his threats. The economies of many European countries—especially those most reliant on Russian gas—would have been badly battered.

One might have thought that existing—but unutilized—European LNG capacity provided a meaningful security blanket to Europe's consumers at this time. In 2013, Europe was only utilizing a quarter of its existing LNG regasification capacity, leaving idle capacity roughly equal to the volumes it was importing from Russia. But, while in theory, Europe could have revved up those facilities to import more gas in the face of Russian obstreperousness, in practice doing so would have been very difficult. For starters, there was the matter of Europe's overall interconnectedness. At the time, a third of Europe's regasification terminals were in Spain or Portugal. Even if those terminals had been fully employed, the dearth of pipelines to carry that gas over the peaks of the Pyrenees and beyond would have made it highly challenging, if not impossible, to replace curtailed Russian gas on the other side of Europe.

Markets also have to be factored into the equation, and 2014 saw intense global competition for LNG. Asian consumers, particularly after Japan forsook nuclear power following the Fukushima tragedy, were thirsty for LNG and willing to pay prices much higher than Europeans. Should Europe have needed to dramatically increase its LNG imports in 2014, it would have been bidding against Asia for what, in the short term, was a finite quantity of LNG. Who is to say that Europe would have won this bidding war? Had it, one can be sure it would have been at a price that was appreciably higher than the $16 per mmbtu drawn by LNG in Asian markets on average between 2011 and mid-2014. The uncertain ability to replace Russian gas at a potentially debilitating price is really not much of an insurance policy.

Now, thanks to the natural gas boom and invigorated European energy policies, this situation is changing, and Europe's options are more real and much larger in scope. As discussed in Chapter Three, until recently, it was common practice for many natural gas contracts to include destination clauses or "take or pay" provisions that require a consumer purchase a certain amount of gas regardless of its needs. Such clauses allowed price differences to persist between Western and Eastern European countries and hindered efforts to create a common energy market among all members of the EU. The surfeit of natural gas and the move toward a global market helped Europeans push back against these once common practices, opening the door to closer integration of EU states and a resultant increase in energy security.

Even more important in increasing European resiliency has been the overall growth of the LNG market. Whereas the tightness of this market in 2014 would have made it extremely expensive for Europe to substitute LNG for piped Russian gas—even if the infrastructure to do so had existed—the well-supplied markets of today put Europe in a much more advantageous position. The shale boom has both directly and indirectly expanded the volume of available natural gas—and the U.S., Australian, and other LNG producers will be adding even greater amounts of gas to markets in the years to come. At least for some time, the essential dynamic will no longer be one of Europe competing with Asian consumers for LNG, but producers competing with one another for markets.

In this more flush market, Europe may still find imported American gas more costly than Russian piped gas when conditions are calm. But in a crisis situation, Europe would be better able to substitute American or other LNG for Russian gas, even at a somewhat higher price, thereby considerably downgrading the threat of a Russian embargo. The ability to withstand a curtailment of Russian gas exports should alleviate the constraints European countries feel today when contemplating a confrontation with Russia and in taking action to circumscribe Russia's ability to challenge the parameters of the existing international system.

This changed dynamic not only helps insulate Europe from Russia's use of natural gas for political reasons, but it also brings benefits in a noncrisis environment. Both the creation of options to secure alternative sources of gas and the increase in the fluidity of markets have given European consumers new leverage over prices and the way they are determined. The most obvious example is, again, that of Lithuania. Before the arrival of the *Independence*, Lithuania paid the highest price for Russian gas of any country in Europe, despite being on Russia's doorstep. Not coincidentally, Russia decided to drop the price it was offering to Lithuania for its gas by a fifth only months before the *Independence* arrived and during negotiations between Lithuanian companies and Norway's Statoil for LNG. In the end, Lithuania received Norway's LNG for less than it had paid for Russian gas in 2013, and Russian gas for even less than the Norwegian gas.

The benefits to European consumers go well beyond those secured by Lithuania. The pressures created by new resources flowing into European markets have caused Gazprom to give in on pricing. Despite insisting publicly that maintaining the price Gazprom gets for its gas sold abroad is the company's top priority, Russia has already made significant accommodations in this area. By directly and indirectly introducing more natural gas into markets that was sold at a spot price (rather than at an oil-indexed one), the unconventional boom gave European buyers leverage to renegotiate the price stipulated in their long-term contracts. This pressure was particularly acute in 2009 and 2010 when much of the LNG exports that were intended to flow to the United States suddenly sought a new home, as burgeoning domestic production of shale gas in America essentially eliminated the need for the United States to import.

As Gazprom points out, new pricing mechanisms may not always work in favor of European consumers; the spot price for natural gas could rise above the oil-indexed one, depending on the price of oil. But in general, Europe can count on paying somewhat less for the same amount of natural gas bought from Russia.

Fly in the Ointment

While Europe is reaping many of the benefits of the unconventional boom in the realm of security, it will find itself at a disadvantage in the domain of economic competitiveness. Today, Europe has some of the highest costs of electricity found worldwide. Such costs are the result of Europe's dedication to decarbonization and to fighting climate change. Europe has in fact been very successful in bringing down its level of carbon emissions. Although the population of Europe is 60 percent larger than that of the United States and its economy is nearly the same size, Europe uses one-third less energy than America and emits 40 percent less CO_2. But Europe has curbed emissions somewhat at the expense of its competitiveness. Extensive commitments to renewable energy sources, in the absence of a truly functioning system for putting a price on carbon, have limited demand for natural gas and its ability to play the role of a "bridge" from a fossil fuel based economy to a decarbonized one.

This imbalance between the competitiveness of Europe and the rest of the world will likely grow wider, as inexpensive American natural gas continues to entice investments to the United States from other parts of the world, including, possibly, Europe. Energy-intensive European companies could be tempted to relocate across the Atlantic as long as energy prices vary so dramatically. The disparity is most evident in comparing natural gas prices in the United States with those in Germany; as of 2014, natural gas prices in Germany were three times higher than such prices in America. The implications of this gap for competitiveness will become even more evident should efforts to negotiate a trade deal between Europe and the United States be revived. In allowing greater access to one another's markets, an American-European trade partnership could make such energy price differentials more obvious as American companies with lower energy costs gain greater access to the EU market.

———

The annexation of Crimea in 2014 signaled to the world that it was dealing with a different sort of Russia—one that was focused more on at least creating the appearance of restoring former glories than it was on maintaining good economic and political relations with its neighbors to the west. Although countries such as the Baltics and some in Eastern Europe had been acutely aware of their vulnerability, Russia's actions in Ukraine provided a wake-up call for the continent as a whole. The threat posed by Russia to the security and well-being of Europe came into sharper relief than it had been at any time since the Cold War. On the security front, a lagging NATO was rejuvenated; on the energy front, Europe snapped to attention from its meandering, lethargic pursuit of energy security.

Europe still depends heavily on Russian energy, but it is much more resilient to potential fluctuations of those flows, whatever the reasons for them. This achievement is in part due to the efforts of European countries and institutions to change regulations, build more infrastructure, and coordinate more closely. Yet, the new energy abundance also deserves some credit for Europe's enhanced fortunes, although not necessarily in the ways initially anticipated. European shale gas production—while originally promising—has and will continue to be a bust. U.S. LNG exports, made possible by the shale boom, will bring some relief, but will unlikely be truly transformative. What will end up mattering most is the greater resilience and more options Europe now has as a result of the new energy abundance. In transforming global natural gas markets, the unconventional boom helped Europe remove impediments to the development of a common energy market and provided Europe with greater options for the sourcing of its energy needs—both at peacetime and in times of crisis.

The new energy abundance does not and will not enable Europe to disentangle itself from Russia and its gas. But just as the quest for energy independence is a distraction for the United States, a focus on eliminating Russian gas exports to Europe would be watching the wrong screen. Rather than obsessing about the need to sever the gas link with Russia, Europe and its allies should appreciate that the new energy environment does something almost as useful: it makes that trade much harder to politicize. In transforming the natural gas markets from favoring the seller to advantaging the buyer, the energy boom has helped make Europe less vulnerable to one of Russia's long-standing foreign policy tools—the political manipulation of natural gas markets.

NINE

Russia

More Petulant, Less Powerful

A podgy academic from Moscow and son of a children's book author, Yegor Gaidar was an unlikely figure to marshal Russia through the tumultuous years immediately following the collapse of the Soviet Union. As the chief architect of Russia's "shock therapy," Gaidar became to many Russians the face of the crippling near hyperinflation of the 1990s. Yet to others, he simply had to manage an impossible situation. Russia was in a severe political and economic crisis when the reform-minded Gaidar took control. To his credit, the nation did not collapse into civil war, widespread starvation, or bloodshed during his tenure as finance minister and then acting head of government.

Gaidar died in 2009 at the age of fifty-three from a blood clot, having been in poor health after he was suspiciously poisoned while eating breakfast in Ireland three years earlier. Before his death, Gaidar wrote a powerful and revealing book claiming that the real cause of the Soviet empire's demise was not the arms buildup spurred on by the United States under President Reagan. Gaidar attributes the collapse—as well as the peaceful manner in which Moscow let events in Eastern Europe unfold—to persistently low oil prices. The earlier collectivization of agriculture meant that the Soviet Union needed foreign exchange for grain imports to feed its people. The low oil prices throughout the 1980s, combined with stagnant Soviet oil production, created a regime-threatening crisis because the government could not finance these essential food imports. If Gaidar were alive in today's era of energy abundance and stubbornly lower prices, he would almost certainly predict gloom and doom for his homeland.

Gaidar's prediction would be one of many prophecies about how the low-price-energy environment will damage Russia. Some have suggested that economic calamity will force the removal of Putin in a palace coup. One expert, Dmitri Trenin of the Carnegie Moscow Center, has in effect compared Russian president Vladimir Putin to Nicholas II, Russia's last czar, who was executed by the Bolsheviks in 1918. Other analysts see the plunge in energy prices as a natural curb on Russian foreign policy adventurism. Optimistic commentators see the period of low prices as "an opportunity to set democratic governance frameworks" for Russia.

These possibilities are intriguing, if a little simplistic. Yet, as in other parts of the world, the most anticipated consequences of the unconventional boom are not necessarily unfolding as predicted and, if they are, they are less significant than expected. In contrast, many of the most meaningful implications of the boom turn out to be surprising or even unforeseen. Russia is no exception.

The new energy abundance is, in fact, creating significant economic troubles for Russia, particularly in the context of sanctions. These difficulties will also have a real impact on the trajectory of domestic events. But they do not seem poised to rock the status quo in the short term. Instead, the new energy abundance is shaping and will continue to shape Russia's foreign policy more than internal politics. The previous chapter showed how new energy dynamics are making it harder for Russia to politicize its energy trade with Europe. But the affront to Russian power does not end there. While increasing Russia's impetus to meddle beyond its own borders as Putin seeks to deliver psychological benefits to the population in place of economic ones, the new energy abundance also dampens the country's prospects for becoming an Asian power and further weakens its standing in Central Asia.

None of this suggests that Russia can be ignored in the coming years. Russia's nuclear arsenal alone, especially when coupled with an increasingly risk-taking leader, will ensure that Russia absorbs considerable time and attention from Western policymakers. But it does imply that Russia will not be in a position to remake the liberal international order on its own—despite its clear desires to—and has less leverage to bully others into joining it to do so.

Stirring the Domestic Pot

Yekaterina, from the Siberian city of Omsk, was the first caller to get through to President Putin during his fourteenth annual *Direct Line* television and radio broadcast in April 2016. She lamented the poor state of infrastructure in her hometown, complaining "it's just one pothole after another." Over the course of the next four hours, Putin answered questions and took petitions from eighty of the more than 2.5 million Russians who sought to speak to the president. In addition to many questions about foreign affairs, Putin addressed queries about inflation, low wages, consumer fraud, property taxes, bank loans, product quality, the cost of medicine, wage arrears, and the decline of traditional industries—just to name a few. At one point, the host of the show asked a man to wrap up his question by reminding him that "brevity is the sister of talent." The questioner responded, "I can't stop. I need to unload, really." Although Putin certainly knew Russia was in economic dire straits before the show, his interaction with average Russians could have only reinforced that sense.

By the time of the broadcast, Russia was in its third year of economic slowdown or decline. The economy had shrunk by 3.6 percent in 2015 and was on track for another year of contraction, albeit more modest. Gross national income per capita had dropped by nearly a quarter in just two years. The number of Russians living in poverty had risen by 20 percent, or by more than three million people, undoing all of the gains achieved in this area over the previous eight years. Some of this decline can be attributed to the international sanctions put in place against Russia in the wake of its annexation of Crimea and interventions in Ukraine. But sanctions are estimated to account for significantly less than half of the economic decline in 2015; the bulk of the economic hit came from declining energy prices. Russian finance minister Anton Siluanov agreed, emphasizing, "First and foremost, the decline of oil prices has an impact"; he estimated that Russia loses up to $100 billion a year as a result of a 30 percent decline in the price of oil.

This assessment comes as no surprise, given the extent of the price collapse and the extreme dependence of the Russian economy on oil and gas. In 2015, the energy sector accounted for nearly a third of the value-added in Russia's GDP and for more than half of its export revenues. Energy sales and taxes provided one out of every two dollars funding Russia's nearly half-a-trillion-dollar 2014 federal budget. Few economies outside the Gulf Cooperation Council (GCC) are more dependent on oil and gas revenues than Russia.

What may, however, come as a surprise is that the adverse effects of the unconventional energy boom on Russia's economy began to be felt well before the price of oil began to waver. According to Russian economist Tatiana Mitrova, increasing supplies of American tight oil and shale gas bear much of the responsibility for ending the energy-driven growth model that Russia enjoyed for more than a decade. In 1999, when Putin first became president, Russian per capita income, as measured in current U.S. dollars, was not quite $13,000; by 2013, it had doubled. The near continuous rise in Russian GDP and GDP per capita over the first thirteen years of the century was fueled by the almost steady increase in global oil prices.

The "effortless" growth slowed dramatically with the stabilization—not even the decline—of the price of oil from 2011 to 2014. As discussed in Chapter Two, throughout these years, roughly one new barrel of American tight oil was produced for every one that was taken off the market due to geopolitical or other disruptions; the net effect was a "stabilization" of the price. Without the propeller of steadily rising prices, Russia's energy-dependent economy began to sputter, tumbling from 4.3 percent growth in 2011 to less than 1 percent in 2014. Had the unconventional energy revolution not occurred, and oil prices been as high as $150 in 2013 as one study calculated they could have been, Russia would likely have been able to continue its energy-driven growth for some time.

Figure 9.1: Russian Prosperity and Oil Prices

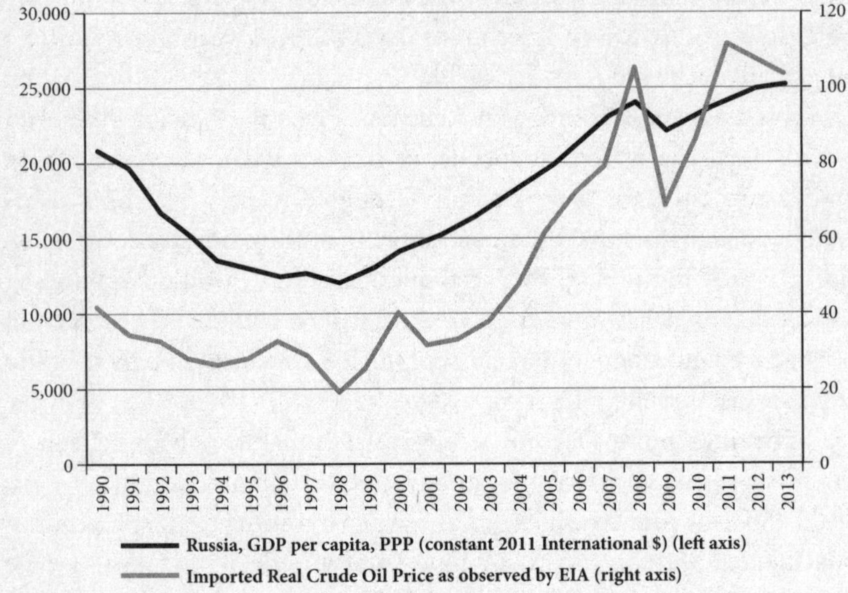

Russia, GDP per capita, PPP (constant 2011 International $) (left axis)

Imported Real Crude Oil Price as observed by EIA (right axis)

Source: "GDP per capita, PPP (constant 2011 international $)," World Bank,
data.worldbank.org/indicator/NY.GDP.PCAP.PP.KD; "Real Prices Viewer,"
U.S. Energy Information Administration, www.eia.gov/outlooks/steo/realprices/.

For all the economic dislocation of 2014–2016, Russia has actually fared better under the double whammy of sanctions and low oil prices than most experts anticipated—and certainly better than Yegor Gaidar might have expected. Russia had several advantages that other energy export–dependent economies did not have, helping ease what was still an abrupt economic slowdown. First, Russia had the ability to increase its production and export of oil, which helped mitigate Russia's "dangerous revenue losses," in Putin's own words. There are, however, evident limits to this strategy without significant further investment, which has not been occurring due to sanctions and low energy prices. Developing Russia's own vast unconventional oil deposits more aggressively also could be an option, but it would require the lifting of sanctions and the application of foreign technology given the country's difficult tight oil geology.

Russia was also able to draw on its Stabilization Fund, which was established by President Putin in 2004 to even out spending through times of low and high oil revenues. One of the successor accounts to the

Stabilization Fund—called the Reserve Fund—proved critical a decade later in keeping sanctioned banks and companies afloat when they struggled to get access to financing. Yet, again, the fund does not provide a long-term security blanket for the Russian economy. Its size dropped by approximately $70 billion or more than four-fifths between mid-2014 and the end of 2016, and is expected to be exhausted during 2017. Finally, and most important, the government moved up the launch of its original plan to allow the Russian ruble to float freely from 2015 to the end of 2014, at least partially in the hopes that it would provide a shield against falling oil prices. The flexible exchange rate did cushion the shock of dropping prices. Almost instantly, the ruble weakened to an all-time low, owing to both sanctions and falling oil prices, thereby yielding more rubles for every dollar earned by exports and mitigating the hit on government revenues that a lower price of oil and natural gas brings.

Figure 9.2: Russia's Reserve Fund (billions of US dollars)

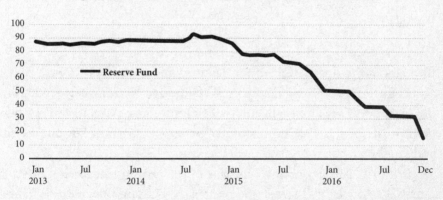

Source: The Economist Intelligence Unit, "Sovereign wealth fund declines sharply in December," January 11, 2017.

Just because Russia's economy did not implode under the weight of dual handicaps does not mean that the country's economic future is bright. In January 2016, Finance Minister Anton Siluanov stressed that the budget needed to be cut a further 10 percent. He warned that a failure to do so could lead to a repeat of the 1998 crisis, when the economy went into free fall, triggering a rush on failing banks by panic-stricken Russians. With the 2016 budget predicated on $50 oil, only an increase in oil prices from where they were in mid-2016 could quickly help Russia get its fiscal situation back on firmer ground. Putin, speaking in December 2015, assured Russians that the worst of the economic crisis was over—although it was qualified with the big caveat if oil prices improve.

Given the new energy abundance, rising energy prices are a thin reed upon which to base the economic future of a country. However, as Andrey Movchan, a Russian economist, emphasized to me at an early morning breakfast in Moscow in October 2016, Putin still has many tools to deploy to keep the Russian economy from free fall. He could raid the country's welfare fund for pensions, resort to expensive international borrowing, further rein in budget expenditures, or print money. Such moves could buy Putin more economic space, but none of them promise to return Russia to positive growth absent a more robust price of oil. This uncertainty about Russia's economic future raises critical issues about both the country's political stability and its prospects for needed economic reform.

Putin's Longevity

Putin's popularity and Russian economic growth were strongly correlated throughout the first decade of the century. The reversal of Russian economic fortunes therefore inevitably raises the question of whether Putin's popularity will sag alongside empty Russian wallets. Putin would not be the first, nor the last, ruler to stumble under the strain of low oil prices and sanctions.

In April 2016, *Foreign Affairs*, one of the preeminent foreign policy journals in the United States, sought to shed light on this question. The publication asked dozens of American and non-American specialists in international affairs—most of them experts on Russia—whether Putin "will still be in control of Russia five years from now." The results must have made Vladimir's heart warm. Of the twenty-five who responded to the survey, four-fifths either "strongly agreed" or "agreed." Only three—interestingly, all of them Russian—disagreed with the statement.

I found a similar confidence in Putin's political future during my visit to Russia in late October 2016. In meetings with dozens of Russian activists, academics, businessmen, and others, I found no one who described Putin to be in a precarious political position. None imagined that a popular uprising such as that seen in the Arab world or in the peaceful uprisings in former Soviet states known as "Color Revolutions" posed a real threat to Putin and his regime. Economic discontent will increasingly rankle the average Russian, and protests particularly around the issue of corruption will sporadically occur. But most experts seem to agree that there is no real opposition to harness this discontent into a political movement that could pose a challenge to Putin's grip on power. Certainly, the September 2016 elections to Russia's parliament, the Duma, pointed to a flagging opposition. Such leaders have either been effectively marginalized or eliminated, or have emigrated. There is no effective alternative to Putin's regime right now, most seem to agree.

At the Pushkin Café in Moscow, Dmitri Trenin explained to me, "Russians do not have a habit of changing their leader through the ballot box every 4–6 years, but they have a record of bringing down the entire state twice in the last 100 years." If Putin were to face a challenge, experts in 2016 seemed largely in accord that it would come not from a popular movement, but from within the establishment leading to a "palace coup." But even this scenario appears unlikely, given how Putin has consolidated control and established himself as a person with unique powers.

Putin plays three critical roles in maintaining the current Russian system. He is the distributor of patronage, the arbiter of opportunities and disputes, and the protector of the status quo. True, as low energy prices eat into revenues, Putin's inner circle—many of them sanctioned by the European Union and America—has come under stress. According to *Forbes*, the number of billionaires in Russia dropped from 111 in 2014 to 77 in 2016. But the nontransparency of the system provides a variety of mechanisms for Putin to create and share largesse, even in declining economic fortunes. Over dinner at Moscow's Hermitage Gardens in the summer of 2014, a journalist shared with me what he deduced from unusual trading in advance of significant comments from Putin. The Russian stock market had plunged by more than 20 percent between the start of 2014 and early spring, causing Putin's cronies to lose great wealth. But trading patterns suggested that Putin was possibly able to offset these losses by informing associates in advance of when he would make a conciliatory statement regarding Ukraine. Given the close correlation at the time between Russia's activities in the Ukraine and the stock market, clever investors were able to make a healthy fortune with the right tip.

According to David Szakonyi, a professor specializing in Russian politics at George Washington University, other methods of keeping Putin's elites satisfied are found in laws and regulations not geared to promote economic growth, but rather to "partially compensate insiders for their losses due to the crisis." The execution of import substitution policies and the awarding of contracts—such as a $4 billion deal to build a rail bridge that was allegedly granted to Putin's judo partner—are two examples. One diplomat serving in Moscow shared another, even more egregious, example. A new mega-gas-pipeline project from China to Russia is being built in stages, not because there is inadequate financing, but because to have one company bid for the entire pipeline would require the bid be put out to public tender; constructing the pipeline in small pieces is slower, but allows the contracts to be awarded to politically favored companies.

More importantly, the politically powerful and the outrageously wealthy stand by Putin not only for what he can bring to them, but also out of fear of their fate in a different system or under another ruler. Although their situation may not be as comfortable as it was before sanctions or the oil price plunge, jettisoning Putin in favor of a less certain successor would entail grave risks. It is more likely that Putin's elites are eager to extend his rule as long as possible. Stephen Kotkin of Princeton University likens the situation to the death of Soviet leader Leonid Brezhnev, whose inner circle "kept him in power for the sake of their own power even after his clinical death." Experts who could at least imagine Putin gone from power in the next five years cautioned democrats to hold their celebrations. The departure of Putin—who "is human," as one specialist felt compelled to point out—would likely be followed by someone who would head a regime very similar to Putin's.

The confidence with which a wide array of experts predicts Putin can and will survive the economic trials posed by low oil prices and other challenges does not necessarily mean that Russia will be politically stable. Although Russians may have a high tolerance for adversity, continued economic hardship—and the budget cuts that persistent low oil and gas prices will force—could have even broader implications than rankling the Russian population. For instance, Moscow could lose some of its ability to maintain a firm hold on the scores of regions that make up Russia. Spanning eleven time zones, Russia is composed of eighty-three federal, regional, and local entities—plus two new "federal subjects" since Crimea was reabsorbed into Russia in March 2014. A very important source of budget revenues for many of these entities is shared federal taxes (which include taxes on oil and gas extraction) and transfers from the federal government. Many such entities saw a cut in transfers from the center even before the price plunge. Now, with less largesse to distribute, Moscow will find it harder to buy off some actors and support others. These limitations could result in security problems in places such as the Northern Caucasus, where three- to four-fifths of the regional budgets consist of subsidies from the federal government. A loss of control of these areas would be gradual and identifiable, at least giving Moscow and other powers whose interests would be threatened by such developments time to react and stave off the worst eventualities.

Putting the Brakes on Reform

If a poor economic situation is unlikely to dislodge Putin from power, it could at least theoretically spur his regime to take new policy initiatives. In the absence of Russia's crisis with the West and the rise of Putin's new politics, reform might have been a predictable outcome of the unconventional boom. A sampling of Russian experts just before the oil price collapse yielded a strong consensus that the economic pressures created by the unconventional energy boom would force Russia to reform its energy industry. In a world of more competitive prices, many predicted the Russian government would be compelled to improve its investment climate in order to attract more outside capital and technological know-how. Gazprom would need to become leaner and more efficient, something perhaps only conceivable if greater competition were introduced within Russia's internal gas market. Energy reforms, some anticipated, would be a precursor to wider economic reforms in the country.

Several developments in recent years *did* suggest a movement toward reform in the ever-critical Russian energy sector. "Independent" oil and natural gas companies, not owned by the Russian government but still heavily influenced by Moscow—such as Novatek and Rosneft—became significant players in the internal market, competing with Gazprom for domestic consumers. In addition, the Russian government began a gradual liberalization of domestic gas prices, although that effort was suspended in 2013. In the final months of that year, Putin instituted a significant change to the long-standing policy that gave Gazprom a monopoly over the sale of all natural gas abroad; now other companies are allowed to export LNG, at least when its destination is in Asia. In a lower-price environment, Russian production (and resultant government revenues) might depend even more on greater efficiency, less corruption, and foreign technology. Reforming the system to be more transparent, less arbitrary, and more welcoming of investment from abroad seemed essential.

Unfortunately, such hopeful signs of economic reform now appear to be relics of the past. On a sweltering Moscow day in the late spring of 2014, an advisor to the Russian government told me, "The government will reform Gazprom when there is no risk to doing so"; I took the statement to be the equivalent of "never." Despite the potential benefits of liberalization in a new low-price environment that were earlier anticipated by many, the pressure Putin and his government is under from international circles and domestic realities actually diminishes the prospects for reform in the critical energy sector. Here's why: in a neutral or positive international environment, Putin and Russia could be confident of reaping the fruits of such reform—increased international investment, capital, and technological know-how—that are vital to the modernization of the Russian economy. In more positive times, Russian leaders might consider this outcome worth the political risks required to reform organizations— such as Gazprom—that have underpinned Russia's power structure for decades. But in the political environment of 2017, such reforms would undermine Putin's mechanisms for delivering patronage to his networks and could eliminate the need for Putin to remain in power as the arbitrator of a corrupt and capricious system. Moreover, in the current international climate, these reforms might fail to bring benefits from external actors, who may be unable to respond to a more positive investment climate due to sanctions or other punitive policies. Remembering the unanticipated events and eventual collapse first set in motion by former Soviet leader Mikhail Gorbachev's reforms, Putin is unlikely to opt for dramatic changes in the energy sector in today's political environment.

The evidence against any real movement toward reform is powerful. Rising economic stagnation is translating into even deeper collusion and entrenchment between Russian business and political power. As a result, Russia's state-owned giants, chiefly Gazprom and Rosneft, are gaining even higher status and political clout.

Nevertheless, the resilience of Putin's power structure and his ability to resist serious reforms in Russia's largest and most critical industry will have its limits over the long run. Some in Russia interpreted the May 2016 welcome of Alexei Kudrin, a former finance minister and constant proponent of economic reform, back into government as deputy head of Putin's economic council as recognition of that reality. But proof will be more in Kudrin's ability to convince Putin to make changes than in his simply assuming a new title. Reportedly, previous ministers submitted plans for reform to Putin, only to have them ignored or fundamentally altered. The proposals Kudrin has advocated—such as increasing taxes, raising the retirement age, and reforming the police—would all threaten Putin's lock on power. According to Kirill Rogov of the Gaidar Institute for Economic Policy, Putin's promise to consider Kudrin's plans will amount to little more than tweaking at the margins. In the *Moscow Times*, Rogov wrote, "Imagine owning a horse-drawn cart: Someone brings you plans to convert it into a car and you promise to 'take it into consideration.' At some point, you attach side mirrors, replace the two rear wheels with alloy rims, apply metallic paint to the sideboards and install an air conditioner for the driver. The result is that you have ostensibly achieved 35 percent fulfillment of the plan. The Russian government takes a similar approach."

Russia Acts Out

If the pressures of the new energy abundance fall short of promoting dramatic changes in the domestic realm, they are having wide-ranging implications for Russia's behavior abroad. In 2010, with the world's economy still fragile, historian Niall Ferguson and I made a bet at the weekly board lunch of the Belfer Center at Harvard University's Kennedy School. The stakes were an ounce of gold, and the bet was whether Russian forces would cross into another country in the next five years. It had been two years since Russian forces had overrun Georgia, a country in the region of the Caucasus. I thought yes: Niall thought no. Although long settled, the bet is a reminder to me how reasonable people can disagree over whether and to what extent economic factors can spur foreign policy quiescence or belligerence.

One might naturally turn to past Russian responses to international crises during times of domestic economic woes to gain insight into how today's economic fragility will affect future Russian behavior abroad. In 1999, when Russia was struggling to recover from financial collapse, Moscow limited its opposition to NATO air strikes in Kosovo to fierce rhetoric. Similarly, at the end of 2004, when the Russian economy looked as if it was slowing down again, Moscow reacted tepidly to the Orange Revolution in Ukraine. Economic fragility at home, one might conclude from these two episodes, will temper Russian meddling abroad.

Drawing this conclusion, however, can be misleading. The Russian government has a very different domestic orientation today than it did during those historical episodes. In 1999, 2004, and even in the aftermath of the Soviet collapse, those in power in Moscow were either liberal reformers convinced that the West held the keys to Russia's restoration, or moderate nationalists (including some of today's leaders at the time) who were pragmatic about the need to cooperate with the West where possible. Today is very different.

Over the past decade, Russia's ruling class has narrowed considerably and those espousing close cooperation with the West have been marginalized. Putin, himself a moderate nationalist in the early years of his rule, has become more of a Russian conservative. His focus is primarily on the protection of traditional Russian society from Westernization and the restoration of Russia's greatness through the reestablishment of a Russian sphere of influence in the post-Soviet space. Putin frequently quotes Russian philosophers of the late-nineteenth and early-twentieth century such as Nikolai Berdyaev, Ivan Ilyin, and Vladimir Solovyev on Russia's great destiny and the need to preserve Russia's historical borders. The Kremlin has even reportedly instructed regional governors to read the works of these thinkers in their free time.

Meanwhile, the Russian government continues to close the space for civic freedoms, placing curbs on "nontraditional sexual relations," press freedoms, and other civil liberties. In this domestic environment, economic and political stresses—combined with perceived international provocation—have encouraged the emergence of a more petulant Russia, rather than a more docile one. In fact, Putin needs a state of tension in order to gain and maintain popular support for his rule.

Certainly, the 2014 crisis over Ukraine and Crimea suggests that domestic economic and political strains have spurred foreign policy adventurism in today's climate of more conservative politics. Some believe the annexation of Crimea was part of a long-standing Putin plan to reestablish the influence of the old Soviet Union. This interpretation of events, however, gives Putin too much credit as a strategist and overlooks the fact that Putin may have felt compelled to react to events in Ukraine as he did because of growing economic weakness at home.

During my May 2014 visit to Moscow, only a week after Russia officially annexed Crimea, I listened as many Russian elites recounted how Europe had rebuffed Putin's numerous efforts to draw closer to it. As evidence, some even pointed to a 2000 BBC interview of Putin in which he answered a question about Russia joining NATO by saying, "Why not? I do not rule out such a possibility." These elites also lamented the perceived European indifference to articles Putin wrote about the need for greater Russian integration with Europe. Moreover, they perceived the levying by the EU of the "Third Energy Package" to be directed at Gazprom specifically. Even more importantly, they believed that Putin saw the United States and the West as aggressively seeking to bring more of Eastern Europe into its orbit, at the expense of Russian pride and national security. Although the Obama administration was not interested in Ukrainian accession to NATO, Putin and Russians generally interpreted events otherwise. In the face of slowing economic growth and declining political popularity, Putin may have found the prospect of the ouster of a Kremlin-aligned president and the embrace of Ukraine by Europe and its institutions as intolerable.

Subsequent Russian intervention in Syria can also be at least partially understood in this light. Putin certainly felt compelled to rescue his ally, Syrian president Bashar al-Assad, when opposition forces were gaining momentum in the summer of 2015. For one, Syria is home to Russia's only naval base on the Mediterranean. And Putin's fears of radical Sunni Islamist forces gaining the upper hand in Syria are genuine, given the potential for such developments to drive further instability in Russia's nearby Caucasus region. But Syria also provided Putin with another opportunity to provide psychological benefits to the Russian people at a time when he could offer nothing but economic pain. Russia's Syrian escapade allowed Putin to show up a dithering United States and to re-establish a physical presence in the Middle East, something it had not had since Henry Kissinger's diplomatic efforts drove the Soviet Union out of the region after the 1973 Arab-Israeli War. As one leader of a Gulf country told me as we sat in his royal court in early 2016, "Now Putin—he knows how to be an ally!"

As Russia's economic situation continues to languish because of sanctions and the new energy abundance, Putin is likely to continue to seek alternative ways of gaining legitimacy and maintaining popularity at home. So far, this approach has served him well; even as the Russian economy contracted, his popularity ratings stayed above 80 percent. As the average Russian increasingly feels the pinch of budget cuts to social services, Putin will continue to seek to project strength in the face of events that challenge Russian power and make him look weak. Harkening back to the glories of the Soviet Empire, or confounding the ambitions of the United States, allows Putin to compensate for his inability to deliver the sort of economic growth that characterized the previous decade. With Putin no longer expecting the West to provide the key to Russian prosperity, as he and others did during earlier foreign policy affronts, we should expect Russia's impulse to meddle abroad to continue.

Two regions in particular will attract extra attention from Russia. First, Putin will likely take advantage of opportunities to interfere in bordering countries with significant Russian populations or in the former Soviet sphere of influence. Second, the Middle East will remain an area of particular concern. While Moscow's influence in that region has many drivers, policymakers from other parts of the world should not overlook the energy dimension of that interest.

Like other powers, Russia has to balance its economic and energy interests with its strategic and security ones in the Middle East. Doing so is not always straightforward. At one level, continued instability in the region boosts the price of oil, which is firmly in Moscow's favor. According to Yegor Gaidar's book, in 1974, then-chairman of the KGB

Yuri Andropov wrote to Leonid Brezhnev, recommending the Politburo act to support the plans of Palestinian terrorist groups to disrupt oil infrastructure and storage in the region. Putin may not resort to such measures today, but he and his inner circle are clearly aware that the quickest path to economic relief is a higher oil price. This reality is not the only factor in shaping Russian attitudes toward the region. Russia's generally constructive approach in the P5+1 negotiations with Iran demonstrates that it will sometimes opt for strategies to stabilize the region. Yet, the desire for a robust geopolitical premium on the price of oil will unquestionably factor in to Russia's willingness to help defuse crises from Iraq to Syria to Libya. Libya, in particular, will be a place where Russia's economic and strategic objectives are at odds. Although Moscow may want to take credit with Europe for resolving a stubborn conflict that feeds directly into European insecurity, a stable, unified Libya could precipitate a large drop in oil prices as the North African country brings more oil quickly to market; one study by the Oxford Institute for Energy Studies suggests that a full recovery of Libyan oil production to pre-conflict levels could drive the price of oil down to $38 per barrel by the end of 2017.

In weighing energy factors in its overall foreign policy strategy toward the Middle East, Russia will not only think about shoring up prices, but also thwarting potential competitors. Ongoing turmoil in the Middle East will keep Iraq and Iran from realizing their full oil production potential and limit the competition to natural gas suppliers that could emerge in a more stable political environment. Both Iran and Iraq have ambitions to export significant quantities of natural gas to Europe through the proposed "Friendship" or "Islamic" Pipeline running from the world's largest gas field in Iran, through Iraq and Syria, to Europe. With the removal of many sanctions on Iran, the Syrian civil war now seems to be the biggest obstacle to that project. Russia worked hard to ensure that Nabucco— another large pipeline intended to bring natural gas west to Europe— never got off the ground. Helping an alternative east–west pipeline to succeed is clearly not high on Moscow's agenda.

Finally, Russian interest in cooperation with the big oil producers of the world—Saudi Arabia in particular—will continue. Russia's participation in, and general adherence to, the 2016 agreement reached between non-OPEC and OPEC members was striking given Russia's past record on this front. But Moscow sees a new opportunity for a grand coalition with Riyadh, as two of the three most important oil producers in the world (the third being America). While Saudi Arabia does not necessarily show this enthusiasm, it will seek to keep its options open with Moscow.

Figure 9.3: Middle East Pipelines

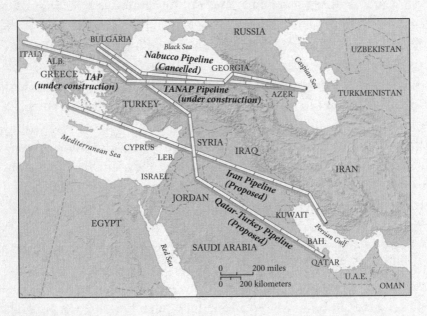

Source: Steve Austin, "Russian gas pipelines," Oil.Price.net, January 17, 2017.

Hastening Russia's Desire to Pivot East

In 2012, long before Russia's confrontation with the West over Ukraine, Putin spoke to a group of Russian energy officials and urged them to look east. Like American policymakers, Russians increasingly see the future unfolding in Asia, where populations and markets—not to mention demand for energy—are booming. Moscow looked at a stagnating Europe that was growing more suspicious and hostile toward Russian ambitions even before Crimea, and sensed the need to turn eastward. Russia has good standing to be a significant Asian power. After all, three-quarters of Russia's landmass is in Asia, where it borders China,

Japan (maritime-boundary), North Korea, Mongolia, and Kazakhstan and therefore directly abuts one-fifth of the world's population. Similar to how Soviet planners saw the industrialization of Siberia as key to the prosperity of the empire, contemporary Russian leaders have focused on the Russian Far East as a gateway for progress. Russia's strengths in many ways seem complementary to Asia's needs and vice versa. Countries such as China, Japan, and South Korea are thirsty for energy, whereas Russia needs capital and technology.

Russia sees not just opportunity in a "Eurasian" identity, but also a need to bolster a flagging population. A great power should not be so unbalanced, with "a massive elephant head in the west, and a mouse's body in the east" as one Russian analyst described it to me. Even the mouse's corpus seems to be withering, with the inhabitants of the sparsely populated Russian Far East dropping by 13 percent since the dissolution of the Soviet Union. Yevgenii Nazdratenko, the recalcitrant former governor of a Far Eastern federal region named Primoskii Krai, called for a "repopulation program" in 2000. He envisioned five million "European" Russians moving to the Far East, in a quest to nearly double the Russian population there.

To make matters worse from Moscow's perspective, while Russian populations dwindle, ethnic Chinese are moving in, according to Alexander Shaikin, the man in charge of the Russian-Chinese border. Shaikin likely exaggerates when he claims that 1.5 million people from China illegally entered the Russian Far East in the year and a half preceding the summer of 2013. But the Russian Federal Migration Service has forecast that in two to three decades, ethnic Chinese will be in the majority in the Russian Far East. Such trends reinforce the reality that, with a seven-hour time difference from Moscow and nearly four thousand miles separating Moscow and Vladivostok, Russia's Far East is much closer to Beijing than to Russia's capital.

This population shift rekindles Russian concerns about a possible Chinese demographic conquest; significant parts of today's Russian Far East were deemed to be Chinese territory for more than a century following the 1689 signing of the first treaty between Russia and China. It is perhaps no surprise that when, at the St. Petersburg International Economic Forum in May 2014, Chinese vice president Li Yuanchao called for combining "the program for the development of Russia's Far East and the strategy for the development of Northeast China into an integrated concept," his Russian hosts chose not to comment. More recently, a 2015 deal in which a Chinese company was to lease 280,000 acres of Siberian land to grow crops and raise cattle caused a firestorm over social media in Russia. Russians declared, "China's creeping expansion in Russia has begun" and "the Motherland is being sold out piece by piece."

If domestic politics, visions of untapped prosperity, a fraught history, and fears of Chinese encroachment were not enough to inspire Putin's pursuit of Asian power status, two more recent developments spurred additional moves in that direction. First, the crisis over Ukraine and Crimea intensified the sense of Russia's leaders that they needed to expedite this transition to the east, in part to demonstrate to the world that Russia was not isolated. Second, the new energy abundance itself provided a jolt. Ten years ago, Russia had the luxury of seeming to be able to sell as much gas as it could produce. Moscow's primary concern at the time centered around producing enough natural gas to meet internal needs and export commitments. Today, Russia is desperately searching for markets for its natural gas; the unconventional boom has curbed the prospects for non-Asian markets that Moscow once thought looked more promising. As we have seen in the previous chapter, while Russia is unlikely to lose its significant foothold in the European natural gas market, it can no longer look to that market for growth.

Even more dramatically, the unconventional boom has eliminated other markets upon which Russia was counting. In 2007, with great fanfare, Gazprom announced plans to develop Shtokman, a massive natural gas field in the Barents Sea containing an estimated 138 trillion cubic feet of gas reserves—more gas than is believed to be in the entire Norwegian shelf. Taking on the challenges of developing these resources—six hundred kilometers north of Russia's northern coastline, one thousand feet below floating icebergs, and in an area in darkness half the year—was justified given the expected robust growth of U.S. LNG imports. In 2009, Gazprom anticipated that it would supply up to 10 percent of the North American market for LNG in ten years' time. These grandiose plans were put on hold in 2012, as Gazprom and its partners faced the reality of spiraling costs and a rapidly disappearing market for gas exports to the

United States because of the booming shale gas production there. According to Gazprom's spokesman, Sergei Kupriyanov, Shtokman will only be revisited "when conditions on the market change: either prices should rise, or costs should go down." With Europe and America offering no prospect for export growth, Moscow's pivot to the burgeoning economies of Asia is more critical than ever to Russia's future.

At the same time that the energy boom is heightening the Russian desire to pivot east, it is also making that objective much more difficult to achieve. As we will see, the unconventional boom has narrowed the time frame in which Russia needs to capture eastern markets, has prevented it from reaching deals on the terms it has sought for decades, and has potentially undermined the commercial viability of projects that once were seen as critical to Russia's strategy to go east.

The Dragon and the Bear

The room in Shanghai seemed far too large for the occasion. The applause of the small number of observers echoed off the ceiling, which was painted with the same pattern that adorned the large brown carpet. Looking distinctly more nervous than his counterpart, Gazprom CEO Alexei Miller signed the agreement and handed it to China National Petroleum Corporation (CNPC) chairman Zhou Jiping. Only after the two stood up from the table, shook hands, and exchanged copies did the cameras home in on Russian president Vladimir Putin and Chinese president Xi Jinping, who had been standing in the background. A woman appeared, offering each leader a tiny glass. In a moment captured on film and tweeted around the world, Putin and Xi looked each other in the eye, clinked their petite glasses, and threw back the drink, sealing what many were calling "the gas deal of the century."

The Sino-Russia gas deal signed on May 21, 2014, was more than a decade in the making. Chinese and Russian negotiators met, haggled over details, and signed promising memoranda of understanding that rarely seemed to lead anywhere. Over the years, the officials disagreed about pipeline route, price, and whether China could acquire equity stakes in Russia's energy fields. The deal centered on the development of greenfield natural gas fields in remote Eastern Siberia and the construction of a nearly two thousand mile pipeline and gas transmission system, coined the Power of Siberia, to the centers of greatest energy demand in heavily industrialized northeast China. Under the thirty-year arrangement, beginning in 2019, Russia agreed to deliver 1.3 trillion cubic feet of natural gas a year to China—a volume that represented about a fifth of Chinese gas consumption in 2014 and could constitute roughly a tenth of it in 2020. Building on the momentum of this deal, the two countries signed an additional memorandum of understanding for a second natural gas deal in November 2014. Although this agreement, in contrast to the May one, was not binding, it signaled Russian intent to ship another 1.1 trillion cubic feet of natural gas to China, this time from already developed Siberian fields via the Altai route entering China in its northwest corner.

Undoubtedly, the global political environment—and Xi's close relationship with Putin—played a role in making this spring visit successful when other efforts had fallen short. Several experts with whom I spoke while in Beijing a few months after the deal was sealed suggested that Xi, sensitive to how a failure to close the deal would damage Putin at a time when he was seeking to project strength in the crisis with the West over Ukraine, tapped his hard-driving negotiators on their shoulders and told them now was the time to sign on the dotted line. However, a look beyond the politics of the moment also reveals the central role of the shale gas boom in heightening Russia's sense of urgency and convincing Moscow to make key concessions that would have been hard to imagine a couple of years earlier.

Time was once on Moscow's side, but the shale gas boom shifted that advantage to Beijing. For a decade, Moscow had been content to let negotiations drag on, in the hopes that China's crushing demand for energy would drive its eastern neighbor to cut a deal on terms more favorable to Russia. During this time, China's demand for natural gas increased nearly one-fifth every year—and China moved from being self-sufficient in natural gas to importing a third of its consumption in 2013. In 2012, CNPC estimated that Beijing's plans to spur growth and to bring millions more Chinese out of poverty would require the annual use of more than 18 trillion cubic feet of gas by 2030—or just somewhat short of the amount consumed by all of Asia in 2010. For most of the past decade, Russia seemed to be the most logical and cheapest way to secure these critical natural gas resources. Not only did China prefer piped gas over LNG (whose delivery comes via waterways patrolled by the U.S. Navy), but China had not yet begun piping natural gas from other markets.

At the turn of this decade, however, China's prospects for meeting its future natural gas demands improved in a fundamental way. Suddenly, China had more choices for how it might meet its gas needs. Piped gas began to flow from Turkmenistan in 2010, and from Myanmar and Kazakhstan in 2012. China—intrigued by the 2011 U.S. EIA report positing that China held the largest reserves of shale gas in the world—began planning to produce more of its own resources. Talk began of how the United States could become an exporter of LNG in the years ahead thanks to its booming shale gas production. More natural gas of both the conventional and unconventional varieties was being discovered and developed in places such as Australia.

"Whether Gazprom slept through the shale revolution or not, it's a difficult question. There is no answer to it yet." These words of President Putin, uttered in April 2013, must have rankled Gazprom officials and made them determined not to miss the narrowing window for a natural gas deal with China. In the run-up to the signing of the first gas pipeline deal in May 2014, Gazprom and Russian government negotiators made some critical concessions, such as abandoning their desired Altai route in favor of the more eastern route preferred by the Chinese. In the final hours, Russia conceded on price as well; although the exact numbers are not public, what is available suggests Russia settled for a price comparable to what it gets for its gas in Europe. This figure is lower than Russia had hoped to achieve, but high enough for outside commentators to value the deal at $400 billion. Moreover, because sanctions have made it difficult to finance the development of new fields, Russia has allowed for Chinese equity investment in Russian oil and gas companies and projects—something that Moscow had long resisted due to the fiercely held belief that Russian resources are the patrimony of the Russian people only.

While Chinese comments about the Power of Siberia gas agreement were subdued, Russian officials were quick to herald it as a harbinger of future strategic cooperation. The comments of Alexei Pushkov, the head of the foreign affairs committee of Russia's lower parliament and an ally of Putin, were representative. He was quick to tweet, "The 30-year gas contract with China is of strategic significance. Obama should give up the policy of isolating Russia: It will not work." Pushkov's counterpart in Russia's upper house, Mikhail Margelov, similarly called the deal a "major step toward a strategic partnership of the two nations."

International reaction to the May gas deal ranged from nervousness to alarm. One commentator asked, "Is anyone else scared to death about this week's announcement out of Beijing that China and Russia have agreed to a 30-year natural-gas deal?" He later likened the deal to a basketball dream team: "when LeBron James and Kevin Durant sign with the same team, you'd better start doing an urgent talent upgrade, or prepare to get buried." Others saw no occasion for levity, with one claiming that Russia and China were "changing the world by shaking the foundations of an order that has ensured the absence of major conflict and paved the way for unprecedented prosperity." The size of the deal heightened fears of an emerging strategic partnership that could challenge the American-led international system. One newspaper subsequently asked if the natural gas agreement was part of a "new superpower axis?" in a not-so-veiled reference to Germany, Japan, and Italy during World War II. Predictions that "the U.S. could find itself facing a formidable Sino-Russian alliance in another multi-decade geopolitical struggle" were frequent.

Reinforcing this sense of a burgeoning strategic partnership is the "bromance" reporters speak of between Presidents Xi and Putin, both of whom share a common outlook. Insiders describe bilateral meetings in which the two discuss concerns about domestic unrest and challenges to their rule. Both leaders see the United States as meddling in their spheres of influence and encroaching upon their legitimate interests. Both see the value of warmer relations between their two countries. In drawing closer to the dragon to its south, Russia is able to counter the narrative of its international isolation, as well as to mitigate the pain of sanctions and locate alternative sources of finance, markets, and technologies. China sees cozying up to the bear to its north as critical to its military modern-ization and important in diversifying its energy sources, especially those that do not flow through the Strait of Malacca. Beijing also sees Moscow as a willing partner in Chinese efforts to use its currency internationally and considers its acquiescence as crucial in its plans to further expand Chinese influence in Central Asia.

Russia and China are making progress in overcoming deep historical memories and insecurities in order to bridge their long-standing divide. The collapse of the Soviet Union back in 1991 offered the chance to resurrect the Sino-Russia relationship. But the two countries were slow to capitalize on that opportunity and there were several false starts. They finally settled most of their border disputes soon after the signing of a Treaty of Friendship in 2001, but it took a decade longer before the Eastern Siberia–Pacific Ocean pipeline began to deliver Russian oil to China.

The pace and scope of this rapprochement, however, increased markedly after the Ukraine crisis of 2014. In addition to the natural gas deals, Chinese and Russian officials penned more than a hundred agreements related to energy, finance, infrastructure, and technology in three different settings in the months from May to November 2014 alone. Unofficial bans on selling Chinese advanced weaponry, allowing Chinese participation in bids on large infrastructure projects in Russia, and permitting Chinese equity investments in Russian natural gas fields were dropped. During President Xi's 2015 visit to Moscow to attend Russian festivities marking the seventieth anniversary of the end of World War II, Xi and Putin signed a joint communiqué pledging to develop a bilateral strategic relationship and underscoring how the two countries would work together to ensure global stability. Later the same month, Russian and Chinese naval forces concluded their first joint exercises in the Mediterranean; even more meaningful joint naval exercises followed that autumn in the South China Sea. Russia became the third largest shareholder—after India and China—in the Chinese-led Asian Infrastructure Investment Bank, launched in mid-2015. Perhaps most significantly, China has stepped in to to help fill the financing gap created by the sanctions and thaw the freeze they have placed on Russia's ability to raise money from western markets.

Thwarting Russia's Asian Ambitions

Despite these outward signs of rapprochement, the relationship between China and Russia remains more opportunistic than strategic. Its future character is still in question. Of the many factors determining its trajectory, energy will loom large. But contrary to many expectations, energy is more likely to frustrate the potential for a meaningful alliance than boost it. This is largely thanks to the new energy abundance, which will exacerbate the already imbalanced nature of the relationship between the two countries. Already, Russia is relying on China to be a source of revenue, growth, and finance in the face of enduring sanctions from the West. The combination of continued sanctions and low oil prices make Russia even more constrained in realizing its pivot to the east due to insufficient access to Western capital and technology. As a result, Russia's shift eastward will be increasingly dictated by China, will happen on China's terms, and will ultimately reduce Russia's capability to diversify its eastern ties. Ambitious Russian plans to meet Japanese and Korean energy demands have slowed, although not been entirely shelved, as the expensive LNG infrastructure projects anticipated to feed these markets have become less feasible.

The delay or potential cancellation of the Vladivostok LNG project is one example of Russia's doubling down on the Chinese market. Once envisioned to supply as much as 15 million tons of LNG a year, the $15 billion project was halted in 2015. The gas once expected to flow as LNG from the eastern Russian port city of Vladivostok to Japan will now eventually be directed to the Chinese market via pipeline. Gazprom's CEO, Alexei Miller, claimed that the venture was no longer "on the list of priority projects, and isn't on the list of projects that will be carried out in the near future." With large volumes of LNG anticipated to come online from Australia and the United States, and technological and financing challenges arising from sanctions, Vladivostok no longer made commercial sense.

In mothballing this project, and potentially others, Russia may well miss its opportunity to claim a significant portion of the global LNG market, given emerging global competition from North America, Australia, and North Africa, as well as smaller producers such as Papua New Guinea and Indonesia. If this window shuts without Russia claiming a share of the global market, Russia's dependence on China as a source of growth will be even more significant.

While Russia's reliance on China will deepen, the new energy abundance will ensure that China becomes less and less tethered to Russia. China will have more and more options to meet the needs Russia seeks to satisfy. The glut in LNG means China will have no shortage of potential suitors seeking access to its market. Moreover, the surfeit of LNG has pushed down its price, eroding to some extent the cost advantage that piped gas—and Russia—has over LNG and other suppliers. Specifically, the new price environment has already led to the freeze of the Altai project and created question marks around the viability of the Power of Siberia pipeline. Finalizing the deal only months before the oil price plunge, Gazprom insisted that the price of gas be linked to that of oil—an arrangement clearly no longer in Russia's favor. Scholars from the U.K.-based International Institute for Strategic Studies argue that, if oil prices remain low beyond 2020, the Power of Siberia project would no longer have a small net present value, but could sustain a loss as large as $17 billion.

Russian gas will maintain some advantages. Because it comes overland, it helps mitigate what former president Hu Jintao called "the Malacca Dilemma" and therefore, as mentioned, will maintain a preferred status with Chinese buyers. Moreover, China will use the possibility of cheaper Russian gas to negotiate better prices for the LNG it receives. But the fact remains that if China receives the other piped gas and LNG for which it has contracted, it will not need both the Power of Siberia and the Altai natural gas pipelines from Russia to materialize. Rather than weaving the fabric of the ever-deeper mutual dependence that Russia envisions, China will be seeking to diversify its supplies and balance its reliance on as many sources as possible.

This dynamic will accentuate what in the past has been a persistent impediment to a closer Sino-Russian relationship: Russia's sense of inferiority. The widening gap between a rising China and a declining Russia will strain the concept of a strategic partnership. As Russia becomes weaker and more dependent on China, the perception and reality of this imbalance will expand beyond Russia's having one energy partner in Asia and China having many. If China gets involved in constructing expensive infrastructure, for example, it will likely only provide finance if Chinese companies and workers are the ones to build the projects. In this situation, Russia's sparsely populated east could experience a major influx of Chinese workers and companies, laboring to deliver the resources that Russia only recently yielded to Chinese equity investments. Not only will this create friction on the ground, but it could further exacerbate

Russian insecurities over alleged Chinese territorial ambitions in Russia's Far East. Moreover, while Russia has long resented the idea that it could become simply an "energy appendage" to China, all signs point to the realization of that vision.

Declining Leverage over Central Asia

Moscow's continued close ties with Central Asian countries after the collapse of the Soviet Union had political and historical bases, as well as practical and economic ones. In the decade after the Soviet Union's demise, Gazprom was dependent on the vast quantities of energy produced in Kazakhstan, Turkmenistan, and elsewhere to meet the commitments it had made to provide Europe with energy. Gazprom would purchase gas from Turkmenistan and Uzbekistan at very low prices, often not paying in hard currency and sometimes not paying at all. Gazprom would then sell the same gas for a fatter price in Europe. This monopsony was buttressed by Soviet legacy infrastructure, which had all pipelines from the Central Asia republics flowing one way: north to Russia.

Any efforts to buck this system were squashed. Just ask the Turkmen, who halted gas exports to Russia in the summer of 1997 in an effort to secure better terms for the sale of their gas. Negotiations between the Turkmen state company and Gazprom followed, with Gazprom CEO Rem Vyakhirev declaring that Turkmenistan would be "forced to eat sand" if it did not sell gas under Russia's conditions. For nearly two years, the gas trade was suspended, wreaking havoc on the Turkmen economy, which had depended on its sales of gas to Russia for 85 percent of its revenues. In 1999, gas trade resumed, although the question of long-term price remained unsettled.

The new energy abundance was one factor in changing this dynamic. Russia is now looking for markets, not for gas. It no longer truly needs Turkmen gas or Kazakh oil to meet commitments elsewhere. But if Gazprom is less interested in buying Central Asian gas, even at low prices, it is now focused on ensuring that such gas does not squeeze into its fiercely protected European markets. This is another lesson that the Turkmen discovered the hard way ten years after the first confrontation. After a tense 2009 summit between then–Russian president Dimitry Medvedev and Turkmen president Gurbanguly Berdymukhamedov in which the possibility of building a pipeline from Turkmenistan to the west was to be discussed, a mysterious explosion damaged the pipeline linking Turkmenistan and Russia. Two days later, Gazprom asked Turkmenistan to reduce its gas sales to Russia by 90 percent. Eight months of negotiations followed, in which the Turkmen economy contracted by 25 percent. An agreement was then struck in which Gazprom bought considerably less Turkmen gas than it had before. The Russian-Turkmen gas trade resumed toward the end of 2009, and Turkmenistan has refrained from any negotiations indicating an interest in sending Turkmen gas west, despite a period of intense interest from the West related to the doomed Nabucco pipeline.

The shift from energy scarcity to abundance changes the dynamics of regional competition in Central Asia in subtle but important ways. Russia remains interested in forging closer links between itself and the former Soviet republics, as evidenced by the launch of the Eurasian Economic Union in June 2014. Energy will help it consolidate such links with smaller, energy-poor countries, such as Armenia, Belarus, Kyrgyzstan, and Tajikistan, all of which had either joined, or were contemplating joining, this union as of 2016. But the new energy realities weaken the pull Russia has over Central Asia's big energy producers and the tools it has to integrate their economies into its own.

At a strategic level, this change will have consequences for the orientation of the whole Central Asian region. Whereas Russia and China used to be in competition with one another for Central Asian gas, today it is Russia and Central Asian countries that are vying for Chinese markets. Russia, in the words of Tatiana Mitrova, "has completely lost control of Central Asia to China." Tension may grow between Russia and China over the implementation of the One Belt One Road initiative—Beijing's big push to link China physically, culturally, and technologically with more than sixty countries to its east. However, there is little competition between Russia and China for the energy molecules of Central Asia—which many had anticipated to be the next Great Game in that part of the world. Since 2009, relations between China and Central Asian countries have burgeoned.

Given Russia's historical relationship and its reliance on Central Asian natural gas to meet its contractual commitments in Europe,

China worked carefully to build closer ties with Central Asia in a way that included, rather than excluded, Russia. For example, in 2001, China became a founding member of the Shanghai Cooperation Organization, a body aimed at increasing cooperation primarily on security and political issues between China, Russia, and Central Asian countries. China used this organization as well as its bilateral ties to promote bilateral trade, infrastructure, and financial assistance to the countries to its west. Critical—and delicate—energy deals followed, moving Central Asian oil and gas to China's thirsty markets. Perhaps today, with considerable amounts of Turkmen gas flowing to China since the 2009 dispute between the Russian and Turkmen governments, Gazprom regrets not fighting eastward-flowing pipelines as well as westward ones. While not posing a challenge to Gazprom's traditional markets in Europe, the Turkmenistan–China gas pipeline and Kazakhstan–China oil pipeline complicate Russia's goal of capturing eastern energy markets. They also tie Central Asian countries formerly in the Soviet sphere physically and economically to China, a trend that will only deepen as China's One Belt One Road effort unfolds.

In 1997, while Russia's economy teetered, Vladimir Putin defended his dissertation at the St. Petersburg Mining Institute. Whether he actually wrote the thesis is an open question, but the core topic highlights his interests—how Russian natural resources are critical to reestablishing Russia as an important economic power. Since Putin was awarded his degree, he has made it his personal mission to restore Russian greatness. No doubt, since he was elected president in 2000, he has seen energy as a key tool in that pursuit. When, in the 2000s, the world was wringing its hands over rapidly diminishing energy supplies, Russia stood strong as the world's largest producer of oil and gas, and the holder of the largest natural gas reserves and the eighth largest oil reserves. Released in May 2009, Russia's *National Security Strategy to 2020* clearly states: "Russia's resource potential and a pragmatic policy of using it have broadened the Russian Federation's capabilities for strengthening its influence on the world stage." For years, Russia's energy prowess afforded it all kinds of geopolitical leverage and was one important factor in giving the country global stature. In 2009, when Fiona Hill, a Russian expert and former official at the U.S. National Intelligence Council, came to speak to my students at Harvard, she rightly referred to Putin as "Lord of the Gas."

The new energy abundance had thrown a very unwelcome wrench into the plans of Russia, making strategies to reassert itself and reorder the international system much more difficult. Russia still has ambitions, and still has serious capabilities, to thwart the United States and to create global difficulties and disruptions. But its tools for doing so are both fewer and less potent given the changed energy environment. Although lower oil and gas prices will not likely unseat Putin from power, nor spur radical reforms of Russia's most important economic sector, they will create persistent and significant problems for the government, forcing it to make tough decisions, unwelcome trade-offs, and what will be increasingly difficult excuses to the Russian people for the hardships they will endure. Moreover, as we explored in the previous chapter, the structural changes in energy markets, particularly natural gas ones, are making it harder for Russia to use its natural gas trade to advance its political agenda.

Unfortunately, economic duress at home will not translate into a more quiescent Russia. In fact, the result will likely be just the opposite as Putin and his inner circle look for emotional ways to placate impoverished citizens. But although Russia in the next five or more years will be more petulant, it will also be less powerful and less able to achieve its other foreign policy goals in many domains. On account of the new energy dynamics, Russia's influence is decreasing in Europe. Perhaps more importantly, Russia is also incapable of reaching its potential in the Far East and in Central Asia.

Russia will and should hold the attention of policymakers in the coming years, if only because its weakness can be as problematic as its strength. In fact, its economic weakness will likely be the cause of puffed-up shows of military strength. Energy is not the only issue determining Russia's political future; Russian nationalism, country demographics, European politics, American leadership, and even Putin's physical health will be other determinants. But it is an absolutely critical factor given Russia's great resource wealth and its traditional reliance on energy to bolster its international stature. In today's low-price environment, the trajectory is clear: the new energy abundance is a bane to Russian brawn.

TEN

China

Greater Degrees of Freedom

I n October 2000, Xinhua News Agency ranked "China's Top 10 Figures in a Century." Mao Zedong, Deng Xiaoping, Sun Yat-sen, and
others on the list are familiar historical figures, even to foreigners.
But one name is almost unknown outside China. Born into poverty in the northwest of China in 1923, Wang Jinxi was a shepherd and
coal bearer before he went to work at an oil field at the age of fifteen.
In February 1960, in a drive for self-sufficiency, Chinese leader Mao
Zedong called for "a massive battle" to dramatically expand Daqing, a
remote swampland on a bleak plain believed to have mass amounts of
oil. Soviet technical advisors had just left China because of the worsen-
ing Sino-Soviet split, and the Chinese Communist Party decided to use
the Daqing project not only to meet China's oil needs, but to demonstrate the country's resilience in the face of the crisis with its northern
neighbor.

Wang Jinxi and his No. 1205 Drilling Team responded to Mao's call
and endured extreme elements to drill the first well at Daqing. As depicted in a 2009 Chinese film celebrating Wang, they braved blizzard
conditions and temperatures of minus 20 degrees Centigrade. Legend
holds that, without any other infrastructure, Wang and thirty coworkers
transported sixty tons of equipment from railway stations to the oil field
by means of long human chains. Equal quantities of water were moved in
the same fashion. Wang himself became known for plunging into huge

containers of muddy water needed for well operation and stirring the frigid flows with his own body to keep the water from freezing. Foreshadowing his own early death from cancer, Wang is said to have declared upon his arrival in Daqing, "I would give up 20 years of life so China can produce oil on its own land."

Wang's enduring celebrity in China attests to the premium the country has put on self-sufficiency and to the great lengths it will go in order to meet its energy demand. Yet despite Daqing's success, China was eventually forced to turn to the outside world to meet its energy needs. The country's energy self-sufficiency ended in 1993, when China began to import oil again after a period of self-sufficiency. But it was still more years before China's energy trajectory became a source of alarm for Chinese officials. The robust economic growth that followed China's accession to the World Trade Organization in 2001 spurred massive increases in Chinese energy consumption. Whereas it had taken more than two decades for China's total energy consumption to double after 1979, it did so again in just the first seven years of the new millennium. By 2010, China had become the world's largest consumer of energy. As of 2014, it accounted for more than a fifth of all energy used in the world each day.

Figure 10.1: Primary Energy Demand for the United States, China, and European Union (million tons of oil equivalent)

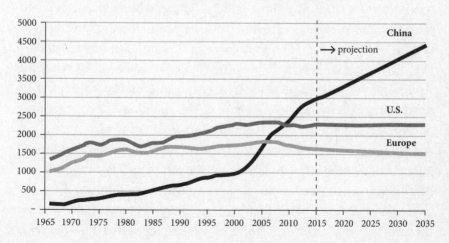

Source: Derived from BP, *BP Statistical Review of World Energy 2016* and BP, *BP Energy Outlook 2017.*

China's rapacious appetite for energy has had major geopolitical consequences, as the need to secure energy has been one of the single largest drivers of its foreign policy over the last two decades. The Chinese government would have preferred to focus on domestic affairs and, in the words of Chinese leader Deng Xiaoping, "hide [China's] strength, and bide [China's] time." But the pace of growth and its intense demands on energy helped push China into the wider world to build deeper relationships in the Middle East, Africa, Latin America, and beyond. A sense of urgency permeated all these endeavors. In 2009, two Chinese officials from the National Development and Reform Commission, the central planning department of the Chinese government, predicted that China would need to import nearly two-thirds of its oil by 2020; the following year in 2010, the IEA projected that four-fifths of China's oil needs would have to be met by imports in 2030. These expectations, coupled with the perception of global energy scarcity—and a looming peak oil—that existed throughout much of the 2000s, primed Chinese leaders to take extraordinary steps to acquire the energy the country needed.

In few places are the potential and actual benefits of the new energy abundance as significant as in China. As we have seen in other parts of the world, the most expected advantages—while real—require some significant qualifications. Yet in China, the downsides are relatively few. The new energy abundance has provided the *opportunity* to the Chinese government to shore up its legitimacy. To some extent—but less than might be expected— the boom and low energy prices have also helped at a time of slowing growth and an uncertain transition to a different economic structure. Moreover, abundant natural gas offers China the possibility of extending this growth with fewer of the devastating environmental costs associated thus far with economic progress—perhaps addressing the pollution that has become yet another challenge to Chinese Communist Party rule.

Although less appreciated, it is in the realm of geopolitics where the benefits of the boom to China are now the most consequential. Given the link between Chinese foreign policy and the pursuit of energy, the new energy abundance will inevitably result in changes in Chinese behavior toward the rest of the world. Some of these changes are already apparent. The new and wide range of options for securing energy supplies has quietly shifted the balance of power between China and its energy-producing neighbors in Beijing's favor. In addition, by loosening the rationale for certain Chinese foreign policy approaches, the new energy realities have at least given Beijing the option to scale back on policies that have produced domestic and international blowback. Finally, the new energy abundance frees Beijing to embrace foreign policy objectives other than securing energy at a time when China's leadership is interested in gaining a larger international footprint.

In this instance, what is good for China has largely been good for the rest of the world and the United States in particular. As discussed in Chapter Six, perhaps no other geopolitical issue has generated more handwringing from Washington to Tokyo than the question of whether a rising China intends merely to tinker with the current international order, or instead to completely revamp it and, in doing so, trigger a great power conflict. Unlike international orders in eras past, the current one is not determined by one country and geared for the enrichment of that single nation. Instead, while U.S.-led, this order depends on coalitions and cooperation and places a premium on integrating countries that wish to rely on the market. China—although not one of the order's original architects—has been perhaps its biggest beneficiary. Participation in an integrated global economy has helped lift hundreds of millions of Chinese out of poverty.

The new energy abundance is by no means the only determinant in the complex calculation of China's rise to global prominence. Other factors, such as China's strategic thinking, nationalism, and territorial disputes will all have their roles to play. But the current global energy situation does figure prominently, and weigh definitively, in favor of China's continued adherence to the broad contours of existing institutions. The new energy abundance has increased the comfort of the Chinese government with the market—a key element of today's order—as the ultimate arbiter of energy resources. It has eroded the rationale for some Chinese nonmarket approaches to securing energy and, in doing so, has given China the opportunity to back away from certain behaviors—such as supporting rogue regimes—that the West has found problematic.

Moreover, the new energy abundance offers an opportunity to both U.S. and Chinese leaders to ameliorate tense bilateral relations. Should leaders from both sides choose, they can use energy as the basis for reimagining a more constructive and, in some places even strategic, relationship between Washington and Beijing. Again, many factors will determine whether a China rising against the backdrop of a dominant United States will lead to conflict or coexistence. But the new energy abundance will matter in this reckoning, not only because it diminishes the possibility of fierce competition over resources, but also because it provides a multitude of avenues for U.S.-Chinese cooperation. Working together to address climate change, an unstable Middle East, and even China's own energy demands will yield benefits on their own—and could also provide critical paths for dialogue and models of cooperation that extend beyond these issues.

Hear China Roar

The transformation of China over the last four decades is one of the most remarkable and consequential tales in the history of the human race. In 1978, when China began to open up its economy to the world, China's GDP accounted for only 1.8 percent of the world's economic output, and China's trade with other countries amounted to only $20 billion a year. In less than forty years, China has grown its economy more than twenty-six-fold. It now constitutes approximately 15 percent of the global economy. In the mere three years from 2009 and 2011, China laid down substantially more cement than the United States did in the entirety of the twentieth century.

As recently as 2005, China's economy was less than half the size of that of the United States. Yet China is now the world's largest economy by the standard of purchasing power parity, if not in absolute measures. It is also the planet's biggest manufacturer, merchandise trader, and holder of foreign exchange reserves. During the first decade of the twenty-first century, nearly 700 million Chinese left behind lives of poverty; more than 200 million attained middle-class status. The scale of this transformation is unprecedented and apparent to the casual eye on the streets of China. In 2005, U.K.-based singer-songwriter Katie Melua released a hit song about Beijing's "Nine Million Bicycles" after visiting the Chinese capital. Had she visited five years later, she might have crooned about the less romantic sounding "Five Million Cars." She would have noticed how many Chinese had swapped their baskets and bells for steering wheels, a point she could have contemplated extensively had she gotten stuck in one of the worst traffic jams in history—a sixty-two-mile-long snarl on the Beijing–Tibet expressway that lasted for twelve days in 2010.

This growth has provided the Communist Party with the legitimacy to continue to govern China. No longer the purveyor of Marxist-Leninist ideology, the party today is more pragmatically oriented to staying in power. Officials are aware that the party has maintained its mandate to rule because it has been able to deliver continued advances in the standard of living to the Chinese population. As long as their material well-being was improving, few Chinese felt compelled to challenge the Chinese government on other grounds. In the words of China scholar Susan Shirk, the party considers "rapid economic growth a political imperative because it is the only way to prevent massive unemployment and labor unrest."

Energy has been a critical component of this story—at least since the turn of the millennium. The growth that occurred in China in the two decades after Deng Xiaoping launched his economic reforms in 1978 was actually *less* energy intensive than previous growth. Just twenty years earlier, China had suffered the Great Famine, in which an estimated 45 million Chinese died. Concerned about the possibility of the return of widespread starvation, Beijing relaxed constraints and strictures on farming collectives. Motivated by more market incentives, Chinese farmers began to produce more. With more disposable income in the rural areas, investment started to flow into local labor-intensive light industry, and away from inefficient, energy-intensive heavy industry—a development that actually brought down the energy intensity of the Chinese economy over the 1980s and 1990s.

No one—in China or in the international energy agencies—seemed prepared for the reversal of this trend in the early 2000s. In 2002, both the Chinese government and the IEA expected economic growth to run in the 7–8 percent range for the rest of the decade and for China's energy intensity to continue its downward trajectory. They also predicted annual energy demand growth would be 3–4 percent between 2000 and 2010. They were wrong—dramatically so—on both counts. China's economy grew rambunctiously—more than 10 percent a year over the course of this decade. Overall energy consumption grew four times faster than had been predicted. But what was even more shocking was the uptick in the energy intensity of this growth, at least in the early part of the decade. Chinese growth was no longer driven by garment factories, but by investment in manufacturing, real estate, heavy energy-intensive industry, and the build-ing of cities for urbanization. By the mid-2000s, almost half of the world's cement and flat glass production came out of China, as did more than a third of the globe's steel and more than a quarter of its aluminum.

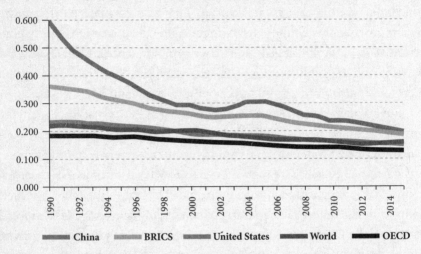

Figure 10.2: Energy Intensity of GDP at Constant Purchasing Power Parities (kilogram of oil equivalent per 2005 U.S. dollar)

Source: "Energy intensity of GDP at constant purchasing power parities," *Global Energy Statistical Yearbook 2016*, https://yearbook.enerdata.net/energy-intensity-GDP-by-region.html.

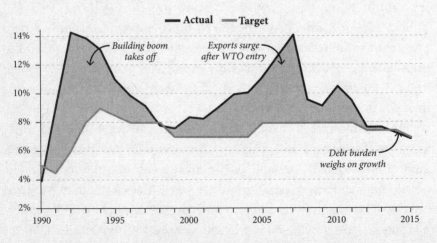

Figure 10.3: Chinese GDP Growth Targets vs. Actual GDP Growth

Source: Bloomberg, "Why China's Economy Will Be So Hard to Fix," February 29, 2016.

Just-in-Time Energy

In many ways, the boom in American oil and gas arrived at just the right moment for China, given the scale and energy intensity of its growth. As China grew at an average growth rate of more than 8 percent from 2011 to 2014, plentiful American unconventional production kept oil prices stable when they otherwise would have been considerably higher. During this period, China consumed an average of 10 million barrels of oil per day, roughly half of which it imported at global prices. One study estimated that, in the absence of the unconventional boom, global prices for oil could have been as much as 36 percent higher during these years. If we simplify things to assume that a higher oil price would have *not* influenced China's demand significantly, China may have needed to devote several hundred more millions of dollars a day, or close to $100 billion a year, to importing oil from 2011 to 2014 had the boom not occurred.

Moreover, the actual price plunge came at a time when China's leaders were—and are—struggling to maintain growth. Xi Jinping assumed the presidency in 2012 with a new economic vision for China. The global recession of 2008 had underscored the fragility of an economic model that relied heavily on government-driven investment and exports to the rest of the world. Xi and his party counterparts pledged to transform this Chinese economy into one that was more consumer-oriented. Future Chinese growth would rest more in the hands of China's 1.4 billion citizens than in the appetites of consumers around the world or in the initiatives of the Chinese government. The benefits of this transformation would be twofold. Not only would future Chinese economic fortunes be less hostage to the health of other economies, but a domestic demand-driven, service-oriented economy would be far less energy intensive than one dependent on heavy industry.

On paper, accomplishing all that sounds straightforward enough, but in reality this transition is proving to be a complex and high-stakes one. What's more, China is seeking to complete it over a much shorter time frame than other countries have done. The United States, for instance, took decades to achieve a similar transition. For China, the shift involves not only changing the drivers of the economy as well as its structure, but also accepting lower overall rates of growth. In March 2015, Premier Li

Keqiang set the target for China's growth for the coming year at 7 percent, the lowest in fifteen years. The Chinese government wanted to orchestrate a "smooth landing," in which fiscal and monetary policies ease the rate of growth and the inflationary pressures that go along with it, while avoiding a slowdown that could have destabilizing effects politically and socially. The unexpected contraction of Chinese exports in the spring of 2015, and the crash of the Chinese stock market later that summer, fueled speculation among experts that the real growth in the Chinese economy was significantly below the 7 percent per annum reported by officials and deemed to be the minimum rate at which China must expand to avoid social unrest. By 2017, there was less speculation about growth rates being misrepresented, but attention had shifted to the ways in which the current growth rates were being sustained. While consumer demand in China was growing at a healthy pace, economic reforms seemed to have stalled and the balance of growth was primarily generated by increases in government consumption.

What is inarguable, though, is that low energy prices helped make a difficult financial situation significantly better. For every one-dollar drop in the price of oil over the course of 2014, 2015, and 2016, China saved $6.8 million every day, or $2.5 billion annually. Put another way, given that the price of oil dropped by $59 from June 2014 to December 2016, China's *daily* oil import bill at the end of 2016 was roughly $400 million less than it was eighteen months earlier for the same amount of oil.

However, despite such benefits, energy abundance has scarcely been a panacea. As has been the case in America, low energy prices have not actually provided an obvious stimulus to the Chinese economy. Low oil prices have also wreaked havoc on China's own sizable oil industry. As high-cost producers, Chinese oil companies struggled, pumping oil domestically at a significant loss. This has led to stress on the industry, which risks adding to the growing pool of unemployed workers in China.

True, the low price of energy has been a welcome development for much of China's society and economy. Yet, the large structural problems of China's economy cannot be addressed or resolved with the modicum of relief that the new energy abundance has brought to most sectors; serious economic reforms will be required. While creating some space for

Chinese policymakers, China will need much more than the serendipity of low energy prices to resolve its enormous economic challenges.

Lubricating Legitimacy

Fabled to have lived thousands of years ago, Da Yu—or Yu the Great—is a legendary Chinese ruler credited with taming the devastating flooding of the Yellow River and thereby enabling agriculture to emerge along its banks. This remarkable feat allegedly won Yu the adoration of the people and their consent to be ruled by him. The story of Da Yu cements in every Chinese mind the delicate but vital link between the environment, civilization, and the mandate of the ruler. At no time in modern history have these connections been so clear—and under such threat—as today. In recent years, the woeful state of the Chinese environment has become a major preoccupation of Chinese citizens. In February 2015, a video titled *Under the Dome* about China's environmental degradation went viral, and was reportedly viewed by 117 million Chinese within twenty-four hours of its release. Simple scenes seemed to resonate with the Chinese. At one point, a six-year-old girl named Wang Huigin tells the filmmaker that she has never seen "a real star," blue skies, or white clouds in her life. The stature of many of the individuals interviewed suggested the documentary was originally approved by the government. But the chord it struck with the Chinese public clearly unsettled authorities, who banned it only a week after it was released.

No longer content with economic progress at all costs, Chinese citizens have increasingly mobilized to demand more sustainable growth. Using social media, ordinary citizens have challenged government plans to build chemical and other potentially hazardous plants. In 2012, in Shifang, a western Sichuan town that suffered badly during the 2008 earthquake, high school students researched the potential adverse health effects of a molybdenum copper plant planned for the town. Through sites such as We-Chat and weibo (China's equivalent of Twitter), they posted information about their findings, galvanizing large-scale protests that turned violent in confrontations with the police. Although journalists had little access to the area, pictures of bloodied protesters were widely disseminated over the Internet, contributing to the decision by

Chinese authorities to suspend plans for the copper plant after two days of protests.

Such incidents are increasingly common. Yang Chaofei, the vice chairman of the Chinese Society for Environmental Sciences, claimed that the number of protests related to the environment rose by nearly a third each year from 1996 to 2011, and increased by 120 percent from 2010 to 2011 alone. A 2013 survey conducted in China found that four-fifths of respondents wanted the government to prioritize the environment over economic growth and nearly the same percentage vowed to join protests if a polluting facility were planned near their residences.

Under the right conditions, the new energy abundance should help China with this very real challenge—and the threat it poses to the legitimacy of the Chinese Communist Party. More abundant natural gas—whether from within China or outside it—could be an important ingredient in China's drive to grow in a more environmentally sustainable way. China's current energy mix is heavily dependent on coal. This dirtiest of the fossil fuels has met an overwhelming 70 percent of China's overall energy needs each year between 1980 and 2014, although the last several years have seen a gradual decline in this percentage. According to the U.S. EIA, "China consumes and produces almost as much coal as the rest of the world combined." Curbing pollution, as well as carbon emissions, will require the realization of the much heralded, but as yet undiscovered "clean coal," or, more realistically, the switch away from coal toward cleaner fuels such as natural gas, renewables, or nuclear power. Even factoring in slowing economic growth today, the scale of this challenge is massive, given China's current energy needs and its expectations for future demand.

The Chinese government has worked hard to ensure that natural gas will meet an ever-growing portion of Chinese energy demand. Multiple obstacles have existed to the greater adoption of natural gas, which met 6 percent of China's overall energy demand in 2015. For starters, the fuel has no bureaucratic champion in a complex, state-run system. Moreover, China's reserves are far removed from likely demand; the Tarim Basin in the northwest province of Xinjiang is roughly the same distance from gas-thirsty Shanghai as New York is from Los Angeles. Perhaps most important, the pricing system for natural gas has provided few incentives for producers, given that it was often sold for less than the price of production. Nor has natural gas been particularly attractive for consumers, given the much cheaper alternative of coal.

Recognizing many of these hurdles, and interested in promoting natural gas for environmental and energy security reasons, the Chinese government has taken steps in recent years to help increase the role of this fuel. Its goal is for gas to account for 10 percent of overall energy consumption by 2020, by both meeting new demand and squeezing out some of the current demand for higher-polluting coal. This may sound like a small amount of natural gas, given that, on average, the rest of the world uses natural gas for 25 percent of its energy mix. But China's targets would equal more than twice what Russia exported to Europe in 2014, or about the same amount that Russia's Gazprom produced for both domestic consumption and export in the same year. To keep the country on track to meet its 2020 targets, the government in 2015 lowered natural gas prices in some sectors and will permit industrial providers to charge up to 20 percent more for natural gas than government benchmark prices. Partially in response to such efforts, natural gas has begun to make inroads into the energy mix, broadening beyond its traditional use in fertilizer and chemical plants to be used in industrial, residential, and even the transportation sectors.

Despite these and other ongoing efforts, it is not certain that natural gas will meet its potential and help China manage the enormous challenges it faces in managing its economy and its environment simultaneously. One possibility is that natural gas may continue to lose out to coal. The global glut in natural gas and China's price reforms notwithstanding, coal still remains three times cheaper than natural gas for the generation of electricity in China. Without additional policy interventions, a large-scale shift from coal to gas—as was seen in the United States—will not occur. China could continue to bring down the price of natural gas to make it more competitive with coal—or China could increase the relative price of coal. The latter could be achieved through China's plans to develop the national carbon emissions cap-and-trade system that was part of Beijing's pledge in the Paris climate agreement. However, if MIT researchers Sergey Paltsev and Danwei Zhang are correct, such a system on its own could actually curb natural gas use, given that resource's own related carbon emissions. Instead, if natural gas is to make a positive contribution to China's energy mix, Paltsev and Zhang conclude a cap-and-trade system will need to be accompanied by a subsidy for natural gas.

Another possible future for natural gas in China is that its expansion gets squeezed out by renewables. China spent $90 billion on renewables-based electricity generation in 2015, making it number one in the world; this amount was more than double the sum that China invested in electricity generated by fossil fuels in the same year. According to the IEA, electricity generated by certain types of solar energy is now cheaper than some forms of natural gas–generated electricity. Dropping costs are expected to make renewables even more cost competitive in the coming decade. Particularly given the need to build more infrastructure to support the greater use of natural gas, investors may come to see putting money into natural gas today to be similar to investing in floppy disks in 2000. Nevertheless, the cost differentials that favor renewable energy to some extent depend on China's continued subsidization of such forms of energy—and assume that policy support for natural gas does not materialize.

Ultimately, China will need both natural gas and renewables to successfully transition to a more sustainable energy mix and economy. Even acknowledging China's slowdown in economic growth and energy intensity, China will need vast amounts of new energy from diversified sources in the future. Moreover, it will need to have non-coal energy sources that can generate power around-the-clock—a need renewables cannot yet meet until there are better options for storage.

"Expect the unexpected" is a common saying in Chinese. It is only in this way that President Xi and the Chinese Communist Party could have anticipated the boon they have received from the new energy abundance. At a time of significant challenge to the legitimacy of the Communist Party and the government, a dramatically changed energy landscape whose roots lie outside China's borders opens new opportunities to help manage some strikingly difficult domestic challenges. The strategic benefits to China of the new energy abundance, however, are not limited to the homefront domain. They extend well beyond the Great Wall to shape China's emerging role in the international arena.

An Opening to a Different Kind of Foreign Policy

The line of angry motorists snaked along the steamy streets of Shenzhen, a large city in China's southern Guangdong Province, for more than two kilometers. Fuel rationing and closed service stations suddenly became the norm, as the province battled shortages of gasoline in the dead heat of the 2005 summer. Concerned about the possibility for social unrest as hot and frustrated drivers jostled to fill their tanks, the Chinese government sent thousands of public security officers and paramilitary police to more than five hundred gas stations across the province.

Analysts attributed this particular crisis to pricing policies that made it unprofitable for refiners to process crude, given rising oil imports and increasing global oil prices. Nevertheless, such scenes rattled the nerves of Chinese leaders, who are ever vigilant about sparks that could ignite larger social discontent. As Zheng Bijian, a senior advisor to then–Chinese president Hu Jintao, explained in 2005, "China has a population of 1.3 billion. Any small difficulty in its economic or social development, spread over this vast group, could become a huge problem." The energy shortages of that summer reinforced the already acute sense of the iron-clad link between access to energy, economic growth, and social stability—and ultimately, the legitimacy of Communist Party rule.

It was exactly to prevent such moments of public turmoil that Chinese officials had turned their attention a few years earlier to securing energy resources abroad. Shortly after assuming office in 2002, President Hu Jintao and Premier Wen Jiabao decided that securing reliable supplies of oil and other resources should be considered not simply an issue for development, but also an element of national security. In 2004, Li Junru, the vice president of the Communist Party's Central Party School, named competition for energy resources as the most important factor shaping China's "peaceful rise," placing it ahead of Taiwan. And one year later, Zheng Bijian, Hu's advisor, cited the shortage of natural resources as a growing obstacle to be overcome in China's development.

The pressure to secure resources from abroad led China to massively expand its diplomatic ties over the first decade of the twenty-first century, with a clear focus on energy-producing countries and regions. It

led to a transformation of China's historical ties with Africa and a re-
juvenation of relations with the Arab world after a lull of hundreds of
years. Energy needs also spurred at least a partial thaw in affairs with its
northern neighbor, Russia, and a careful and deliberate construction of
ties between Beijing and the countries of Central Asia. And, over time,
China—through investment and its appetite for commodities—became
one of the primary drivers of growth in Latin America.

As helpful as it has been, the new energy abundance has not freed
China from the quest to secure resources from abroad. True, the prospects for greater Chinese energy self-sufficiency have improved with the
discovery that China has the greatest technically recoverable shale gas
reserves in the world—almost twice those of the United States. But difficulties in extracting this gas on a large scale dampened the initial enthusiasm of the Chinese government and caused it in 2014 to scale back its
expectations by half for how much shale gas will contribute to the Chinese energy mix over the rest of this decade. While China may well become a shale gas powerhouse over time, it will first need to contend with
tough geological realities, find ways to increase the incentives Chinese
national oil companies (NOCs) and private companies have to produce
shale gas, and address obstacles such as water scarcity, lack of infrastructure, discouraging property rights, and insufficient environmental regulation. China's shale oil resources—also believed to be substantial—are
even further from development.

Even if the new energy abundance will not change China's massive
need for resources from abroad, it *will* dramatically change the context in
which that pursuit occurs; China will seek to satisfy its more modest energy
demand growth in a context of relative plenty, not scarcity and cutthroat
competition. Chinese and other leaders now know that huge quantities of
oil and gas can be produced at prices that are higher than the troughs of
2016, but lower than the $100 a barrel that seemed "reasonable" to both
consumers and producers just a few years earlier. Oil prices will rise and
fall depending on a range of factors, but the scenario of great powers competing for limited oil resources—so prevalent only a few years ago—now
seems like the discarded storyline of outdated movies or novels.

The price plunge beginning in late 2014 made the oversupply of oil evident to the world, and, a few years later, a surfeit in natural gas emerged. U.S. shale gas production alone has had positive knock-on effects for China. LNG exports once destined for the United States ultimately made their way to Asia, where they added to the liquidity of the market and helped keep prices from rocketing even higher at a time of growing Asian gas demand. As discussed in Chapter Three, pressure mounted on suppliers to revise their contracts and move natural gas prices away from a near total linkage with very high oil prices to reflect spot prices to some degree. In this positive economic environment, China emerged, seemingly overnight, as the third largest LNG importer in the world, increasing its appetite sixfold from 2008 to 2013. The oil price plunge of 2014–2015 provided some added relief, as the combination of adequate supplies and lower oil prices drove down once sky-high natural gas prices to Asia significantly.

China and other consumers of imported natural gas look ahead and see a natural gas landscape even more suited to their needs for the next decade or longer. Australia and the United States together could bring nearly 50 percent more LNG to global markets by 2020. Looking out two more decades, LNG supply appears set to increase by more than two-thirds. Lower than expected increases in demand growth—owing to dampened Chinese growth or the resumption of nuclear power by Japan—could further deepen this glut. China now finds itself in a surprisingly comfortable position: it has contracted for more natural gas in the coming years than it expects to use, even if it meets its objective of transitioning to a more natural-gas-intensive economy.

The critical link between energy, growth, and the legitimacy of Chinese Communist Party rule is still strong, and as a result Chinese leaders will not become complacent about securing needed energy from abroad. But much of the urgency and anxiety around procuring these resources has been moderated. This change in temperature has subtle, but extremely important, repercussions for China's foreign policy, given the central role that securing energy has played in the past. As we will see, the new energy abundance gives China an opportunity to rethink and revise core elements of its foreign policy, as well as provides more leeway for China to focus on objectives other than securing energy. In doing so, The new energy realities further align China with the current international order, rather than pitting it against it.

Downsizing the Gallery of Rogues

After months of careful consideration, extensive study, and vigorous debate within the U.S. State Department, Secretary of State Colin Powell was ready to issue a verdict. The date was September 9, 2004, and he was seated in front of the Senate Foreign Relations Committee, testifying on recent events in Sudan. After detailing the situation there and the administration's response, Powell wound his way to the key moment: "We concluded—I concluded—that genocide has been committed in Darfur and that the government of Sudan and the Janjaweed bear responsibility— and that genocide may still be occurring." This was the first time the executive branch of the U.S. government had *ever* used the word "genocide" in relation to an ongoing conflict.

Nine days later, the paltry fruits of American efforts to get the United Nations to pressure Khartoum to protect civilians and cooperate with African Union monitors were announced to the world. The title of the U.N. press release sums it up: "Security Council *Declares Intention to Consider* Sanctions to Obtain Sudan's Full Compliance with Security, Disarmament Obligations on Darfur" (italics added). China, which ultimately abstained, managed to dilute the resolution referred to in the press release so as to make it almost meaningless, and threatened to veto any subsequent efforts to sanction Sudanese leaders.

China's position toward Sudan, which it maintained for much of the rest of the decade, had two main drivers. Beijing sought to shield Khartoum's domestic behavior from international scrutiny, in the hopes that it could similarly dissuade the world from peering into China's own internal affairs. Equally important, China aimed to protect its substantial investment in Sudan's oil industry, even after shareholders had forced out Western companies on moral grounds. By 2004, China had invested substantially in Sudan's oil industry and was importing more than four-fifths of Sudan's oil output. China's deputy foreign minister, Zhou Wenzhong, was unapologetic about China's stance. "Business is business," he stated. "We try to separate politics from business. Secondly, I think the internal situation in the Sudan is an internal affair, and we are not in a position to impose upon them."

While China's approach to Sudan drew more international attention and outrage than its policies toward other countries, it was not one of a kind. In fact, at the time, Beijing's policy toward Sudan was fairly representative of China's long-standing "going out" strategy. In essence, the "going out" approach was and is an effort by the Chinese government to encourage Chinese companies—state-owned ones in particular—to seek investment opportunities in a wide array of resources beyond China's borders. This encouragement was not just rhetorical but came in many forms. These included arranging high-level visits by government officials to countries where Chinese companies were seeking deals, tying development aid and other financial assistance to resource-wealthy countries to the completion of deals, and extending preferential credit to Chinese companies seeking to compete with international firms. Adding to the attractiveness of such deals was the Chinese government's indifference to the domestic policies of the country in question. Investment, finance, and aid were predicated on ownership of oil and other resources. Human rights, fiscal policies, domestic subsidies—none of these things appeared to matter to prospective Chinese investors and their government backers with deep pockets.

First articulated in 1997, China's "going out" strategy originally seemed geared toward diversifying China's investments and introducing competition to some of China's largest state-owned firms. But by the early 2000s, a sharper objective had come into focus. The Chinese government and China's NOCs joined forces to seek and obtain equity oil investments worldwide, particularly in Africa and Latin America. Chinese presidents, premiers, and development ministers crisscrossed the vast continents of South America and Africa in the company of Chinese oil executives. Almost without fail, they left in their wake dozens of signed trade, investment, and even military cooperation agreements as well as billions of dollars of investments, many of them in equity oil. In 2010 alone, Chinese NOCs spent nearly $30 billion on acquiring oil and gas assets globally, with nearly half that sum invested in Latin America. In the decade following 2004, China invested more than $45 billion in exploring and producing oil and gas acquisitions in continental Africa. At least initially,

China seemed to believe that equity ownership of African or Latin American oil would help insulate it from growing international competition for resources. Anticipating an ever-more-competitive energy landscape, the duo of the Chinese government and the country's NOCs aggressively sought equity investments with the expectation that, in some future time of crisis, China would be able to ensure this oil flowed in its direction.

In seeking out such investments, Chinese NOCs initially felt disadvantaged and unable to compete with international oil companies that already had substantial presences on both continents. Seeing themselves as late to the game of overseas resource development, Chinese NOCs targeted countries that—for a variety of reasons often related to sanctions or high political risk—did not have access to Western capital or markets. From Venezuela to Sudan, China provided much needed capital to at-risk governments in return for actual ownership over their resources, or claims to guaranteed flows of oil.

The result—intended or not—was that Beijing appeared to be cultivating its own gallery of rogues across Latin America and Africa. Such relations often contained an anti-American or anti-Western tinge. For instance, between 2007 and 2015, Beijing loaned Caracas approximately $50 billion, most of it to be repaid in shipments of oil. For years, these loans-for-oil arrangements have brought with them close political ties. In late 2004, Venezuelan president Hugo Chávez told a group of Chinese businessmen, "We have been producing and exporting oil for more than 100 years. But these have been a 100 years of domination by the United States. Now we are free, and place this oil at the disposal of the great Chinese fatherland."

The "going out" strategy has also been perceived as a challenge to Western efforts to promote good governance, environmental protection, and economic reform in Africa and Latin America. Western governments generally view the investments made by international companies domiciled in their countries as entirely separate from development aid and finance to countries in need, much of which is administered by multilateral development banks. China, in contrast, saw the two as complementary. When given the choice between Beijing's unconditional assistance and the highly structured and conditional aid packages of the World Bank, IMF, and Western governments, few developing countries chose the more stringent option. Perhaps the most oft-cited example of how this Chinese approach undermined Western efforts to promote certain liberalizing economic and political policies occurred in Angola.

In 2002, Angola emerged from twenty-seven years of civil war with its countryside and cities devastated and half a million of its people dead. Once an exporter of food, by the 1980s, Angola had become an importer of grains and a country in which nearly half of the children were suffering from malnutrition. Like other countries in dire need of basic infrastructure and development, it turned to the IMF to provide a significant loan for such purposes. As was common practice, the IMF required an accompanying "stabilization program" through which Angola would agree to greater transparency and certain economic reforms; Angola was already a significant oil exporter, and according to the IMF, at least $8.5 billion of public funds had been unaccounted for over the previous five years. Abruptly, in 2004, Angola broke off these negotiations with the IMF and soon afterward announced it would be the recipient of $2 billion in soft loans from Beijing, granted at very favorable interest rates. Angola would repay the loan in shipments of oil over the following twelve years. Around the same time, negotiations were concluded in which Sinopec, one of the Chinese NOCs, secured a 50 percent stake in Angola's Block 18, which had belonged to Shell before being sold to the Angolan company Sonangol earlier that year. Two years later, in June 2006, when the prime minister Wen Jiabao visited Angola, Angolan president José Eduardo dos Santos summed up the relationship, "China needs natural resources, and Angola wants development." Soon thereafter, China's EximBank announced a $2 billion loan for Angolan infrastructure projects. As of 2017, Angola was China's second largest supplier of crude oil imports, providing more oil to China than did Russia, Iraq, or Iran.

"Going Out" Blowback

The new energy abundance is one of several factors that have called into question the value of this "going out" strategy. Beyond doubt, China's "going out" strategy has brought benefits to Latin America and, especially, Africa. Over the last decade and a half, China has built critical infrastructure within countries and across the continents. By far the largest financer of African infrastructure, Chinese institutions have supported the construction of dams intended to boost hydropower-generating capacity and rehabilitated or constructed from scratch roads, railways, and airports across the continent. Africa's telecommunications sector has also been transformed by China's investment and financing. Little wonder that China remains generally popular in Africa; a 2014 Pew poll found that, except in South Africa, the overwhelming majority of those polled in five African states had a favorable view of China.

There is, however, evidence of growing discomfort with China in some parts of Africa. Writing in 2013, Nigeria's central bank governor, Lamido Sanusi, called for Africans to "wake up to the realities of their romance with China. . . . China takes our primary goods and sells us manufactured ones. This was also the essence of colonialism. The British went to Africa and India to secure raw materials and markets. Africa is now willingly opening itself up to a new form of imperialism."

Sanusi wasn't alone in his concern. African civil society groups and others have protested the effects of Chinese practices on human rights, governance, labor conditions, and the environment. The more than one million Chinese expatriate workers living in Africa in 2015 have also been an irritant to some local populations; African governments have promulgated new local content laws in reaction to this influx of Chinese workers. The Zambian presidential election of 2011 served as a further example of the backlash against China in some parts of Africa. In a country where banking can be done using the renminbi, the Chinese currency, Michael Sata, the successful candidate, ran a campaign highly critical of Chinese influence. Sata stirred up crowds at his rallies with declarations such as, "Zambia has become a province of China. . . . The Chinese are the most unpopular people in the country because no one trusts them. The Chinaman is coming just to invade and exploit Africa." He reviled the Chinese for "bringing in their own people to push wheelbarrows instead of hiring local people." Rumors circulated that his opponent's campaign was funded by China. Speaking in Beijing in 2012, South African president Jacob Zuma warned that China's relationship with Africa was "unsustainable," delicately but clearly referencing African concerns about the emergence of a neocolonial relationship.

More recently, China's "going out" strategy garnered scrutiny within

China itself. Over time, the NOCs executing the strategy became more and more driven by commercial motivations, particularly after they were partially privatized. Increasingly, observers concluded that NOCs—not the government—were the main drivers of the "going out" strategy, using government largesse and influence not to augment Chinese energy security but to advance their own corporate objectives. While the activities of Chinese NOCs have increased global oil supply and, therefore, the energy security of all consumers, relatively little of this production has found its way back to China. One industry insider shared with me his private estimate that only 10 percent of the "going out" investments generated oil eventually consumed in China; the rest was sold on the open market. Moreover, the oil that does flow from these investments in Latin America and Africa to China in most cases is transported via waterways protected by the U.S. Navy and therefore considered vulnerable. While the equity oil investments certainly provided an economic hedge against high oil prices, these realities made it difficult to argue that they directly and materially enhanced China's energy security.

Other critics have focused on the fact that in commercial terms many of these equity oil endeavors proved to be poor investments for the Chinese; indeed, as many as two-thirds have been estimated to be unprofitable. For example, Chinese NOC Sinopec made investments of $10 billion in Angola between 2008 and 2015, yet reported no returns on its investment. Corruption has flourished in deals of such great size, with so many players, in nontransparent environments. One of the most powerful senior officials or "tigers" targeted by President Xi's corruption probe was Zhou Yongkang, a former head of the country's domestic security and an earlier head of China National Petroleum Corporation (CNPC). Zhou had been an aggressive promoter of CNPC's international expansion and had personally spearheaded the company's entry into Sudan, among other places. Sentenced in 2015 to life in prison, Zhou was convicted of leaking confidential documents and accepting $118,000 in bribes, including those from Jiang Jiemin, a former head of CNPC who was also jailed for corruption. Zhou and Jiang were not alone in their convictions; more than a dozen senior officials from Chinese NOCs lost their jobs and were investigated for corruption.

The revelations behind these corruption probes raised eyebrows both internationally and domestically. Chinese consumers especially chafe at the thought that NOCs might have pressured the government to offset their losses overseas with subsidies and higher domestic oil prices, contributing to domestic inflation. Some African and Latin American governments may be equally dismayed to learn how Chinese officials enriched themselves in such deals, which could potentially affect their ability to strike similar deals in the future.

China's oil-for-loans programs also look less attractive in light of new energy abundant realities. Take Venezuela. The loans-for-oil arrangements between Beijing and Caracas mentioned earlier have meant that Venezuela receives little payment for as much as 1 mnb/d of its roughly 2.6 mnb/d of exports, if highly discounted exports to the Caribbean are added to the Chinese flows. As a result, Venezuelan revenues are already well below what might be expected from the OPEC exporter, even before the dramatic price plunge of 2014–2016 is taken into account. As a result, few funds exist for reinvestment into or development of Venezuela's vast reserves, leading to wobbly production projections. Although Beijing has provided Caracas additional financial assistance, Chinese officials might well wonder about the wisdom of further propping up a very unpopular regime struggling to deliver on its commitments over the long run. Dr. Xue Li from the Chinese Academy of Social Sciences and one of his colleagues articulated this view clearly as early as March 2015: "Considering [Venezuela's] political and economic situation, it is inappropriate [for China] to further increase the amount of investments and loans. In short, China should guard against bad debts." In today's new energy environment, China can be much more confident in the market to deliver energy security than in a brittle, insolvent regime in Venezuela.

For all these reasons, China's "going out" strategy was overdue for a reexamination and revision. The new energy abundance gave Chinese leaders—in government and in the NOCs—space for undertaking such review and reform. In a world of oil and gas plenty, China can be more comfortable in relying on the market as a mechanism to secure needed energy resources. Protecting and strengthening oil-rich governments—and enduring international opprobrium for doing so—seems unnecessary in a world where the Chinese can be confident the market will deliver its energy needs. The changed energy environment gives Beijing more leeway to choose another, less confrontational, less easily misunderstood path. If the costs of the "going out" strategy were deemed high in a world of energy scarcity, they are even more unreasonable—or unnecessary—in a world of abundance.

There is already some evidence this shift is occurring. In recent years, China has been willing to impose sanctions on Khartoum for egregious domestic behavior, and support for Venezuela may be more qualified in private than it appears in public. Commercial considerations seem to be getting the upper hand in deciding on investments, as the profile of Chinese overseas investment is shifting to more mature economies—such as Canada and the United States—from developing Africa and Latin America.

Ceding Space to New Drivers of Foreign Policy

Chinese officials insist Xi Jinping's One Belt, One Road plan is an "initiative," not a "strategy." They also emphatically reject any reference to it being "a Chinese Marshall Plan," as they are eager to divorce this big idea from notions of geopolitical competition and expansion. Yet however one refers to it, Xi's grand scheme demonstrates China's readiness to step away from its reluctant role in the world and instead to be a driver of major change in its region and beyond. In November 2013, Xi presented this proposal to create new corridors of connection and integration between China and more than sixty countries by land and by sea. A bit confusingly, the "belt" refers to efforts to build on the old Silk Road, which extended overland from China, through Central Asia, to Europe. The "road" loosely retraces the maritime voyages through Southeast Asia to the east coast of Africa made by Zheng He, a eunuch admiral of the Ming Dynasty whose ten-thousand-mile journey unfolded almost a century before Christopher Columbus traveled to the Americas.

The modern revival of these two routes underpins the grandest conception of the One Belt, One Road initiative. Although there is considerable uncertainty about the exact details of the effort, hundreds of billions of dollars already appear devoted to it through a Silk Road Fund, the new China-led Asian Infrastructure Investment Bank, and the China Development Bank. These funds are supporting the building of roads, railways, pipelines, and ports. Rather than being limited to tangible infrastructure, however, the initiative also encompasses efforts to promote greater financial integration through the region-wide use of the renminbi; it also entails an "Information Silk Road" of optical cables and other communication nodes to facilitate greater people-to-people exchanges and interactions.

The One Belt, One Road initiative could be seen by some as simply the "going out" strategy on steroids. Part of the appeal of China's western neighbors is, no doubt, their significant energy resources; energy infrastructure is certainly central to the success of the initiative. China would unquestionably welcome a diminished dependency on oil and gas traveling through the Strait of Malacca. Nevertheless, unlike in previous decades, energy is no longer the main animating feature behind China's most dramatic and consequential foreign policy initiative.

Two important factors have allowed other foreign policy objectives to take a seat alongside secure energy in China's list of top priorities. First, as noted, the new energy abundance means China is more comfortable relying on the market to secure its needs and thus can place less of a premium on establishing equity oil arrangements, particularly given their many drawbacks. Second, efforts to rebalance China's economy away from energy-intensive industries means that China's officials are less preoccupied with sating rapaciously growing energy demands. This reality has opened the door for China to place a new, pressing priority at center stage in its foreign policy: the need to create demand for Chinese goods, services, and companies to soak up the excess capacity that has emerged now that the economy is no longer booming at the same rates of earlier years. The lower growth necessarily accompanying the transition to a less investment-driven and more consumption-led economy has left Chinese steel, gas, and concrete makers with far too little demand to satisfy. Booming infrastructure projects supported by the One Belt, One Road initiative outside of China's borders could at least temper the pain of the transition.

Africa neatly illustrates how Chinese economic restructuring at home is replacing energy security as a significant driver of foreign policy. The combination of the downsides of the "going out" strategy and the diminished need to seek equity oil investments in an energy-abundant world appear to be opening the door to a new Chinese approach to Africa—one in which oil assumes a less prominent role. President Xi Jinping signaled this new approach in December 2015 when speaking to the triennial Forum on China-Africa Cooperation in Johannesburg. Nervousness about the trajectory of China's relationship with the continent preceded his speech, stoked by a 40 percent decrease in Chinese investment over the previous year. Chinese imports from Africa—more than four-fifths of which are commodities and crude material—had fallen in value by a third due to the crash in prices and weaker Chinese demand growth. Xi's speech was intended to squelch anxiety that China was a fair-weather partner for Africa. His pledge that China would invest $60 billion in the continent surprised even African optimists. At previous summits, China had consistently doubled the figure from three years earlier; the 2015 commitment, in contrast, tripled it. But what is most remarkable about Xi's speech, and most telling, is the virtual absence of any mention of natural resources.

Why, then, such generosity toward Africa? Because, rather than acquiring equity oil, China is focusing on markets into which it can export the excess industrial capacity it has built up at home during what Daniel Yergin calls "the great build out of China." Now that the age of breakneck industrial expansion and infrastructure development is slowing, China has many firms that are underutilized and would benefit from new markets and contracts overseas. As China seeks to move to a more service-driven economy, it is eager to shift its labor-intensive (and energy-intensive) industries to Africa. "Industrial capacity cooperation" and "strategic complementarity" are not just new words for describing an old relationship. They signal China's intent to broaden its engagement with Africa to better meet its own needs beyond the energy equation.

An Opportunity for U.S.-Sino Cooperation

China is emerging as a critical global actor. After decades of insisting it is only a developing country that needs to focus only on its internal challenges, President Xi Jinping now speaks determinedly of promoting "the Chinese dream." While China's intention to assume a larger role on the international stage is clear, the country and its leadership do not yet seem to have constructed a grand strategy to guide its emergence. One can discern elements of China's desired *regional* order, but the leadership's thinking on the *international* order seems more nascent. Given China's size, the way in which China develops itself as a global power will have massive implications for countries around the world.

How and to what ends China seeks to wield its influence in the world will be of great consequence to the United States in particular. Many factors will determine whether the United States and China emerge from this historical juncture as partners or adversaries. Graham Allison, a renowned professor at Harvard, describes the momentum toward conflict between the United States and China as "the Thucydides trap." According to the Greek historian, it was the rise of Athens, and the fear it inspired in Sparta, that triggered the start of the Peloponnesian War in 431 BC. Allison studied sixteen historical cases in which a rising power challenged a ruling one and found that in a dozen of these circumstances, the outcome was war.

Energy will not be the only factor weighing in one direction or the other as the United States and China chart a new course. But it is an important one, and one that—in a dramatic shift from the predictions of less than a decade ago—leans squarely in favor of the two powers successfully negotiating some sort of peaceful coexistence, or at least finding common cause in some strategic arenas. There are several spheres in which these two countries, thanks to energy, can find mutual interest if their leaders so chose. These areas have value in their own right, but are even more important as they represent possible points of collaboration in an otherwise increasingly fractious bilateral relationship, which happens to be the most significant relationship in the world today.

Notably, the new energy abundance will align China and the United States more closely on the question of the appropriate international order. China is likely assessing whether it can satisfy its global ambitions under some modification of the current international order or if it needs to challenge and remake the international order entirely to serve its interests. While the jury is still out on this hugely consequential question, the new energy abundance at least increases the chances China will find a comfortable home in some version of the current order. If, as predicted, the new energy abundance contributes to China revising its "going out" policy, that, too, should diminish China's *rationale* for supporting rogue states and therefore potentially remove one arena in which China has been at odds with established norms and institutions. More sanguine about their ability to meet energy demands, China's leaders need not feel the same impulse to cultivate recalcitrant regimes—and shield them from the concerns of the international community—in order to maintain access to their resources.

As discussed, the new energy abundance has also increased the confidence of the Chinese leadership in one of the main principles of the current international order: the primacy of the market. A decade ago, China mistrusted the market as the mechanism to deliver energy to the country. Washington's denial of the 2005 attempt by China National Offshore Oil Corporation (CNOOC) to purchase the American oil company Unocal, the U.S.-led war in Iraq—which Beijing viewed as being motivated by energy—and the perception of diminishing global energy resources collectively pushed Beijing to look to nonmarket means of acquiring the energy so essential to China's prosperity and the Communist Party's legitimacy. A decade later, Chinese companies have invested more successfully in North American energy companies, and the United States looks chastened from its robust military action in the Middle East. But most significantly, the dramatic shift from scarcity to abundance greatly diminishes the chances that the market will be unable to manage the allocation of resources and keep China from getting the resources it needs.

The new energy abundance also provides grounds for recasting ties between the United States and China. For starters, it helps deflate the "China conflict narrative" that had become the dominant lens through which nearly all Chinese overseas activity was viewed by the United States and other Western countries. In April 2005, Mikkal Herberg, the director of the Energy Security Program at the National Bureau of Asian Research, told a group of academics that he could not foresee any future scenario where there would not be confrontation between the United States and China over energy. This perspective became common among U.S. policymakers and the main framework for evaluating China's rise within the international system.

According to U.S. defense officials and members of the American Congress, competition for energy was one of the most commonly cited potential flash points for the inevitable conflict between a rising China and a declining West. In 2006, the Pentagon's annual report *Military Power of the People's Republic of China* broadened its analysis of drivers of Chinese militarization beyond Taiwan to include the need to potentially use force to secure resources. Two years later, a new version of the same Pentagon annual report noted that growing Chinese resource needs could spur China to develop more robust defense capabilities and a "more activist military presence abroad."

The advent of the new energy abundance has shifted the framework from one of inevitable competition and possible conflict between the United States and China, to one of potential and actual cooperation around energy. Neither country has to see the energy-inspired actions of the other in a zero-sum light. China's foreign policy need not be dominated by its desire to secure resources at all costs. The possibility to pursue other aims—ones not so likely to create friction with the United States—opens up.

Of course, it is conceivable that a China with a wider foreign policy agenda will challenge U.S. interests in other, perhaps even more fundamental, ways. Growing tensions over the South China Sea suggest, at least in that part of the world, this will be the case. But the new energy abundance not only removes the old lens through which American policymakers viewed Chinese actions abroad, but it also offers new grounds for cooperation.

Indeed, some of the most direct and obvious new avenues for fruitful U.S.-Sino cooperation are in the development and trade of energy. For instance, U.S.-Sino cooperation could dull Russia's heavy-handed use of energy to shape politics to its advantage. Rather than seeking to curb China's efforts to extend its energy influence in Central Asia through initiatives like the Asian Infrastructure Investment Bank, the United States and its allies might pursue just the opposite tack. Yes, it would be nice if the energy-rich Central Asian republics built closer ties to their west rather than to their east, but energy realities—and geopolitics—simply do not make Europe a realistic alternative to the Chinese market for these countries in the coming decades. Central Asian oil and natural gas, however, will be the best counterweight to Chinese overdependence on Russia, especially given that this energy comes overland, not via sea lanes

Beijing views as insecure. Pipelines from Myanmar and, potentially, via Pakistan will also have this advantage.

In addition, the United States also should be vested in helping China meet its natural gas needs, as the resource is key to China's ability to realize its climate goals and keep its economy growing—both of which are in the interest of the United States. The United States has made some efforts in this direction; as discussed in Chapter Six, the U.S. government engaged early on with China in an effort to help the country assess its unconventional resources and to provide advice on how to develop them.

Yet, the United States should also do what it can to see that China has diverse, secure, and affordable LNG sources. The U.S. shale gas boom and the advent of U.S. LNG exports will continue to exert downward pressure on the price that China will pay for its LNG, partially eroding the cost advantage of Russian gas. But the United States could be more forward leaning in making it clear that it welcomes China as a customer and investor in U.S. LNG. Early signals were not encouraging. During a Senate committee hearing in 2015, Senators Debbie Stabenow and Angus King expressed reservations that U.S. LNG exports to China were consistent with U.S. interests. That same year, some American companies working on LNG export projects were left with the impression that involving Chinese companies or customers could make their projects politically untenable. Michael Smith, the CEO of Freeport LNG, said there was Chinese interest in Freeport's project, but recalls that "We were advised by the DOE to be careful who our customers were, because this is very political . . . a political hot potato we couldn't take the risk on." As of early 2017, China had received ten shipments of LNG cargo from Louisiana's Sabine Pass facility, but all through third parties. There did not appear to be direct contracts between Chinese companies and U.S. LNG facilities, however, suggesting that both parties saw an interest in keeping a distance.

Instead of seeing Chinese involvement as customers and investors in U.S. LNG as politically undesirable, Congress and the U.S. administration should welcome it. If China wants to create a dependency on U.S. energy in the interest of diversifying its sources—however small it would likely be in the context of China's overall needs—the United States should be more than happy to oblige, particularly if it helps weaken Sino-Russia links.

Finally, members of Congress and the administration could make it clear that Chinese investment in U.S. energy companies is welcome—except under extraordinary circumstances in which most foreign investment would be curtailed. While Chinese investment has quietly flowed into the American energy sector, episodes such as the 2012 forced reversal of a Chinese investor in a wind farm may leave the opposite impression.

Reimagining the Strategic U.S.-Sino Relationship

Looking beyond questions of investment and trade, the new energy landscape has generated further opportunities for the United States and China to recast their bilateral relationship in a new light. President Nixon and Henry Kissinger orchestrated the United States opening to China in 1972 with a big idea in mind. Recognizing Beijing and building a relationship with China would generally provide the United States more flexibility in the international arena, help ease the "pain of an inevitably imperfect withdrawal from" Vietnam, and would specifically give Washington more leverage in its dealings with the Soviet Union. For China, the opening brought equally important strategic gains, from securing military assistance to increasing the international standing of the People's Republic. Both countries were motivated by the need to check the Soviet Union, and were united in that strategic objective, even if they still had major differences in other arenas.

The geopolitical context has now changed, and the agenda between the United States and China is exponentially more complex. Policymakers in both capitals seem in need of bigger, animating ideas to underscore why a closer relationship between the two countries is in their strategic interests. The new energy abundance offers just that by generating opportunities for cooperation in two areas that together offer sufficient grounds to reimagine the bilateral relationship: climate and cooperation on the Middle East.

On November 11, 2014, after the 2014 Asia-Pacific Economic Cooperation summit, President Obama accepted a rare invitation to visit the Zhongnanhai government complex in the Imperial Garden next to the Forbidden City. The heart of the Communist Party, Zhongnanhai remains somewhat of a mystery to Chinese citizens and foreigners alike. In stark contrast to the White House on Pennsylvania Avenue in Washington, D.C., Zhongnanhai is believed to encompass the home of the Chinese president, although his exact address is unknown. President Obama was to meet President Xi Jinping on Yingtai, an idyllic island in the South Lake of the Imperial Garden. The two leaders were not the first to use this spot to address pressing issues; the Qing emperor Kangxi was believed to have formulated his strategies to ease civil strife amidst the beauty of the place. The choice of Yingtai was clearly meant to deliver a message.

The two leaders took some time before getting down to business. Despite the cold and the wind, the evening began with Xi's giving Obama a personal tour of the island and a lesson in Chinese history. A formal meeting between the two leaders, flanked by their advisors, followed and flowed into a dinner banquet. Afterward, the pair retired for a private chat over tea. The private meeting lasted almost twice as long as the presidents' schedulers had anticipated, as the leaders spoke about matters spanning the Chinese economy, Chinese reforms, human rights, and sovereignty. According to the Chinese press, at the end of the night, Obama allegedly remarked to Xi that "this evening has given me the deepest, most thorough understanding of the Communist Party's history and governing philosophy that I have ever had in my life, and allowed me to see your perspectives." While this quote is likely embellished, it demonstrates the cordial tone that set the stage for the historic announcement Obama and Xi would make the next day.

A mere few hours later, at a press conference at the Great Hall of the People, Obama and Xi presented the U.S.-China climate agreement to a surprised world. With the stroke of a pen, the prospects for real international action addressing climate change went from fanciful to possible or even probable. As discussed in Chapter Six, the United States transformed earlier domestic goals into international commitments, pledging to reduce greenhouse gas emissions 26 percent to 28 percent below 2005 levels by 2025. China, for its part, announced its intentions not to reduce carbon emissions, but at least to ensure that they peak by 2030, and to expand its share of zero-carbon energy by 20 percent by the same date.

At least until President Trump stated his intention to withdraw

America from the Paris Agreement, both Washington and Beijing needed one another. President Obama sought to make climate one of his legacy issues, and China's willingness to constrain its carbon emissions removed one of the largest and longest-standing American arguments against U.S. action to address climate change. Since Kyoto, opponents of action to address climate change have argued, with some justification, that it makes little sense for the United States to take potentially costly measures to curb its own emissions if China and other large emitters refuse to do so. For President Xi Jinping, addressing carbon emissions, and environmental hazards more generally, is critical to bolstering the legitimacy of the Communist Party. But doing so will create challenges to vested interests, ones that can be better neutralized if Beijing is adhering to commitments made at the international level, not only the domestic one. Beijing would not be the first capital to use international agreements to force a domestic political agenda. As the United States calculates the cost of jettisoning such goals, it should take the impact of backtracking from these goals on its relationship with China into account.

The Middle East is another matter. To date, there is limited U.S.-Sino cooperation in the Middle East. But it is not difficult to imagine how a closer bilateral partnership could bring a greater measure of stability to the troubled region.

China's ties to the Middle East go back more than two millennia, to Emperor Hu of the Han Dynasty and his decision in 138 BC to dispatch Zhang Qian, a soldier and diplomat, to build economic and political relationships far to China's west. This initiative blossomed into the Silk Road, which, for more than a thousand years, was the route by which endless numbers of merchants, soldiers, nomads, and pilgrims made the journey across deserts and mountains from the Mediterranean to China. In the fifteenth century the expansion of sea trade—which was quicker and better suited to transporting larger quantities of goods—led to the Silk Road's decline, as did China's inward reorientation and subsequent preoccupation with the arrival of European imperialists.

Today, the ancient links between China and the Middle East are being resuscitated. Trade has burgeoned, fueled largely by China's ever-growing dependence on Middle Eastern oil; China's trade with the Middle East increased more than sixfold in the decade following 2004. These links will continue to grow. As North America becomes increasingly self-sufficient, Asia is becoming a more and more important market for Middle Eastern oil. In 2040, the U.S. EIA anticipates that 90 percent of Middle Eastern oil will be destined for Asia. And it is not just trade flows that are increasing, but also investment. The 2017 decision of Saudi Aramco to invest $7 billion to build a massive refinery in Malaysia is indicative of what will be a larger trend as Saudi Arabia and other Gulf states look to ensure markets for their crude in this new energy abundant environment.

China's vision for closer economic and political ties with Arab states was laid out clearly by President Xi in a keynote address to the China-Arab States Cooperation Forum in June 2014. Against a backdrop of pledges of China's commitment to peace in the region and the establishment of a Palestinian state based on the 1967 borders, Xi advocated a new "1+2+3" mode of cooperation between China and the Arab world. Energy cooperation would continue to be the core of the relationship (1), but infrastructure construction, and trade and investment facilitation would build upon it (+2), with an additional focus on three high-tech fields (+3): nuclear energy, space satellites, and renewable energy. Harking back to historical cooperation between the East and West along the Silk Road, Xi predicted that Sino-Arab trade would grow from $240 billion in 2013 to $600 billion over the next decade; such trade had stood at less than $6 billion in 1996.

With this growing economic interdependence will inevitably come greater pressure for Chinese political involvement in the Middle East. China still maintains that its political relations with the region are rooted in the well-worn adage of "noninterference in the domestic affairs of other countries." But there are indications that such a stance is slowly yielding to greater activism, as China's vital economic interests and its concern over its restive Muslim-majority province, Xinjiang, give Beijing a growing stake in stability in the Middle East.

Just compare China's response to Iranian threats to close the Strait of Hormuz in 2008 with that of 2012. Beijing kept quiet when in 2008, heightened tensions in the region led Iran's Revolutionary Guard Corps Commander Mohammad Ali Jafari to warn that "one of [Iran's] reactions [to an attack on Iranian nuclear facilities] will be to take control of the

Persian Gulf and the Strait of Hormuz." This passive Chinese stance is in stark contrast to January 2012 when Tehran reacted to potential new sanctions by once again threatening to shut down the strait. Prime Minister Wen Jiabao, then in the midst of a six-day visit to the Gulf, made it clear that any such Iranian actions would be completely unacceptable to China—and topped off his blunt language by stating that China "adamantly opposes Iran developing and possessing nuclear weapons."

Until recently, U.S. policymakers have tended to see China's involvement with the Middle East as problematic. In speaking with journalist Tom Friedman in the summer of 2014, President Obama smiled wryly and described China's role in this way: "They [the Chinese] are free riders . . . they've been free riders for the last 30 years, and it's worked really well for them." But a close observer of Chinese action in the region would discern a more positive trend: a willingness to limit imports of Iranian oil during the nuclear negotiations, the offer of financial and military support for Iraq in its effort to combat ISIS, serious contributions to U.N. peacekeeping initiatives in Africa, and participation in antipiracy patrols around the Horn of Africa.

The United States and China will increasingly share a common outlook toward the Middle East, one that prioritizes stability over other objectives related to internal modes of governance. The two countries, therefore, are likely to agree more on desired outcomes in the future than at any other point in the recent past. China's useful role in the P5+1 talks with Iran is testimony to this trend. U.S. political and military capabilities in the region are substantial, although such history comes with a certain amount of baggage. American interests will also remain considerable, even though they are marginally declining with diminishing oil imports from the region. (See Chapter Eleven on the Middle East.) In contrast, China's interests in the region are expanding, and owing to its past noninterventionist stance, its baggage is minimal. China's political and military capabilities, however, remain extremely limited.

The shifting U.S. energy stance, however, has convinced many in Beijing that China can no longer be certain that "free riding" will be sufficient to protect its interests in the region. There are signs, in fact, that China's long-standing aversion to the exercise of American military power outside its hemisphere is waning—as Beijing becomes more and more concerned about a power vacuum developing in the Middle East. When asked about Chinese views on the U.S. bombing of ISIS in Iraq in August 2014, a spokeswoman from the Chinese Foreign Ministry declined to make the usual condemnation. Instead, she stated that her country would "keep(ing) an open mind" about operations that would "help maintain security and stability" in Iraq.

Both powers, in short, are in need of a comprehensive, sensible, sustainable strategy toward the Middle East. Devising these strategies in tandem, to the extent possible, could help the U.S. and China advance their interests at more reasonable costs. Large questions remain before such a coordinated approach to the region could be realized. To what extent is China willing and able to devote military resources to the region? And if it does, how might the United States and China cooperate and coordinate in the interest of regional stability? How willing is the U.S. military to yield or at least share operational or other duties with the Chinese? How much intelligence sharing would meaningful cooperation require and what are the associated risks? Are the Chinese willing to use their extensive economic investment in the Middle East to exert political leverage on warring parties and other conflicts? Experiential differences factor in as well.

The Chinese are quick to note that they have only a fraction of America's understanding of the complex dynamics in the Middle East, while also observing that Washington's superior knowledge has seemed to do little to keep it from costly, controversial engagements in the region. Americans cite the difficulty of sharing intelligence, systems, and plans with even their closest allies, never mind with a potential adversary.

Yet, despite these and other inevitable friction points, the trend will be more toward U.S.-Sino cooperation in the Middle East in the years ahead. The problems of the region are growing, not decreasing, and they are likely to overwhelm the ability of any one state to defuse. Moreover, the incentive for closer bilateral cooperation goes beyond the good that might be done in the Middle East. In establishing relationships and modes of cooperation in less prickly areas, such U.S.-Sino cooperation could prove helpful in managing larger points of contention elsewhere.

The rise of China and its consequences for the international system will shape the world in coming decades, or even centuries, more than any other global political development. Policymakers in Washington, Beijing, and everywhere in between are cognizant of certain factors influencing China's trajectory, such as rising nationalism, weakening economic growth, and growing military capabilities. Energy is an equally important factor. Policymakers need to appreciate the many ways in which energy—and the new energy abundance—shape China's possibilities and challenges. In considering the impact of this new energy landscape on China, U.S. and other policymakers would do well to consider the many ways in which seemingly disparate issues are connected by energy. They will find more opportunity than peril in trying to understand China through the prism of energy.

The benefits of viewing Chinese interests and actions through an energy lens, however, extend beyond simply gaining a better understanding of Chinese motives and strategies. It will also open new avenues for Washington to reimagine its relationship with Beijing, to craft approaches that capitalize on common interests, and to offer new possibilities of establishing channels of trust and cooperation in an otherwise tense relationship.

ELEVEN

The Middle East

Trying to Make the Most of a Tough Situation

S ir Mark Sykes slid his finger across the map that was unfurled on a table at No. 10 Downing Street. "I should like," he said to British Prime Minister H. H. Asquith, "to draw a line from the 'E' in Acre to the last 'K' in Kirkuk." Sykes then traced an imaginary boundary from a city on the Mediterranean coast to one near the mountains of present-day northern Iraq. The year was 1915. Sykes had been engaged in a secret mission with François Georges-Picot, a French diplomat and lawyer, to divide up the vast Ottoman Empire into British and French spheres of influence. World War I was to stretch on for several more years. But the colonial powers were already well focused on the gains that could be made, and interests that needed to be protected, in the aftermath of the conflict. The Sykes-Picot Agreement was signed the following May, but it was nearly a decade—and many other deals and treaties later—before the modern borders of the Middle East emerged.

This Sykes-Picot Agreement, decided without the input of the Arabs in the region, broke the pledges of freedom the British had made when enlisting Arab support against the Ottomans. It also conflicted with the vision of "self-determination" promoted by President Woodrow Wilson during the war. The U.S government only learned of the agreement two weeks after it was signed, when British foreign secretary Arthur Balfour revealed it to Edward House, a foreign policy advisor of Wilson's.

Enraged, House wrote "It is all bad and I told Balfour so. They are making it a breeding place for future war."

Figure 11.1: Sykes-Picot Agreement, 1916 (present countries borders are shown in dotted gray lines)

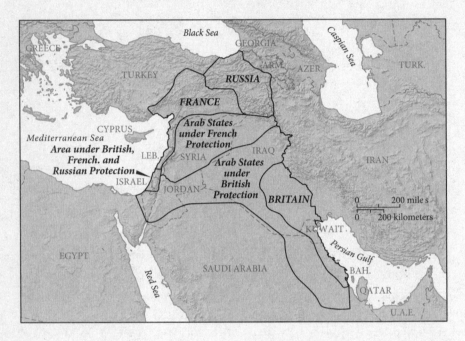

Source: "Sykes-Picot Agreement Map, signed 8 May 1916," Wikimedia Commons, October 7, 2011.

Fast-forward one hundred years to 2016 and Edward House looks prescient. The leaders of the Middle East may agree on few things. But one outstanding point of accord is that the era of Sykes-Picot is over. Walid Jumblatt, the leader of the Druze community in Lebanon, put it succinctly: "Sykes-Picot is finished, that's for sure, but everything is now up in the air, and it will be a long time before it becomes clear what the result will be." Barham Salih, a Kurdish leader and former deputy prime minister of Iraq, voiced similar sentiments: "The system in place for the past one hundred years has collapsed. . . . It's not clear what new system will take its place." Even ISIS leader Abu Bakr al-Baghdadi was focused on the death of Sykes-Picot, declaring in 2014 that "this blessed advance [of ISIS] will not stop until we hit the last nail in the coffin of the Sykes-Picot conspiracy."

Since 2011, the Middle East has been buffeted by multiple upheavals, any one of which might have been sufficient to challenge the post-Ottoman arrangements. It was in that year that political revolutions began to rock countries including Tunisia, Egypt, Libya, and Yemen. Soon these upheavals reverberated into full-blown civil wars in Yemen and Syria, contributing to a near–state collapse in Iraq.

The same period also saw a growing prominence of nonstate actors, some of whom transcend borders—not only ISIS, but also the Tamarod Movement in Egypt, the Muslim Brotherhood, and the Tunisian National Dialogue Quartet. These changes also coincided with the rise of regional powers such as Iran and, for a time, Turkey. Iran was already flexing its muscles in multiple parts of the region despite international sanctions. Once the nuclear deal was completed, its economy started to strengthen. Turkey saw its GDP per capita nearly double within a decade, and it became more assertive, at times aligning, and at times clashing, with the interests of other regional players like Israel, Saudi Arabia, and Qatar.

Amidst these political transformations, the new energy abundance has instigated a second revolution of a whole different sort in the Middle East. Oil and gas are the economic lifeblood of the region—either directly for the producing countries, or indirectly, via large transfers of wealth from them to those countries bereft of the resource wealth of their neighbors. Given this reality, most expect the energy revolution will mean greater tumult, instability, and loss of influence for the region. This view is true to some extent, particularly when seen in conjunction with the other upheavals. Some countries, the most obvious being Iraq, were pushed to the brink of collapse by falling revenues in the midst of security and political crises. Moreover, we have already seen how the new energy dynamics weakened OPEC—an institution that is older than many countries in the region and has wielded enormous global influence for decades.

Nevertheless, as has been the case elsewhere, the realities are more complex than they first appear. Many of the other anticipated effects of the energy boom on the geopolitics of the Middle East will not materialize, or at least not nearly to the extent predicted. For instance, the U.S.'s domestic energy boom will not lead Washington to withdraw from the

Middle East, nor even spur a dramatic diminution of American interests in the region. Nor will the boom necessarily dilute the ability of the region to wreak energy-inspired havoc on other parts of the world, either through embargoes or less politically driven production disruptions. In fact, in a persistently low-energy-price environment, the world could become *more*, not less, reliant on the Middle East.

Many of the less anticipated, but more meaningful impacts of the energy boom on the Middle East are surprisingly encouraging. Perhaps counterintuitively, the new energy abundance has actually opened the door for some positive developments in the region. At this point, such possibilities are just that—possibilities. There is nothing inevitable about their realization and, in fact, considerable odds are stacked against their success. But it is fascinating how the new energy dynamics have worked with other circumstances to create the potential—or even imperative—for countries in the Middle East to pursue paths that would have seemed unimaginable only a few years ago. Ambitious reforms, and even new peace initiatives, are plausible where none had been on the horizon a short while ago.

Protecting the Producers

On Valentine's Day 1945, an American destroyer covered in Oriental rugs pulled up alongside a larger vessel named the USS *Quincy* in the waters of the Suez Canal. With dozens of American sailors standing at attention, Saudi King Abdulaziz Al Saud, also known as Ibn Saud, slowly made his way over a gangplank to where President Franklin Roosevelt waited to receive him.

The two leaders came from very different worlds. Ibn Saud had never been out of his country before this trip. He arrived with a pared-down entourage of forty-eight, including the royal astronomer. He slept outside on the deck of the ship and insisted on slaughtering his own sheep onboard so he could eat fresh meat. Roosevelt, by contrast, had traveled abroad from the age of two and was a champion of progressive reform. He had repealed Prohibition and was successfully leading American forces in the largest international conflict ever.

The differences extended to their countries. To much of the world, the United States was synonymous with notions of accountability, democracy, freedom of speech and religion, and opportunity to people of all backgrounds and races. Saudi Arabia was a hereditary desert tribal kingdom founded only thirteen years previously on a conservative interpretation of Islamic law, with strict limits on personal and political freedoms. The nation was also still relatively unknown—although its vast reserves of oil would soon change that.

Nonetheless, the meeting was very congenial. Both men had trouble walking, and Roosevelt gifted Ibn Saud with his spare wheelchair. Sitting close together, the leaders laughed and smiled over the course of the conversation. During a four-hour meeting in the Great Bitter Lake portion of the Suez Canal, the two leaders set the foundations for the modern U.S.-Saudi relationship and, in many, ways, American foreign policy toward the whole region.

The arrangement was simple. Following World War I, Britain had claimed a mandate over oil-rich Iraq. American companies, searching for additional external sources of petroleum, had discovered large reserves in Saudi Arabia. As late as 1936, the desert kingdom had not exported a single barrel of oil. But, with American investment and expertise, it quickly became a global oil giant. So, in exchange for a steady flow of oil to global markets from Saudi Arabia—and eventually other Gulf countries—the United States guaranteed that no one would threaten the country or its ability to export its commodity.

Over the course of the twentieth century, the relationship between these two otherwise very different nations has been important enough to be the explicit subject of multiple presidential doctrines. In 1980, the United States and Saudi Arabia both viewed the Soviet Union's invasion of Afghanistan as an attempt to gain direct access to the vast oil reserves of the Gulf just three hundred miles to the south. A few years earlier, the CIA had done a secret study called the "The Impending Soviet Oil Crisis," which raised concerns that Soviet oil production would soon peak, causing the nation to forcefully look abroad for new sources of energy. In his last State of the Union address, President Jimmy Carter responded to the Soviet Union with forceful words, "Let our position be absolutely clear: An attempt by any outside force to gain control of the Persian Gulf region will be regarded as an assault on the vital interests of the United States of America, and such an assault will be repelled by any means necessary, including military force." His words were interrupted several times by bipartisan applause.

Just one year later, newly elected President Ronald Reagan affirmed his readiness to protect American interests in the Middle East. "There is no way," he said, "as long as Saudi Arabia and the OPEC nations there in the East—and Saudi Arabia's the most important—provide the bulk of the energy that is needed to turn the wheels of industry in the Western World, there's no way that we could stand by and see that taken over by anyone that would shut off that oil."

A decade later, the United States backed up its promises to defend Saudi Arabia and other Gulf nations with decisive action. In 1990, Iraqi forces overran Kuwait, and threatened to do the same to Saudi Arabia. As discussed in Chapter Five, gaining control of the region's energy riches would have exponentially enhanced the power of Saddam Hussein, Iraq's megalomaniacal leader. In addition to the vast riches from oil sales, Saddam would have been in an excellent position to blackmail the world by orchestrating huge price spikes overnight.

Once again, an American administration made clear that this was an unacceptable situation. "The economic lifeline of the industrial world runs from the gulf," said Secretary of State James Baker, "and we cannot permit a dictator such as this to sit astride that economic lifeline."

Within months, the United States led a coalition of countries in a rout of Iraqi forces, quickly driving them out of Kuwait. The Iraqi invasion of Kuwait marked a turning point in U.S. strategy and military presence in the Middle East. As late as 1989, the United States had fewer than seven hundred military personnel deployed in all of Bahrain, Kuwait, Oman, Saudi Arabia, and the UAE combined. Yet, even a decade after Iraq was ejected from Kuwait, tens of thousands of American troops remained in the region—a number that expanded to nearly a quarter million by the late 2000s at the height of the Iraq War.

U.S. Withdrawal: Not So Fast

Given the centrality that oil has played in U.S. involvement in the region, many Americans view their new energy prowess as a vehicle for deliverance from such costly and controversial engagements in the Middle

East. If a full withdrawal is not in the cards, then, at a minimum, proponents argue, Americans can look forward to a significant retrenchment from the region. According to economist Anders Aslund, "U.S. interests in the Middle East will decline along with U.S. energy imports." Middle East policy expert Daniel Pipes likewise concluded, "Washington will be largely freed from having to kowtow to the oil and gas pashas."

Predictions that the United States may soon be able to scale back its costly Middle Eastern diplomatic and military engagements have also played well with politicians. In 2012, President Obama voiced the expectation that new energy sources would make the United States "less dependent on what's going on in the Middle East." Obama's Republican opponent in the 2012 presidential election, Mitt Romney, largely agreed with his campaign, saying that America's energy wealth would mean "the nation's security is no longer beholden to unstable but oil-rich regions halfway around the world." In 2016, Donald Trump's campaign platform pledged to "become, and stay, totally independent of any need to import energy from the OPEC cartel or any nations hostile to our interests."

These leaders, and many others, will be disappointed. U.S. interests, and therefore engagement, in the Middle East will fluctuate based on the propensity of leaders, America's strength at home, the sort of threats and opportunities emanating from the region, and other foreign policy challenges elsewhere. Energy, however, will remain more or less a constant driver of U.S. engagement in the region—despite new American resource riches.

It is true that the United States may no longer need to import any oil from the Middle East in the foreseeable future even if, as described in Chapter Four, the much coveted status of U.S. energy independence remains elusive. For most of the 2000s, the United States imported on average just over 2.3 million barrels per day—or roughly a quarter of its daily imports—from the Gulf region. Then, beginning with the economic slowdown from the 2008 financial crisis, U.S. imports of Gulf crude oil began shrinking significantly. That trend continued even as the American and global economy rebounded, as the swell of U.S. tight oil kept import levels low. By 2014, the United States was importing less than half the total volume of oil that it had been seven years earlier. U.S.-Saudi crude trade reflected this same trend, falling by nearly two-thirds from a peak in the spring of 2003 to the end of 2015. The IEA and some oil and gas experts such as Harvard fellow Leonardo Maugeri suggest that this trend could accelerate so rapidly that the United States may not import any oil from the Middle East in the coming decades.

Fretting over exact volumes, however, is a waste of time. If history is any indication, the volume of oil flowing between the Middle East and the United States will have little bearing on the quality of the overall relationship. Take the bilateral relationship between the United States and Saudi Arabia. Oil imports from Saudi Arabia fell to almost zero in 1985 on account of the massive Saudi rollback in production in the early 1980s, yet the bilateral relationship remained solid. From the first Gulf War in 1991 to 2009, U.S. imports of total crude and oil products from Saudi Arabia remained stable despite significant ups and downs in the relationship. They were approximately 1.4 mnb/d following the invasion of Kuwait by Iraq in 1990, when President George H. W. Bush told Saudi Arabia's ambassador to Washington that "If you ask for help from the United States, we will go all the way with you." And they were more or less at the same level after September 11, 2001, when tensions were so high that American soldiers who had received medals from the Saudi government in the first Gulf War returned them in disgust—in protest of the primarily Saudi group of terrorists that perpetrated the horror of that day.

In reality, factors other than the volume of oil trade matter much more in determining U.S. interests in the region. Many issues of critical importance have no direct relationship to oil. Fighting terrorism, resisting the proliferation of nuclear weapons, and supporting Israel are all American foreign policy priorities of the highest order that demand U.S. involvement of one sort or the other in the Middle East. ISIS and al Qaeda pose a threat to European and American cities; letting the vicious conflicts between these groups and Middle Eastern governments unfold without U.S. involvement would have adverse implications for regional stability and even U.S. homeland security. Iran's regional ambitions threaten not only to destabilize a precarious balance among Middle Eastern powers, but could shatter a counter-proliferation regime that has helped minimize stray fissile material and technologies which could vastly increase the capabilities of groups opposing Western powers. Both developments could be devastating to Israel, an important partner of the United States.

The importance of the Middle East to American foreign policy—and the relative significance of oil—is captured in an illuminating study by American academics Mark Delucchi and James Murphy. They estimate the costs to the United States of protecting oil supplies in the Gulf by considering a number of different scenarios. Interestingly, they consider one scenario in which the Gulf exists, but has no oil. When they evaluated this scenario in 2004, they found that even were the Gulf to be oil-free, the United States would still spend substantial amounts on security (not to mention diplomacy and economic engagement) to advance its interests there. They reckon that the United States would still spend 25 to 40 percent of current peacetime expenditures in the Gulf in this scenario—all for non-oil purposes.

In addition to having interests in the Gulf unrelated to oil, the United States continues to have *energy* interests there that go beyond petroleum imports. The Delucchi and Murphy study also highlights this nicely in examining another scenario in which the Gulf does have oil, but the United States does not consume it (although other countries do). The price tag for this arrangement differs considerably from the scenario where the Gulf is bereft of oil, illustrating that several energy concerns necessitate U.S. engagement, even if America is not consuming a single drop of Middle Eastern oil.

The reality is that the global nature of the oil market ensures that the United States will be interested in Middle Eastern stability, regardless of the direction of the flow of oil. Even if the United States meets all of its energy needs in the future by relying only on its own resources and those of Canada and Mexico, it will still be connected to the global market; the prices its consumers and companies pay will therefore be determined by global supply and demand. If a major disruption of any sort sharply curtails oil supply from the Middle East, global prices will rise accordingly. The United States will be better insulated from some of the downsides of a price spike thanks to its new energy riches. (For instance, rather than having dollars flow out of the United States to oil-producing states, they will stay inside the country, and accrue instead to oil companies and beneficiaries there.) But most of the dangers of a sudden price spike for the economy will remain the same, making the United States almost as keen as ever to see stability in the Middle East.

More tangibly, while the United States may soon no longer look to Middle Eastern countries to directly supply its consumption, many U.S. allies will be more dependent than ever on that volatile region. This will be particularly true in a persistently low-oil-price environment, as higher-cost producers outside the region are squeezed out. Asian dependence on Middle Eastern oil in particular is set to rise. By 2040, one of every two barrels of internationally traded oil will be destined specifically for China and India. Given China's function as the engine of global growth, a sharp rise in price induced by supply disruptions in the Middle East could have devastating effects well beyond Asia if it slows China's economy. America's energy boom may allow it to curb its own dependence on external sources of oil, but the United States will be in no position to substitute for large volumes of crude that a Middle Eastern crisis might take offline; gone are the days of the 1956 Suez Crisis when booming U.S. oil production enabled America to organize the Oil Lift to replace the Arab oil denied to France, the U.K., and Israel after their invasion of the Suez.

As discussed in Chapter Five, the United States will also continue to have *strategic* interests in the oil of the Middle East, even if it does not have big commercial interests there. As was the case concerning Iraq in 1990–1991, the United States will be concerned about any single leader or country controlling a disproportionate amount of global oil reserves or production, particularly if that entity is an adversary of the United States or its allies. True, to the extent that the production of unconventional oil becomes a global phenomenon, a smaller percentage of overall global production could be located in the Middle East. But, at least for the coming decade and likely beyond, the highest concentration of oil production and proven reserves will still be in the Middle East. For instance, growing Iranian influence in Iraq is worrisome to the United States for many reasons, but high on the list is the fact that if Tehran were to control Iraqi oil policy, it would double the proportion of world reserves, as well as the level of oil production over which it has sway.

Adjusting to a New Balance of Power

In the coming decade, advancing U.S. interests—both energy- and non-energy-related—will require American involvement in the region. But rather than simplifying Washington's relationships with Gulf capitals, the new energy abundance will add fresh complications to the relationship. For decades, Saudi Arabia and the United States had a symmetrical relationship. One side offered to be a swing producer as described below; the other side offered security. The energy glut is changing that relationship. As Joe Nye of Harvard mused, "Power rises from asymmetries in interdependence." No longer a relationship of symmetrical interdependence, Washington's position vis-à-vis Riyadh will be enhanced.

Thanks to the new energy abundance, the United States is far less likely to be the supplicant it has found itself to be so many times over the past decades. Over the course of multiple administrations, American presidents and diplomats have made the fourteen-hour plane journey to Saudi Arabia with a single request. Each time, they asked the kingdom to increase global oil supplies—either in anticipation of a coming disruption, such as the 2003 war against Iraq, or as a means of easing tight markets like as those resulting from unexpectedly robust demand growth in the mid-2000s.

Riyadh's ability to respond to these requests has been a huge source of political power and geopolitical stature for the kingdom, even more so than the large volume of oil it produces or the exports it generates. Being responsive has required the maintenance of spare capacity—the building and maintaining of oil fields and infrastructure that are not generally used for regular production, but can be brought on line in short order. Only a few countries—Saudi Arabia, the UAE, and Kuwait—have been willing to voluntarily incur the expense associated with holding this spare capacity. Doing so, however, has allowed them to react to changes in the global market, bringing new production to market to curb price peaks that otherwise would have only eased through a much slower process of eventually spurring new investment to increase production and dampening consumption to decrease demand.

In an energy abundant environment, there may be little reason to implore the Saudis to increase production. If anything, we may see a repeat of the soft message Vice President George H. W. Bush delivered to the Saudis in 1986 to *decrease* their production in order to protect the U.S. oil industry. American leaders will need to knock less frequently, if at all, on Riyadh's door with a special plea to dip into its spare capacity and produce more oil.

As a result, there will be greater opportunity for the United States to raise other matters in the relationship. As all envoys—from the junior to the presidential—know, there is a limited number of issues that can be raised seriously in any one meeting. When a president meets a king, the president rarely has the luxury of giving serious treatment to more than a few subjects; rarely is a laundry list of concerns rattled off in the hopes that one of them makes an impression. Instead, American government figures carefully select the issues of greatest concern to the United States and seek to impress upon their interlocutors the importance of those few issues. Tumultuous times in the region ensure there will be no shortage of issues for U.S. and Saudi leaders to discuss in the years ahead.

The new energy realities do not, however, translate into all good news for the United States in the conduct of its relationship with Saudi Arabia and other Gulf powers. First, U.S. leaders may wield less heft in the eyes of Saudi sheikhs now that America has lost the special status as Saudi Arabia's largest oil importer. Asian nations have become the new consumers ensuring Middle East security of demand and—as King Salman's month-long trip to Asia in 2017 suggests—the Gulf will lavish more time and attention on the countries to its east. In addition, the perception—whether founded or not—that Middle Eastern countries are less important to the United States has already affected their behavior in ways America may not appreciate. During a trip I made to Turkey with a small delegation in mid-2016, one of Ankara's most senior diplomats summed up the view of many in the region nicely; he said, "Leadership is needed. And that leadership must come from the United States. And when you decide not to lead, it all falls apart. Maybe you think you are living in a new reality now. You are no longer dependent on the region for oil. So maybe in this context you think that you do not need to care about the Middle East. Perhaps this is how you see things now."

Gulf countries in particular were not prepared for their largest customer to become their competitor. Even before King Abdullah's death and the rise of a more assertive Saudi leadership in 2015, the kingdom acted with greater independence in the region, in part due to the perception that the United States was no longer invested in regional outcomes to the same extent as in the past. Whereas Riyadh and Washington had long sought to tackle problems in tandem, the House of Saud has increasingly charted its own course. It provided billions of dollars to the government of Egyptian president Abdel Fattah al-Sisi, at a time when the United States sought to use its more modest economic leverage to orchestrate a return to democracy. And it finalized deals to buy billions of dollars of advanced U.S. weapons at the same time it sought increased military partnerships with China. The perception of diminished U.S. interest in the Middle East may have made a bigger difference to regional affairs than any increase in American leverage in the U.S.-Saudi relationship; surprising some, the United States may have less influence in the region now that it will be less dependent on its oil.

Finally, U.S. leaders will need to face up to the challenge of explaining to Americans why the Middle East still matters in a new world of energy abundance. The average U.S. citizen could resist future efforts to involve the United States deeply in the region now that oil no longer plays the same role. U.S. presidents will need to be sensitive to this perception—however erroneous—and work hard to explain the continued importance of this troubled part of the world. But continuous and mounting political pressure will also lead future U.S. administrations to seek the help of other countries to share burdens America has traditionally shouldered alone. It will take time for these efforts to bear fruit, given Europe's weakness and China's reluctance. But when they do, the usual challenges of coordinating multiple actors in a complex environment will apply.

Middle East Oil: Even More Important in Some Scenarios

David Ottaway's book *The King's Messenger* opens with the recounting of a scene that is bound to make U.S. policymakers squirm. Ottaway describes how, in the waning days of the summer of 2001, Saudi Ambassador Prince Bandar bin Sultan delivered a letter from Crown Prince Abdullah to President George W. Bush. The crown prince, who effectively ruled Saudi Arabia at the time, was outraged that the United States was doing nothing to restrain Israel in the face of the outbreak of the second intifada. The letter allegedly presented an ultimatum to President Bush: either the United States take action to secure a lasting agreement between the Israelis and Palestinians, or the Saudis would use their oil weapon and wreak havoc on the American economy reminiscent of the 1973 and 1979 energy crises.

According to Ottaway, Prince Bandar "watched with enormous relief as Bush and a lethargic White House had jumped into action," crafting a new initiative within days to spur an Israeli and Palestinian peace.

One would expect that, at a minimum, such blackmail would be a thing of the past in an era of energy abundance. Certainly, neither Saudi Arabia nor other countries can effectively threaten to cut exports to American customers if such flows have dwindled or ceased. But, with a number of important caveats, Saudi Arabia and other Middle Eastern producers could still exert significant leverage over the global economy in the medium to long run.

As mentioned, this will be particularly true if the new energy abundance ushers in a long-term era of very low prices. The IEA created a "Low Oil Price Scenario," in which prices stay between $50 and $60 a barrel well into the 2020s, before rising to $85 by 2040. If prices remain this low, for this long, they will inevitably affect global production patterns. In this scenario, low-cost producers in OPEC and the Middle East will account for a larger share of total global output, with high-cost U.S., Canadian, and European producers accounting for less. Correspondingly, the Middle East's share of global crude oil exports rises significantly in this situation. According to the IEA, by 2040, the Middle East would be the source of 57 percent of all inter-regional trade in the low-oil-price scenario, in comparison to just half in the most-likely, reference case (higher price) scenario. Asia in particular becomes more vulnerable to Middle East disruptions, as by 2040, 85 percent of the crude sourced by Asian refiners could come from the region in this low-oil-price scenario.

Theoretically, Asia's increased dependence on Middle Eastern oil could make it susceptible to the sort of political blackmail reportedly threatened in the Saudi letter to President Bush. But the caveats are important. First, any embargo of an Asian economy would not be limited to the target country; because of the global nature of the oil market, all economies would suffer the effects of an overnight embargo if the oil was completely withheld from the market. Second, such a move—and the price increase that would accompany it—would only add urgency to the already ardent search for alternative, non-fossil-fuel energy sources, a push no oil producer should be willing to give. And finally, as discussed elsewhere in this book, tight oil's quick response to higher prices would at least mean that the pain of a price spike would be alleviated much sooner by an increase in U.S.—and possibly other-country—tight oil than has been the case in the past.

While this set of reasoning should be enough to dissuade any Middle Eastern power from taking such steps, the real risk is not in a political embargo of the sort experienced by the world in 1973. Despite the fact that global oil policy has long been animated by the specter of another such embargo, Middle Eastern producers learned from the largely failed effort of 1973 that such moves did not easily translate into political influence; the Arab producers of OPEC were not only unable to drive Israel from the occupied territories, but the world plunged into a period of stagnation that was bad for producers and consumers alike.

The greater risk to the global economy in a continuous, very low price scenario is that global production becomes ever more concentrated in the most volatile region of the world: the Middle East. In this circumstance, any number of wars, terrorist attacks, or collapsed states could send the global economy reeling. Until and unless tight oil production becomes truly a global phenomenon and significantly dilutes the influence of the Middle East on global markets, this scenario remains an all too real global risk.

An Invitation to Reform

In 2006, seized with enthusiasm for a blossoming idea over lunch, journalist Tom Friedman grabbed a napkin and drew two lines on a graph, one a sharp mountain shape and the other the mirror image of it. The lines intersected on the far left and far right, forming a diamond. On the vertical axis was notionally the price of oil, with some index of political freedom on the horizontal. Later dubbing his theory "The First Law of Petropolitics," Friedman proposed an inverse relationship between political freedom and the price of oil. Pointing out that Bahrain is both the first Arab Gulf state to be running out of oil and the first to hold a free and fair election where women could run for office as well as vote, Friedman deduced, "I don't think that's an accident." Although Friedman came under a barrage of criticism for his methodology, history could well favor his theory. The problem is, if it does, neither Friedman nor many of us may be alive to give him credit. As the years since the first hopeful protests dislodged authoritarian rulers in North Africa demonstrated, the road to greater political liberalization will be long and fraught with potholes.

Figure 11.2: Tom Friedman's First Law of Petropolitics

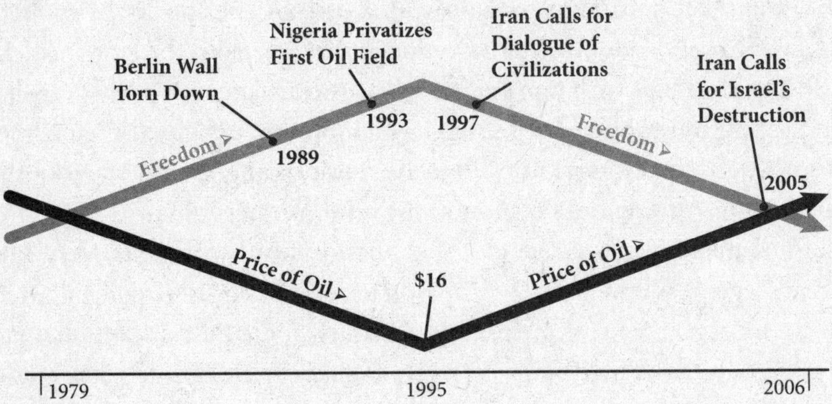

Source: Thomas L. Friedman, "The First Law of Petropolitics," *Foreign Policy* 154, no. 3 (2006), 29.

Whether Friedman was conscious of this or not, the first part of his logic lies in the comprehensively researched and debated idea known as the "resource curse": countries whose economies are heavily reliant on natural resources tend to grow more slowly and to be less democratic than those whose economies are not. Despite some recent improvements, Nigeria was long held up as the poster child of the resource curse. Even though it was the thirteenth largest oil producer in the world in 2012, more than half of its population lived below the poverty line. But the phenomenon of the resource curse is not only about corruption. Resource wealth is also believed to shape the nature of the political system, lending itself to governments that are less responsive to their people (particularly given that they often do not pay taxes) and have invested heavily in repressive institutions. Friedman's theory builds on this idea by positing that when faced with lower revenues, the ruling classes of resource-rich countries cannot maintain the systems that have kept them in power.

Given the political upheaval already experienced by the Arab republics—Tunisia, Egypt, Libya, Iraq, and Syria—the obvious place on which to test the validity of this theory is the single part of the region that has thus far escaped the roil of popular uprisings: the Gulf monarchies of Saudi Arabia and, to a lesser extent, the United Arab Emirates, Qatar, Kuwait, Oman, and Bahrain. Will the kings, emirs, and sultans of the Middle East be able to keep their thrones in the face of the new energy abundance? Or, will the new energy abundance combine with other factors to deliver a new wave of Arab revolutions to the Gulf?

The challenge to existing regimes is greater than it might first appear. In the past, the Gulf states largely needed to manage a price cycle whose details were unclear, but whose broad contours were well known. In the early 1980s, they endured years of low oil prices, and were forced to exhaust their reserves, to resort to debt, and—in the case of Saudi Arabia in 1986—to repeatedly delay the passage of new budgets. But the familiar cadence of rising and falling prices helped these monarchies preserve their power. Lean periods were tough, but they were inevitably followed by fat ones. The new energy abundance, however, is changing this customary cycle. Gulf producers must now think beyond weathering a few months or years of a low oil price. If lower prices are the new normal, these governments must wrestle with the much more difficult challenge of adapting to lower revenues over the long haul.

Moreover, two revolutions are tougher to survive than one. The challenge of managing the effects of the new energy boom have been greatly compounded by the complex political, security, and demographic problems facing the region and the ways in which many regimes initially chose to deal with them.

From their palaces across the desert kingdom, the Saudi royal family watched the events occurring in Egypt, Tunisia, and Libya with alarm. In response, the Saudi government increased its arrest of dissidents and cracked down on suspected instigators of political unrest. But its immediate hallmark reaction to the Arab revolutions in the region was notably softer. On February 23, 2011, less than two weeks after Egyptian president Hosni Mubarak was forced from power by a popular uprising, the Saudi regime announced new pay hikes, bonuses, new housing projects, and other spending for the population; the total price of these perks soon reached $130 billion.

Although this massive expenditure was ostensibly to celebrate King Abdullah's return from the hospital, the lavish outlay had the desired effect. Many Saudis, long citing the disarray the Iraqis experienced in their quest for democracy, focused on enjoying their new wealth, rather than on organizing against the regime that provided it. Such handouts were not the first of their kind, nor the last. Shortly after acceding to the throne in early 2015, King Salman bin Abdulaziz Al Saud showered his citizens with grants, gifts, and bonuses in a giveaway that one economist anticipated will cost more than $32 billion, an amount equal to the annual budget of Nigeria or the amount the United States spends each year on medical research through the National Institutes of Health. Appreciative Saudis created a new Arabic Twitter feed, #two_salaries, to share how they were spending the two months' extra salary the new king had sent their way. Some Saudi men declared the largesse would enable them to take a second, third, or fourth wife, while others paid off loans or planned holidays abroad.

Such financial benevolence is part of a broader arrangement that has underpinned the stability of the Gulf monarchies since their establishment. These regimes have sustained elaborate, if unwritten, agreements with their populations. The rulers provide economic prosperity, physical security, public services, and comfortable government jobs for their citizens, without the undue burden of taxes. In exchange, the people consent to be governed with little or no say in the system, relying instead on the goodwill and judgment of their rulers. Those objecting to this contract can expect harsh treatment. The particulars of these arrangements vary significantly from country to country to reflect different histories, institutions, and peoples; Kuwait has a vibrant parliament wielding actual power, while Saudi Arabia has no elected representative body with real authority. The general idea, however, is similar. The generosity and durability of these compacts was a major factor in explaining the ability of the monarchs to survive, relatively unscathed, the Arab political revolutions that unseated their republican counterparts.

Maintaining these ever-expensive social contracts in the face of declining revenues is the real challenge facing the Gulf monarchies. Young, growing populations and ever more generous benefits have strained budgets. The yawning gap between the "fiscal breakeven price"—which refers to the price of oil at which a country is able to fund its internal budgetary commitments—and the actual price of oil in late 2014 was one indication of the dire straits in which royal families found themselves. In Saudi Arabia, the breakeven price nearly tripled in five years alone, shortly before the price of oil began to drop. At the end of 2014, soon after the oil price began to plunge, everyone from talk show hosts to IMF economists noted that breakeven prices for virtually all the Gulf monarchies (as well as the constitutional republics) exceeded the actual price of oil. Saudi Arabia, Bahrain, and Oman needed oil close to $100 to meet their fiscal needs. In late 2014, only Kuwait seemed on track to balance its budget.

These numbers conjured up memories of the low oil price of the 1980s and the collapse of the Soviet Union, or the days of $10 barrels of oil and the 1990s rise of Hugo Chávez in Venezuela. These historical examples might lead one to predict dark days for many Gulf producers.

Yet the Gulf countries proved resilient, at least in the short and medium term, for several reasons. First, the extreme wealth of these countries provided them with immediate shock absorbers. In 2014, Saudi Arabia, for example, had financial reserves equivalent to its entire 2013 GDP—or the total amount estimated at the time by the United Nations needed to rebuild war-torn Syria, Iraq, and Gaza. Put differently, Saudi Arabia had enough money in its reserves to fund three years of spending at 2014 levels without drawing on any other source or accruing any further revenues. The reserves of the emirates of Abu Dhabi and Dubai were even more impressive, totaling nearly a trillion dollars, or roughly twice the annual U.S. defense budget or four times what it would cost to buy all American professional sports teams. For a small country with more expatriates than citizens, Kuwait's financial reserves amounted to approximately $150,000 per person or nearly half a million per citizen. Even the few nations that did confront immediate fiscal dangers—such as Bahrain and Oman—still had the option to raise debt.

In the twelve months after the oil price began to falter in 2014, all of these countries cut their budgets. Big capital projects were targeted first, allowing countries to reduce their breakeven point without touching spending on social contracts. In the three months between October 2014 and January 2015, Saudi Arabia—like many other Middle Eastern countries—revisited its budget and effectively slashed its breakeven price from $102 to $87 by the estimates of the IMF. Bahrain, Kuwait, Oman, and to some extent the UAE, did much the same thing.

Such measures, however, did not seem adequate to allow all of the Gulf states to maintain their social contracts over the long run. Bringing down fiscal breakeven prices from $107 to $87 is easy to do when fat can be cut. But getting down from $87 to a more realistic $70 or even $60 will prove much harder in the face of growing demands and the continued political imperative to defend—and even sweeten—the social contract.

Saudi Arabia: A Giant Learns New Tricks

Of all the Gulf monarchies, Saudi Arabia will find this adjustment the most difficult. Behind Saudi Arabia's staggering reserves lurk warning signs about the fiscal future of the kingdom. While Saudi financial reserves are abundant, they have been drained rapidly by the combination of high spending in 2014 and early 2015 and low oil prices. From 2014 to the end of 2016, Saudi financial reserves decreased by more than a quarter, or $200 billion. In addition, Saudi Arabia needs to be cognizant of its growing population. At more than thirty million people, the number of inhabitants in the desert kingdom far outweighs the tiny populations of Kuwait, Qatar, and Bahrain, and still dwarfs that of the UAE, which has fewer than ten million people.

Moreover, unlike other Gulf countries, Saudi Arabia faces significant demographic pressures. Birth rates in the kingdom have begun to decline, but in the thirty years before 2010, the population grew by three times the global average, or by 180 percent. As a result, a surge of young people born in previous decades is now testing the ability of the Saudi state to provide health care, education, and jobs. Such demographic growth would be viewed favorably in many societies seeking growth, from Japan to Italy. But in Saudi Arabia, where almost three-quarters of those employed work for the public sector and where no taxes are levied on income, this demographic explosion portends more burden than boom. More than a quarter of a million young people enter the work force in Saudi Arabia every year. In 2012, a report by Alkhabeer Capital, an asset management and investment bank firm based in Jeddah, claimed that 27 percent of Saudi youth and 78 percent of university graduates were unemployed. Providing health care and education for Saudis each year already takes up a whopping 44 percent of the government's budget.

Low oil prices are not the only worry. The kingdom's finances are also under pressure from the growing domestic energy demand that is eating away at the volume of oil Saudi Arabia can export. In 2015, gasoline was just 45 cents a gallon. It has been heavily subsidized, as have other forms of energy. In 2013, electricity prices were roughly one-fourth the average paid in OECD countries. As a result, domestic demand for energy soared; demand for electricity—more than half of which is generated inefficiently from burning fuel oil or diesel—was growing at 7.5 percent a year in 2014. In 2009, overall Saudi energy demand required the domestic use of 3.4 million barrels of oil-equivalent a day. Extrapolating that year's demand growth rates forward, in 2010 Khalid al-Falih, the then-CEO of Saudi Aramco, warned that by 2028 Saudi Arabia might need to divert nearly a third of current crude oil production away from exports to satisfy domestic demand. This would place a major squeeze on government revenues at a time when the Saudi social contract was getting more and more expensive to sustain.

Finally, the kingdom's finances are even further strained by the responsibilities of being a regional leader. Given the growing insecurity in the region, Saudi Arabia would be wise to anticipate continued—even larger—handouts to friendly regional governments in order to stabilize them. Aid to Bahrain, Egypt, Jordan, Oman, Yemen, Palestine, Morocco, Sudan, and Djibouti reportedly cost the Saudi government nearly $23 billion from 2011 to 2014. Moreover, during a visit of Saudi King Salman to Egypt in late 2016, the kingdom agreed to supply Egypt with 700,000 tons of refined petroleum products each month for five years on favorable terms, a deal also valued at $23 billion. Expensive, but a Saudi investment in regional stability.

Amidst such staggering simultaneous economic and political challenges, some positive developments have emerged. Rather than simply hoping for a revival in oil prices to spare them tough choices, many governments in the Gulf have embarked on programs of economic reform. As mentioned, they initially focused on cutting the less politically sensitive capital projects. But eventually they were forced to revise some of the privileges their citizens have long enjoyed. Kuwait, Saudi Arabia, and the UAE scaled back generous fuel subsidies in place for generations. Some Gulf states have begun requiring their inhabitants to pay for water, in a major departure from past practices. The countries of the Gulf Cooperation Council are deliberating whether to impose a region-wide value-added tax and a corporation tax.

Of all the Gulf states, Saudi Arabia launched the most ambitious reform efforts, partially out of necessity and partially due to the emergence of a new, young crown prince, Mohammed bin Salman—or MbS as he is called for short by many. The favorite son of the elderly and possibly ailing king, MbS has emerged as the most powerful prince his age since President Roosevelt's counterpart and the kingdom's founder, King Abdulaziz Al Saud, Ibn Saud, according to longtime expert on Saudi Arabia Karen House.

Since his father inherited the throne in January 2015, MbS has assumed the positions of crown prince, deputy prime minister, minister of defense, chairman of the council for economic and development affairs, and the head of Saudi Aramco. In short, he holds the reins for the defense, economic well-being, and oil policy of the kingdom. At thirty-one years of age, MbS has been viewed by many in the royal family as too young and too brash to be in charge of the kingdom, particularly at such a precarious time. But his energy and his willingness to work hard and hold people accountable have been welcomed by others, especially Saudis tired of a ruling class that seemed to do little, yet reap so much of the country's wealth.

MbS has anchored his own personal ambitions to become king to the development and execution of *Saudi Vision 2030*, a national plan to move the kingdom away from its dependence on oil to a modern, diversified economy. While the document is filled with quirky, seemingly peripheral goals—such as getting Saudis to exercise more—at its core is a bold economic transformation. It invokes familiar prescriptions, such as increasing foreign investment, better utilizing female Saudi labor, privatizing much business done by the government, and incentivizing Saudis to spend their disposable income in the kingdom, rather than in neighboring Dubai or Bahrain. But the plan also extends to the controversial, including an initial public offering of part of Saudi Aramco, the country's massive oil company. The proceeds will provide added capital to a revamped Public Investment Fund, which is envisioned to eventually contain $2 trillion and generate a nonoil source of income for the kingdom similar to the sovereign wealth funds of other Gulf countries. When he was still deputy crown prince, the young prince declared that by 2020 Saudi Arabia will no longer be dependent on oil.

Despite the vigor with which the new plan was launched, those championing it face an uphill battle. The Saudi public sector is massive, but its real capacity to enact dramatic reform on a compressed time scale is very much in question. An even greater obstacle is the Saudi labor market, which does not yet have the talent and expertise needed to transform the economy into a more dynamic one where the private sector is a driver of growth and provider of jobs. The energy boom in the United States is not only an impetus for reform, but could also be an obstacle, as cheaper natural gas in North America deprives Saudi Arabia of some of its relative global competitiveness in manufacturing and petrochemicals; this competitiveness will be further eroded by the lifting of subsidies on energy feedstocks, which will be inevitable under the new reforms.

Yet the fortunes of MbS, *Saudi Vision 2030*, and Saudi Arabia as a whole do not rest solely—or even primarily—on the matter of economics. The success of these economic reforms will ultimately require Saudi Arabia to make simultaneous political and social reforms. The connection between the two is obvious in some areas, such as attracting foreign direct investment, bringing women into the workforce, or encouraging Saudis to spend their leisure time in the kingdom. Changes to lubricate these reforms will require deft handling and the expenditure of political capital, particularly as they will encroach on issues of importance to the powerful religious establishment, whose backing the royal family relies upon as a source of its legitimacy.

Even more precarious, however, are reforms that will be required as MbS essentially renegotiates the contours of the social contract. As mentioned, the reform plans have already chipped away at subsidies that most Saudis feel are their entitlements. Additional planned steps, such as the adoption of a value-added tax and the reduction of public wages as a proportion of the budget, will further put to the test the support of Saudis. MbS and others may hope the promise of more jobs and continued public services will be enough to convince Saudis to adjust to the new burdens. But if my conversations during a 2016 trip to the kingdom are any indication, Saudis will look for more than maintaining privileges in a renegotiation of the social contract. While not every Saudi with whom I spoke asked for the same changes in return for losing subsidies and paying taxes, many expressed a desire for greater transparency into the finances of the royal family, more equity, and less corruption. If MbS can deliver on such expectations, he will have a wider berth for realizing his vision of Saudi Arabia in 2030.

The range of outcomes for Saudi Arabia is not binary. The plans may eventually succeed, fail, or end up at a thousand points in between. However, if reform plans do not achieve at least a moderate level of success, the future of the kingdom will be bleak. While popular uprisings are not impossible to imagine, a failed effort to renegotiate the social contract is more likely to invite more government repression and create a more disaffected population. A poorer, more repressive, more radicalized Saudi Arabia would not only be bad for Saudis, but it also would be destabilizing for the region and the world.

Iraq: Reinforcing the Unity of the State

Ashti Hawrami, a brilliant British-educated petroleum engineer from Sulaimaniya, a Kurdish town near the Iranian border, was the mastermind behind the swift emergence of the Kurdish region of Iraq as an energy force in its own right. While other Kurds fought Saddam's forces in the mountains as their contribution to Kurdish freedom, Hawrami's input was developing the economic options of the region. Over the years since 2007, under Hawrami's watchful eye, the Kurds opened their previously unexplored territory to international oil companies, discovered significant oil reserves, began producing them, and constructed infrastructure to sell their oil to global markets through Turkey.

All of this was done without Baghdad's involvement and often over its objections. In fact, the development of the energy wealth of Iraqi Kurdistan has been a point of contention between Baghdad and Erbil, the Kurdish capital, since the ouster of Saddam in 2003. Partially in an effort to repress the Kurds, Saddam never allowed for the exploration of the Kurdish region for oil and gas. While many other parts of Iraq were found to be awash in these resources, the Kurdish region seemed to be bereft of them.

In complex negotiations of which I was a part during my time in Iraq, the Kurdish region decided to resist the long-standing pull of declaring independence in 2003 and to remain part of the Iraqi state at least conditionally. American pressure and other geopolitical realities played an important role in this decision. But also critical were new arrangements with Baghdad that made Iraq a federal country (and the Kurdistan Regional Government a federal region), gave the Kurds considerable autonomy, and stipulated a budget arrangement in which the Kurdish region would receive 17 percent of all the revenues accruing from the sale of Iraqi oil and gas produced elsewhere in the country. These arrangements also meant that Kurdistan would have the ability to explore and develop whatever oil and gas resources it might have.

While assurances of a significant portion of Iraq's revenue were agreeable to the Kurds, their long, troubled history with Baghdad made them reluctant to be dependent on the center for their well-being. For decades, Kurds had suffered—including genocidal acts and the use of chemical weapons—at the hands of Saddam and a powerful Iraqi state. While there was a new beginning in Iraq, the Kurds were wary that Baghdad could again become more authoritarian, with devastating consequences for them. For this reason, the efforts of Hawrami and others to develop Kurdistan's own resources were critical to the Kurdish project. In a few short years, the Kurdish region of Iraq was able to boast its own considerable production—dramatically changing the outlook for this little corner of the world.

Never disguising his disdain for the power center of Baghdad, Hawrami openly and frequently "did the math." In spring of 2014, sipping a glass of French wine in the guesthouse of the Kurdish prime minister, and speaking over the rush of a water fountain in the indoor courtyard, Hawrami asked me a rhetorical question, "How many barrels of oil does Kurdistan need to produce and sell in order to exceed the revenues it is owed under the budget agreement from Baghdad? At what point would it make greater sense for Kurdistan to simply rely on the sale of its own oil, rather than to be dependent on Baghdad for its economic well-being?" He paused, then confided, "We are closer than you might think."

In early 2014, with oil prices at a steady and healthy high, Hawrami could probably taste Kurdish "economic independence." With oil prices north of $100, it would not be long before Kurdish oil production was sufficient to garner as much or more than the revenues the regional government was supposed to receive while staying in Baghdad's orbit.

Rarely are so many damp cloths put over a dream at the same time. Over the course of six short months in 2014, the idea of an independent Kurdish entity lost its air of inevitability and imminence as events took unexpected and discouraging turns. Pressuring governments from Italy to Morocco, Baghdad waged a largely successful international campaign to dissuade buyers from purchasing Kurdish crude. The Iraqi government even enlisted American courts in its efforts, leading a U.S. court in Houston to interpret Iraq's constitution and determine what should happen to an oil tanker filled with Kurdish crude that had spent more than five months circling a navigational buoy sixty miles off Galveston, Texas, while awaiting permission to dock and unload. Turkey refrained from helping Syrian Kurds under siege in the border town of Kobani, stoking wariness among Iraqi Kurds already nervous about putting their fate in the hands of Ankara, a former adversary. Most dramatically, the Islamic State—ISIS—steamrolled across western and northern Iraq, suddenly threatening the sparkling malls and restaurants of Erbil and leading to a mobilization of the Peshmerga, the Kurdish regional security forces whose last military encounters had been with Saddam.

Perhaps most significantly, the price of oil fell by more than half, punching several holes in Hawrami's careful math. International oil companies shifted from being eager to being reluctant to be part of the Kurdish energy scene as prices fell and political risk rose. The volume of exported oil needed to meet the region's budgetary needs without Baghdad's help skyrocketed from within reach to far from it.

Both Baghdad and Erbil felt the financial squeeze at the same time, making rapprochement between the national and regional governments attractive—perhaps even essential—to both sides. Iraq could no longer send the oil produced in its vast southern fields north for export through Turkey; ISIS controlled the territory through which the pipeline ran. With its exports already constrained by insufficient infrastructure in the south, Baghdad suddenly saw value in a Kurdish pipeline able to carry crude to

Turkey without traversing territory held by ISIS. In a low-price environment, Iraq needed to export every barrel of oil possible to fill dwindling coffers. An arrangement in which increased exports of Kurdish oil would contribute to the national kitty was also attractive for the same reason. Lubricated by new and more visionary leadership in Baghdad, the two governments struck a deal in December 2015. Although the deal did not last, it marked the first meaningful energy cooperation between Baghdad and Erbil in years, and elements of it persisted, such as Baghdad's acquiescence to the export of oil by the Kurds through Turkey.

The push for Kurdish independence from Baghdad is of course about much more than economics and is set again to intensify in the aftermath of the ejection of ISIS from the northern Iraqi city of Mosul. After enduring decades of repression under Baghdad, and having preserved their own culture and language for centuries, the Kurds are eager to be masters of their own destiny. Steps taken by the United States and others to fight ISIS have inadvertently strengthened the Kurds and their quest for a separate state. Many would be willing to pay an economic price if that was the only one to be incurred for political independence. But given the spate of additional political and security calculations that must be made in the drive for independence, the dramatically changed energy and economic environment created a pause for cooperation. It opened the door for one more chance for Iraq to maintain itself as a unified state. Even if Iraqi Kurdistan eventually breaks with Baghdad, the more sober energy environment has cooled the fervor of those wishing to hastily make such a move—and may have created the possibility for interim arrangements that could lead to a more peaceful and sustainable outcome.

Israel: Gas Monster

Further west, the new energy abundance has presented intriguing possibilities for peace between Israel and some of its neighbors. In 2010, American and Israeli companies working in tandem discovered Leviathan, a large natural gas field off the shore of Israel. Named for the biblical sea monster in the Old Testament, Leviathan is not far from the Tamar field that had been found one year earlier. Together, the two fields are believed to hold 32 trillion cubic feet of natural gas, or twice as much as the U.S. Geologic Survey says is recoverable from the National Petroleum Reserve in Alaska. This amount is more than enough to meet Israel's domestic needs and offers Israel the prospect of becoming a natural gas net exporter by 2020. Despite threats from Hezbollah, a Shi'a militant group based in Lebanon and considered to be a terrorist group by Israel and the United States, Tamar started production in March 2013. After working through a range of antitrust suits and regulatory problems, Leviathan looks set to be developed, with the first gas expected at the end of 2019.

These developments have fueled hopes among many Israelis that natural gas might serve as a platform for peace in the region. There are already indications that at least some of these hopes can become realities.

In May 2010, the world watched transfixed as a flotilla of boats carried Turkish and international peace activists across the Mediterranean to the Gaza Strip. The group intended to break the Israeli and Egyptian blockade of Gaza by delivering aid and to raise global awareness of what it perceived to be an illegal cordon. In a predawn raid, Israeli commandos rappelled onto the Turkish ship *Mavi Marmara* from circling helicopters. Although they sought to verify that there were no weapons or ammunition onboard the boat, fighting broke out immediately, soon leaving nine Turks dead and one fatally wounded.

The incident led to a bitter fallout between Israel and Turkey. The two countries had been close allies until that moment, but the raid on the flotilla precipitated multiple inquiries, the withdrawal of ambassadors, and icy relations for six years. The rhetoric was vicious. Turkey was once Israel's closest ally in the Muslim world, but, in 2014, Turkish prime minister Recep Tayyip Erdogan accused Israel of having "surpassed Hitler in barbarism." In a telephone call orchestrated by President Obama in 2013, Israel apologized for the deaths, but a full rapprochement was stalled by disagreements over compensation and other matters.

It was not until a 2016 meeting in Rome that Turkey and Israel finalized the elements of an agreement and normalized relations between the two countries. Surely, the deteriorating security situation in the region was one spur to the new accord. But energy undoubtedly was also a major impetus. Relations between the two capitals began to thaw in late 2015, not coincidentally after Turkey shot down a Russian plane over Syria.

That action had prompted Gazprom, Russia's natural gas behemoth, to (at least temporarily) suspend construction of TurkStream—a pipeline to deliver Russian gas to Turkey and beyond.

Israeli prime minister Benjamin Netanyahu was explicit about the role that the prospect of natural gas deals played in the reestablishment of relations between Turkey and Israel, both when I met him privately with a small group of former U.S. officials in Israel in August 2016 and in public. At the time of the diplomatic agreement, Netanyahu stated, "This agreement opens the way to cooperation on economic and energy matters, including the gas issue. Gas is so important and contains the possibility of strengthening the Israeli economy and state coffers with vast capital. . . . Leviathan could supply both the Egyptian market we intend to work with and also the Turkish market as well as the supply of gas through Turkey to Europe, and this is a strategic issue for the State of Israel. This could not have come sooner without this agreement, and now we will take action to advance it."

Israel's hopes to showcase the benefits of peace by forging substantial and mutually beneficial economic partnerships with the only two Arab states with which it has signed peace agreements—Egypt and Jordan—are also on track to be at least partially realized. In September 2016, Jordan's state-owned national power company signed a $10 billion deal with the developers of the Leviathan field to provide natural gas to Jordan. Because of both the physical proximity of Jordan to Israel and infrastructure already in place, the sale of natural gas can still be economical in a low-price-energy environment. Israel had similarly sought to sell substantial amounts of natural gas to Egypt; that country's two LNG export terminals have been idle due to lack of gas for the last several years. Domestic debates about Israel's regulatory structure, however, slowed the finalization of agreements to do so, and fate intervened. Late in the summer of 2015, Eni, the Italian energy company, announced it had discovered a "super giant" natural gas field off the shores of Egypt's Mediterranean delta. Zohr, as the field was named, is believed to contain enough gas to both meet Egypt's domestic needs and supply Egypt's LNG terminals. As a result, the prospects of Israeli natural gas providing a lifeline to a gas-thirsty Egypt—and the political benefits that may have accompanied such an arrangement—are now far less likely to materialize.

Syria: An Outside Chance

No country in the Middle East—or perhaps the world—needs a push for peace more than Syria. The death toll of its civil war is staggering, with nearly half a million Syrians killed since the conflict erupted in 2011. Equally astonishing is the number of displaced Syrians. The country is being depopulated. At a time when other Arab states are battling "youth bulges," Syria is becoming a country without children. Its absolute population has declined from almost 21 million in 2010 to 18.5 million in 2015. Fully half its population has been driven from their homes, with a third in foreign countries and two-thirds uprooted within Syria's borders. The economy has been destroyed and infrastructure is devastated. A video clip filmed by a drone flying over the city of Homs—a major industrial center that was at a time Syria's third largest city—shows the once vibrant city to be nearly abandoned and in ruins.

While one can imagine a scenario in which energy plays a palliative role in the war, the pressures of the new energy abundance have not yet worked in favor of resolution. There is one exception that is relatively small given the scope and complexity of the conflict: the new energy realities have clipped the financial fortunes of ISIS. The group had commandeered oil fields in both Syrian and Iraqi territory in 2014 and initially reaped significant financial gains from operating them. But the dual developments of low oil prices and aerial bombing of these locations has limited the contribution these oil sales have made to the coffers of the extremist group.

Although Syria itself was never a large oil producer, nearly all major external actors in the Syrian conflict have significant interests in energy. Russia, Iran, Saudi Arabia, and the United States are five of the top seven oil producers in the world; only the budget of the United States is not dramatically impacted by the price of this commodity. One might therefore think that major budgetary pressures on Russia, Iran, and Saudi Arabia would have the potential to curtail the involvement of these countries in the conflict. Thus far, however, there is little evidence that the cost of intervention is weighing heavily enough to spur any shifts in course, particularly given the enormous strategic stakes all sides perceive to be at risk in Syria.

For the United States, Russia, Saudi Arabia, and even Iran, the cost of intervention in the conflict in Syria pales in comparison to their overall military expenditures. Russia and Saudi Arabia are the third and fourth largest military spenders in the world, with military budgets of $70 billion and $61 billion respectively. While neither country wants an expensive, long-term military engagement in Syria, the costs of involvement are not the primary drivers of their commitment. Iran's official military budget is smaller, at $12 billion, but that figure does not include "paramilitary" spending. The $500 million that the United States used to fund a failed effort to train and equip Syria opposition figures in 2014 and 2015 was less than 1 percent of the overall Pentagon budget.

Perhaps more interesting is the question of whether low oil prices can help overcome the hostility between Saudi Arabia and Russia and Iran. There were at least some rumblings in November 2014 about the possibility of a deal between Saudi Arabia and Russia that would link efforts to shore up oil prices with the end of hostilities in Syria. With oil prices beginning to soften that November, Saudi foreign minister Saud al-Faisal visited Moscow for talks with his Russian counterpart, Sergei Lavrov. In speaking to the press after the talks, Lavrov vehemently denied that a deal over oil would compel Russia to alter its policy toward Assad. Immediately after a news story suggesting just such machinations, Alexei Pushkov, the head of the Foreign Affairs Committee in the Russian State Duma, rejected the idea that any such pact was discussed. He blamed the rumors on fake news, saying, "*The New York Times* distorted information so many times, especially since the Ukrainian crisis started. I wouldn't advise you taking it as a reliable source. There were no talks of such exchange." Speaking anonymously, a Saudi diplomat at the talks was more circumspect: "If oil can serve to bring peace in Syria, I don't see how Saudi Arabia would back away from trying to reach a deal."

Since this time, low oil prices have in fact spurred some modicum of greater cooperation between Saudi Arabia, Russia, and Iran. The economic duress resulting from the price plunge helped the three countries overcome tensions at least enough to agree to an oil production cut in 2017. There is, however, not yet any strong evidence that such limited cooperation will translate into a meeting of minds on Syrian policy. In fact, by agreeing to a production cut, the Saudis gave away some of the leverage they would have in convincing the Russians to alter course on Syria. While Russia appears to be interested in building a more strategic relationship with Saudi Arabia, Riyadh has thus far remained cool, preferring instead to focus on forging a new relationship with the United States under the Trump Administration.

Iran: Hoping for More

Finally, Iran. As described in Chapter Five, the new energy abundance played an important role in paving the way for the international sanctions that eventually brought Iran to the negotiating table—and ultimately to the signing of an agreement with the P5+1 in July 2015 to curb its nuclear pursuits. Yet however instrumental the new energy abundance was in the conclusion of a deal, it is likely to be distinctly unhelpful in upholding the very same agreement. As was already evident by mid-2017, many Iranians feel disappointed over the fruits of the deal, potentially jeopardizing the durability of the accord. The lifting of sanctions in January 2016 did lead to greater oil production and exports and did allow Iran to garner foreign interest in many nonoil sectors of its economy. But central government revenues and expenditures have lifted only marginally, thanks to lower oil prices.

For this reason, many Iranians feel less optimistic about the country's trajectory in general and economic situation in particular in 2017 than they did before the deal was signed. In the words of Ebrahim Mohseni, a research associate at the University of Maryland who conducted a poll of one thousand Iranians one year after the deal, "Iranians were expecting a lot but feel that they have received nothing tangible in return for Iran's nuclear concessions. . . . Unless Iranians see real economic gains from the nuclear deal, the nuclear agreement, those who negotiated it and the foreign policy approach that made it possible will all be in jeopardy."

Iran is likely to continue to struggle to deliver significant economic gains to its population. The new energy abundance will be, at least in part, to blame. It adds to the myriad of existing obstacles to attract greater foreign investment to develop Iran's oil fields—a prerequisite to Iran reaching its goal of producing 4.8 mnb/d by 2021. Corruption, lack of transparency, and continued unilateral U.S. sanctions have tempered the enthusiasm of many outside investors; low oil prices have piled on by cutting the investment budgets of many companies and forcing countries that do want to attract foreign investment to offer more attractive terms. Whereas some international companies were clamoring for access to Iranian oil in the late 2000s, the advent of unconventional resources means that companies can now be highly selective about where they invest their money. As long as prices stay at moderate levels, the international oil companies are capital constrained, not opportunity constrained.

In the realm of natural gas, the story is similarly downbeat. With the second largest conventional gas reserves in the world, many expected that Iran could quickly move from its current position as a marginal supplier of gas to Turkey to be a major global gas player. While Iran had a larger window to realize that status in the 1990s, the advent of unconventional gas calls into question the feasibility of some of Iran's big gas projects, not to mention the possibility of Iran emerging as a global LNG provider in the near term. The combination of U.S. and Canadian shale gas exports, along with the completion of gas projects in Australia, Israel, the countries of east Africa, and elsewhere, could mean that Iran will miss the boat and need to wait until the LNG market tightens many years down the road before making its splash.

In a region beset with bad news and worrying trends, the new energy abundance was expected to spell nothing but doom and gloom for the big powers in the region. In an era of copious oil and gas supplies, many inside and outside the Middle East thought the people of the region could only look forward to more state weakness, political marginalization, and increased conflict to add to their already substantial woes. Such developments would further pile on today's challenges, increasing the possibility of economic collapse or political turmoil. But such negative outcomes are not inevitable, nor necessarily the ones most likely to materialize.

Certainly, the Middle East is in the early innings of coping with the effects of a much altered energy environment. In future years, Middle Eastern governments and citizens could need to grapple with the demand-dampening results of a stronger global push to address climate change. Those with ample oil and gas reserves could face the real prospect of "stranded assets"—where their once valuable resources no longer contain significant worth in a world moving away from fossil fuels.

While it is not too early for Middle Eastern leaders to contemplate and prepare for such eventualities, the news for the next five or ten years is not all bad. Those who view the U.S. presence in the region as a stabilizing one can be heartened by the notion that America will not curtail its engagement there due to its new energy prowess. Moreover, given the possibility that in a continuously low-oil-price environment Middle Eastern oil will be more important to global markets than ever, those in the region can be confident of continued international efforts to build relationships and to better understand the countries and cultures there. Finally, the new energy abundance has created some intriguing possibilities on the positive side of the ledger—even if their realization is far from assured. It has spurred serious and long-overdue efforts to reform economies in the Gulf—initiatives that may, perhaps inadvertently, also lead to political and social change. And it has opened the door for steps toward peace if not across the region, than at least in some parts of it desperate for good news.

Conclusion

From Serendipity to Strategy

Energy has been a key driver of many of the most important events of the past hundred years. For example, in 1913, Winston Churchill, then First Lord of the Admiralty, decided to shift the British naval fleet to run on oil instead of Britain's plentiful coal reserves. The decision gave the Royal Navy much greater range while simultaneously making it dependent on access to an energy source thousands of miles away. As a result, the oil-rich lands in the Middle East, including Persia (now Iran), catapulted to strategic importance. Soon thereafter, World War I emerged out of a complicated series of events, of which the competition between European countries for access to oil was a significant, if underappreciated, factor.

Following the war, oil became more and more important to economies—and militaries—around the world. By World War II, access to oil influenced military strategy as much as any other single consideration. In 1941, for example, the United States embargoed oil sales to Japan. Tokyo, desperately in need of oil to fuel its war efforts and fearful U.S. forces would disrupt its access to the oil-rich East Indies, made the fateful decision to bomb Pearl Harbor. The attack drew the United States into the war on both fronts. A year later, Germany's drive through southern Russia to capture the Baku oil fields stalled, and ultimately failed, at Stalingrad. These two events, both prompted by the overwhelming need for oil, were among the most important ones in determining the eventual outcome of the war. Even the Marshall Plan of the late 1940s and early 1950s—one of the most well-intentioned foreign policy initiatives of the twentieth century—had a major energy component. The plan helped consolidate postwar Europe's move away from coal by providing funds to build a pipeline connecting American-run oil fields in Saudi Arabia to Western Europe.

In 1973, Arab nations embargoed oil exports to the United States as retribution for its military support of Israel in the Yom Kippur War. The global economy, now more oil-dependent than ever, slumped in response to sharply higher prices before moving into an economic malaise that lasted throughout the 1970s and early 1980s. Many countries also attempted to make their economies less vulnerable to an oil price shock, investing in fuel economy and alternate fuel initiatives. Japan accelerated its move toward less fuel intensive electronics manufacturing and honed its automaking until its more efficient vehicles finally cracked the huge U.S. market. In the 1980s, low energy prices weighed heavily on energy producers and ultimately contributed to the collapse of the Soviet Union and the global order that had endured for decades.

In more recent decades, geopolitical events in the oil-rich Middle East have been prominent. In 1990, after accusing Kuwait of "slant drilling" across its border, Iraq invaded Kuwait, seizing its oil fields. Iraq was subsequently challenged by a coalition led by the United States, driven out of Kuwait, soundly defeated, and placed under sanctions and the U.N. Oil-for-Food Program that traded Iraq's oil for humanitarian supplies for years. The decade's low oil prices also created the poor economic conditions in Venezuela that helped pave the way for Hugo Chávez's leftist Bolivarian Revolution in Latin America.

In the first decade of this century, Russia used its natural gas trade to express unhappiness with Ukraine's political trajectory through a mix of coercion and disruptions of energy supplies. In the Middle East, another U.S.-led coalition used force against Iraq; this time Saddam Hussein did not survive. Since then, volatile politics and religious extremism have been contributing factors to the decade or more of turmoil in the region. Alas, it is no coincidence that, a century after World War I, the most oil-rich region of the world is once again the setting for direct military action by some of the world's most powerful militaries, including those of the United States, France, England, and Russia.

Energy will continue to be as important in shaping the world as other geopolitical events that receive a lot more attention. Yet in one important way, the past will not be a harbinger of the future. Whereas, particularly in the last thirty years, energy scarcity and the fear of energy shortages have shaped the relationship between energy and international affairs, abundance, not scarcity, will be the defining feature of the coming years. The combination of technological advances, politics, and policy has already made possible the production of vast quantities of new energy resources. Understanding the new global dynamics created by energy markets awash in oil and gas—and eventually renewable sources—could be as important as discerning the role of radical extremism, infectious disease, nuclear proliferation, or climate change on global affairs.

The comparatively low energy prices of the last several years have been the most pronounced expression of this new energy abundance. But they are not necessarily its most enduring, nor its most important, manifestation. In fact, an energy-abundant world is not necessarily a world free from spikes in the price of oil or gas. Given the many factors that affect price—geopolitics, natural disasters, psychology, technology, investment patterns, policy—some price spikes are still inevitable. The key difference is that such spikes are likely to be less enduring and will tend to have tactical implications—rather than being long lasting and carrying strategic consequences. Think of the eight months in 2008 and 2009 when oil prices seesawed wildly, falling from a peak of more than $140 a barrel to below $40 and back above $100. Then compare that rollercoaster ride to the steady price plunge that began in 2014 and stretched well into 2016. The first period impacted markets and businesses over the short term, but changed little geopolitically; the second had wide-ranging effects that are still unfolding. A price spike in the coming years is plausible, given the hundreds of billions of dollars cut from investment budgets of oil and gas companies in the wake of the 2014 price drop. But if tight oil does its job, responding with greater alacrity to a higher price with increased production, such a spike should moderate within a relatively short period. It will not represent a new price plateau that lasts for years.

In any case, the changes in oil and gas markets induced by the new energy abundance are even more fundamental and consequential than price. The global oil market is in new territory, with the market playing a much greater role in balancing supply and demand than has been the case for more than a century, an anemic OPEC notwithstanding. Global gas markets have been transformed even more radically by new supplies and new technologies. Greater integration among regional gas markets, while not amounting to a global market, has meant more efficient allocation of capital and greater flexibility. As explored throughout this book, these changes in the structure of oil and gas markets—and the shift in power from the producer to the consumer that has accompanied them— have already had profound effects for geopolitics. And the story is, in many way, still unfolding.

A Strategic Boon to America

Nowhere in the world have these fundamental changes in the realm of energy had so many consequences as in the United States. Most evidently, the surge in tight oil and shale gas production has fueled an economic boom that has been good for jobs, government coffers, and the economy as a whole. But it has also brought with it enormous *strategic* benefits. To the disappointment of some, it has not delivered—and may never deliver—the energy independence long coveted by the United States. Yet, even more importantly, the energy boom has reinforced the foundations of both hard and soft American power. The strategic benefits to the United States also go well beyond its own borders, to shaping the international environment in a way that is—on the whole—more conducive to the promotion and protection of American interests. There are many issues affecting what kind of global power the United States will be in the coming decade and beyond. The challenges it faces are significant. Internally, it needs to address its fiscal situation, divisions over race, and the dysfunctional politics that have fueled a rising populism. Externally, the United States must come to terms with a world in which power is distributed more diffusely and determine what it means to be an exceptional country given these new realities. No single issue—including energy—will in itself determine the U.S. trajectory. But there is no question that the balance sheet of American strengths and vulnerabilities has been profoundly altered by the energy boom—and overwhelmingly, if not uniformly, in the interests of the United States.

A deeper dive into the ways in which the new energy abundance has enhanced American power yields some surprising insights. Many of the conventional wisdoms that have arisen around the new energy landscape are—as so often is the case—ill-informed and incomplete. The misplaced fixation on energy independence—and the belief that the United States has or will soon reach this exalted status—is perhaps the most glaring example of such misperceptions. In addition, contrary to popular understanding, the energy boom will not significantly diminish U.S. interests in the Middle East and thereby absolve America from the difficult interventions that have marked U.S. foreign policy in the last decades. Nor will copious American natural gas enable the United States to displace Russian gas in Europe's energy mix and free Europeans from the specter of Russian energy embargoes and the political influence that results from this possibility. At the same time, contrary to what is commonly thought, the new energy realities are working against, not in favor of, a strategic partnership between China and Russia. And, finally, while low oil prices spell tough times for the Gulf states of the Middle East, they do not render inevitable state collapse, revolution, and greater conflict. While these commonly expected scenarios will not arise, as demonstrated throughout this book, the new energy abundance *is* exerting its pressure in other numerous and subtle ways, all of which add up to a changed strategic landscape for the United States and for other countries around the world. In ways that are both surprising and consequential, the new energy trends are exerting powerful influences on how the world works.

This energy-induced geopolitical reordering is just beginning. What has occurred thus far has largely been happenstance—serendipitous for some, unfortunate for others. Now that the new energy reality is established, policymakers must more consciously shape the future in light of these new trends. Some regions, such as Europe, are moving directly toward a future that—with the critical exception of Europe leaving its own shale gas largely undeveloped—harnesses many strategic benefits of the new energy environment. Other countries, where the boom in tight oil and gas has caused unwelcome dislocations, are also clear-eyed about where policy needs to go in the future, even if they have not yet traveled there successfully. Canada, for example, continues to pursue its dogged, if not yet fruitful, efforts to bring energy from its vast interior to its tidewater for export to destinations other than the United States. However, many other countries have yet to embrace the strategic benefits or weaknesses of the energy boom. Regardless of where they are now, countries including the United States, Russia, and China, as well as regions like the Middle East, Africa, and Latin America, all need to consider the new energy environment as they plan for the future.

Thus far, the improved domestic and international position of the United States resulting from the energy boom has been largely the product of markets. Although the United States is the epicenter of this new abundance, American policymakers as a whole have not yet fully embraced a mind-set that labels energy a strategic asset. For decades, they were repeatedly reminded—by the intelligence community, academe, think tanks, and their own daily experience—that the United States depended on imported energy and that energy markets favored producers over consumers. It was gospel that the American need to access to reliable and affordable energy was a major strategic vulnerability. The energy boom has transformed realities faster than policymakers have been able to change their own mind-sets.

There is some evidence that policymakers are beginning to adapt. In terms of rhetoric, it is still possible to find the occasional policymaker warning against the evils that imported oil brings to the United States. Yet, some policymakers have at least identified the foreign policy opportunity presented by the boom. In 2013, National Security Advisor Tom Donilon gave a speech at Columbia University claiming, "America's new energy posture allows us to engage from a position of greater strength." For Donilon, energy was a new foreign policy asset for the United States. Similarly, the National Security Strategy of the Obama Administration, released in February 2015, stated that "America's energy revival is not only good for growth, it offers new buffers against the coercive use of energy by some and new opportunities for helping others transition to low-carbon economies."

While these acknowledgments have been largely limited to generalities or conventional wisdoms—which, as we know, tend to be off the mark—some actions have also been taken to better align U.S. interests with new energy realities. The historic lifting of the forty-year-old ban on U.S. oil exports at the end of 2015 was largely done for economic reasons. Yet in the flood of justifications for the move, some advocates highlighted geopolitical benefits. For example, the Bipartisan Policy Center asserted that if the ban were lifted "foreign crude oil exporters will lose market share and political power."

The United States is still in a nascent state in terms of truly understanding the energy boom as a gift to American foreign policy. U.S. policymakers first need to appreciate the vast strategic benefits that the United States has already reaped from the boom. One way is to imagine the American position in a world without this boom. With much higher oil prices and the absence of a growing energy sector from 2009 to 2014, the Great Recession would have been much harder to overcome. In 2007, the U.S. government expected net crude oil imports to continuously increase. Absent the new energy abundance, the volume of U.S. imports could have reached all-time highs, even if we take into account that higher prices would have led to Americans driving smaller cars and consuming less oil. Moreover, the increase in American energy imports would have been coming at least in part from less-than-savory sources. For instance, in a world where producers still had the upper hand, the United States might have well been heavily dependent on imports of Russian natural gas.

Exactly how this energy picture would have influenced American foreign policy is impossible to ascertain, but some questions about what might have been are worth considering. If Russia had been an important natural gas supplier to America, could have Washington taken the same strong stand against Russia in the wake of its annexation of Crimea? How would Europe have reacted had natural gas markets been tighter than ever, rather than being increasingly flush with supplies from other sources? If Russia was the only viable source of the natural gas needed to sustain Chinese growth, would Beijing have been more vocally supportive of Russia's foreign policy adventurism? If natural gas had not been available to substitute for nuclear power in the wake of the Fukushima disaster, would Japan have maintained its political stability in the face of massive energy shortages? How much stronger would the voice of Venezuelan president Nicolás Maduro be today in the region were oil prices at significant highs? Would President Obama have been able to pursue his opening with Cuba had Havana not been worried about the economic viability of its near only patron, Caracas? Would the United States have been able to marshal support for economic sanctions against Iran with oil prices at record levels? If not, would the Iranians have come to the negotiating table where the nuclear deal was signed? What would have been the eventual fate of American partners in the Gulf had economic pressures not compelled them to take the hard decisions on reform? Would the United States have been in a position in 2015 to credibly present a plan to tackle carbon emissions and motivate the world to follow had natural gas been both expensive and imported? The previous chapters describe in detail my answers to each of these questions, but the macro answer is a generalized "no." Through multiple paths, the new energy abundance—often in conjunction with other forces and impetuses—has produced outcomes beneficial to the United States that might not have otherwise materialized.

We do not and cannot know all the contours of the energy future. But the United States has a real opportunity—and even an obligation—to take advantage of this moment to craft its policies and strategies to ensure it harnesses the new energy abundance to advance its interests and those of its friends and allies. Few countries ever see their strategic situation improve so suddenly and even fewer policymakers are successful in maximizing the opportunity such a shift presents. In his book, *A World in Disarray*, Richard N. Haass, the president of the Council on Foreign Relations, laments how the United States did not fully capitalize on the new circumstances created by the collapse of the Soviet Union in 1991 to build a new global order. Imagine, therefore, how much harder it will be for current policymakers to seize upon a geopolitical moment created by markets, the internal workings of which are often obscure to the foreign policy establishment.

The Trump Administration clearly views the energy boom as having enormous value to the United States and that in itself is a welcome orientation. As evident by both its rhetoric and its early policy moves, the new administration is placing a high priority on removing impediments to greater oil and gas production and putting in place new infrastructure to move these resources once produced. On the whole, these inclinations are reasonable and are a positive departure from a national conversation that was moving in the opposite direction. Yet, the lens from which the administration appears to be viewing the energy boom is far too narrow and, as a result, carries some significant risks.

Rather than seeing U.S. oil and gas production as an end in itself, or at best a vehicle for more jobs in the American economy, U.S. policymakers should adopt a different outlook. They should view the boom as a means of maximizing strategic value and benefit to the United States and its partners and increasing overall American standing in the world. Adopting such a perspective would result in a somewhat different and broader set of policy prescriptions. More specifically, U.S. policymakers should embrace an expansive range of actions that can roughly be grouped into four categories: sustaining the boom at home; using America's new energy position to advance wider, non-energy foreign policy and national security goals; taking advantage of the new strategic landscape created by the energy boom to advance American interests; and anticipating and planning for new challenges that may arise as a result of the new energy landscape.

Sustaining the Boom at Home

While focusing on increased domestic oil and gas production cannot be the entirety of the U.S. policy approach, it is certainly an important element of the equation if America is to maximize the strategic value of the boom. As explored throughout this book, some of the geopolitical benefits of America's new energy prowess come simply from the changes that greater quantities of American oil and gas have instigated through market forces. OPEC's influence over oil markets—and therefore global economic growth—has been much diminished, in part by the quantities of tight oil produced in the United States, but more importantly due to the new business model tight oil has introduced. Similarly, the American shale gas bonanza has been a major factor in changing gas markets and diminishing the geopolitical power of producers such as Russia; these developments are partially a result of how the shale boom both directly and indirectly increased the volume of LNG on global markets and the terms though which consumers can gain access to this energy.

As a result, a comprehensive U.S. strategy to take advantage of the new energy abundance would include measures to boost and sustain tight oil and shale gas production. As discussed elsewhere, some reform of onerous or duplicative regulations would be beneficial. Also needed is a shift in the weight of who regulates oil and gas production from the federal government to the state level, at least where the capacity to do the job well exists. State officials are likely better positioned to regulate local production, given that they are more sensitive to local conditions. Finally, the United States is in dire need of more infrastructure to transport oil and gas to demand centers and export terminals.

As important as these measures are, unfettered oil and gas development may ultimately go against the broader goal of amplifying America's strategic benefit from the boom—and even compromise the more narrow goal of maximizing oil and gas production. This is because producers must be conscious of maintaining the social license to operate; if they lose the trust of the communities in which they work, their efforts to produce will be severely hampered, even if their activities are legal. As demonstrated in Europe, and in locales far closer to home, public support for fracking is already on a thin margin in many places. Environmental or other accidents could tip a fragile acquiescence in favor of allowing oil and gas companies to operate to a stronger consensus against it.

Sustaining the boom at home will also require U.S. policymakers to take a hard look at a range of non-energy national security and foreign policy approaches. Although often overlooked, the fortunes of America's own energy producers are dramatically affected by what happens in other domains. In fact, policymakers may find that the largest impediments to U.S. oil and gas expansion are not regulations, but policies in other areas that, on the surface, appear to be unrelated to energy. For example, the American provision of global public goods in the protection of sea lanes from the Gulf to the Pacific underpins robust global trade in energy. Any suggestion that the United States will or could withdraw from these responsibilities could severely hamper these flows, which now include exports of American oil and gas. In addition, both energy trade and global energy demand will be adversely affected by a retreat from the embrace of free trade—a ballast to worldwide economic growth for decades. And finally, America's own energy fortunes will be affected by policies espoused toward individual countries, Canada and Mexico in particular. Where some may not see the connection, a U.S. approach that destabilizes Mexican politics and creates new barriers to trade and investment between the United States and its southern neighbor will crimp America's own energy boom—and any hope Washington might have of reaching the most meaningful form of energy self-sufficiency: North American energy independence.

Using Energy as a Means to Achieve Wider Goals

The second domain in which U.S. policymakers need to challenge their thinking is one in which America's new energy situation is a tool to attain non-energy pursuits. Here the possibilities are large and span the globe.

First, the United States should prioritize maintaining a position of global leadership on the issue of tackling climate change. The Obama administration leveraged the decline in U.S. carbon emissions resulting from the shift from coal to less emissions-intensive natural gas to reassert American influence on an issue of great importance to a large number of countries around the world. Even putting the importance of mitigating climate risks aside, the United States should treasure this source of soft power, particularly in light of the growing difficulties associated with wielding hard power. Ideally, Washington would take the lead, as the federal government not only has the most influence on policy, but is best positioned to generate and make use of soft power. However, states, localities, corporations, and even individuals can help substitute for federal action if necessary.

Second, the United States should use its energy instruments to further integrate its economy with those of Canada and Mexico. In truth, the three countries have already reached an extraordinary level of interdependence, but energy offers the opportunity to take this integration further, particularly between the United States and Mexico. Not only will this be good for U.S. natural gas exports, given that Mexico is by far the largest importer of this American commodity. But the United States and Mexico have almost a unique economic relationship in which they literally build products together. As a result, supporting a prosperous Mexico, not a failing one, is unequivocally good for the American economy.

Third, America can also reap greater gains by reinvigorating earlier efforts to help other countries with unconventional oil and gas reserves to develop them. While initial endeavors by non-U.S. countries to unleash their own oil and gas potential proved disappointing, such efforts are by no means over. Vast quantities of oil and gas exist that will be tapped in the years ahead, although doing so may require approaches that vary significantly from those pursued by America. The United States should be at the forefront of helping the countries that desire assistance manage this process. Programs under the Obama administration to provide this support were popular, but lost steam in large part due to environmental sensitivities. Now there seems to be little funding for such efforts. But these programs should be revived and, when they are, they should be evaluated on a broader basis than whether the country in question develops its oil and gas. Instead, their assessment should take into account the extent to which they provide critical avenues for advancing U.S. ideas related to transparency, open markets, and anti-corruption.

Fourth, the United States should look for ways in which its good energy fortunes can help ensure that China realizes its objective of increasing the role of natural gas in its domestic energy mix. One may be tempted to think that if China fails to reach such goals, renewable energies will be the beneficiaries. However, if natural gas does not permeate the Chinese economy in a larger way, it is more likely that China will continue its heavy dependence on coal. Whether China accelerates its shift away from coal is critical both to the fate of global efforts to tackle climate change and to the Chinese government's ability to maintain stability in that vast country. It is in the interests of the United States that China is successful in both domains. Helping China realize greater natural gas use will not necessarily require huge policy shifts on the part of the United States; the impact of U.S. shale on global markets is already very positive. But policymakers could reinforce these positive trends in a variety of ways, from publicly welcoming LNG flows between the United States and China, to removing any hurdles to Chinese companies directly importing U.S. gas, and reexamining any policies that could stand in the way of China adapting and utilizing technologies to produce shale gas and tight oil itself.

Some foreign policy analysts prefer that U.S. companies not help

China address its energy vulnerability. For them, this is particularly the case if, as a producer, the United States can leverage China's vulnerability for its own diplomatic ends. Ultimately, however, Americans are better off realizing that the more energy self-sufficient China becomes, the less it needs to engage in problematic policies abroad, such as partnering with Russia. The more confident China is in the current order managing its energy needs—whether in the global market or through its own production capacity—the less need it will feel to build alternative institutions.

The fifth way in which specific energy tools can help the United States advance broader foreign policy goals is by shoring up alliances and building new, more pro-America centers of power abroad. For instance, in Latin America, the energy boom has helped create further opportunities for the United States, this time in the Caribbean. The new energy abundance is but one factor that has pushed Venezuela and its Bolivarian government close to collapse. A destabilized, poor Venezuela has already begun to curtail Petrocaribe, its program that provides cheap heavy fuel oil to countries in Central America and the Caribbean in exchange for political support. As this program fizzles out, the United States and others have the chance to help countries in this region find energy substitutes for Venezuela's oil that are not only cleaner, but also come without political expectations. More generally, America's status as a natural gas exporter is helping strengthen U.S. allies from Europe to Japan—not by fully replacing Russian gas, but by providing another source of diversity of supply and by shifting the balance in favor of consumer countries against traditional producer ones.

Finally, U.S. policymakers should consider the extent to which they want to use certain energy instruments—such as approvals for the export of natural gas or a revamped and reconsidered Strategic Petroleum Reserve—to promote foreign policy interests. Doing so would not be a complete departure from past practices; the United States has been an avid implementer of sanctions against oil exporters in efforts to combat terrorism and constrain proliferation and advance other goals. Nevertheless, the United States will need to carefully balance its commitment to free and smoothly functioning energy markets—which has been so critical to U.S. energy security—with the opportunity to exert geopolitical leverage. While it will be tempting to deploy these new or revised tools in an effort to advance an international agenda, the United States will probably find that maintaining well functioning energy markets delivers greater benefits to it on the whole.

Leveraging the New Strategic Environment

One of the main takeaways to emerge from this book is that the new energy abundance has dramatically transformed the global strategic environment in a way that is conducive to U.S. interests. Crafting a U.S. strategy that truly maximizes the benefits accruing to the United States as a result of the boom also entails taking full advantage of the new opportunity presented by this strategic landscape. Again, the prospects here are both varied and highly consequential.

Most importantly, the United States needs to take advantage of the additional breathing space given to it to shore up the liberal international order under which America and its partners—and many of its competitors—have prospered economically and politically. This order has been under attack and the United States has been lethargic as its defender. However, several developments related to the new energy abundance can help the United States in what should be its main foreign policy objective: fending off challenges to the order and reforming it so as to enhance its sustainability.

For starters, the new energy abundance degrades the power of the most aggressive and influential challenger to that order, Russia. The economic difficulties from low energy prices may, counterintuitively, push Russia to be more petulant in the international arena over the short and medium terms. But the structural changes in energy markets—as much as the price—engendered by the new energy abundance make some of Russia's traditional tools of coercion less powerful and influential.

At the same time, the new energy abundance affects what many experts consider to be an ongoing debate within China about whether it should join forces with Russia to construct a new order in the place of that which has existed for decades. Rather than making that option appealing, the energy boom reinforces Chinese confidence in one of the key elements of the liberal international order: the market. Moreover, existing global energy governance bodies provide the United States the chance to show that international institutions can be changed and adapted to accommodate growing Chinese power; doing so is one way to make the point that China need not remake the liberal international order to get the respect and influence commensurate with its changed size and status.

In addition, the United States needs to take advantage of the new dynamics that have emerged in the Middle East as a result of the new energy abundance. While there is no question that low energy prices place immediate—and in some places like Iraq, existential—strains on countries in the region, the news is not all bad. Particularly in the countries of the Gulf, the new energy abundance has provided policymakers a strong impetus to reform. It has pushed ruling regimes to begin the very delicate—but absolutely essential—process of renegotiating the social contracts they have with their citizens. It is far too early to conclude they will be successful in this renegotiation of reduced benefits for increased freedoms and greater transparency. Even an optimist might be hesitant to bet on such an outcome. But the new energy abundance, in combination with the volatile politics that surfaced in the region during the revolutions in Arab republics, has forced these issues to the fore. What is hopeful is that these countries, and the royal families who rule them, are taking the initiative while they still have sufficient time and the immense wealth to provide some room for maneuver in what will inevitably be a very difficult renegotiation.

When it comes to the Gulf states, and Saudi Arabia in particular, the United States needs to see the situation for what it is: a struggle against time and culture to reform economies and societies before the only options become revolution or greater repression. The United States, and others in the region and throughout the world, have a huge stake in these reform efforts succeeding. Ultimately, success or failure falls on the shoulders of the people in those countries. The United States cannot micromanage the desired outcome. But it should maximize its efforts to encourage the countries to reach some critical threshold of reform. The United States should not hang back as a disinterested party, perhaps with notions about more democratic regimes coming to power if current royal families fail. In some cases, the reform efforts under way *do* essentially amount to a renegotiation of the long-standing social contracts between ruling families and the people of their countries. The United States should feel very comfortable in supporting such efforts. Many Arabs, looking at Iraq and other countries striving for democracy in the region, may not press for immediate elections as part of the renegotiation. But they will almost certainly push for less corruption, greater transparency, and more scope for civil society. These measures can all coexist within a monarchy. Eventually, they can also serve as the foundation for even greater political liberalization.

A third realm in which the energy boom has created new strategic opportunities for the United States relates to greater U.S.-Sino cooperation. These energy-induced openings are of particular value, given the importance of this bilateral relationship and its overall brittleness. For the decade ahead, the relationship between the United States and China will be filled with tensions, as the two powers seek to avoid the so-called Thucydides trap, which suggests that rising and declining powers are almost always bound to clash. While the new energy abundance will not erase many points of friction, it does present the United States and China tangible and meaningful areas of cooperation. The two most obvious opportunities are efforts to tackle climate change and to stabilize the Middle East. Both are challenges of immense proportion, whose outcomes will reverberate throughout the globe. In both arenas—unlike the South China Sea—the United States and China have similar interests, but lack the ability to solve the problems unilaterally. Ideally, these two issues can be a foundation upon which relationships can be built and frameworks created that provide models for bilateral cooperation in other spheres.

Fourth, the new energy abundance has created a fissure in the relationship between China and Russia and, in doing so, provides the United States with the opportunity to ensure that these two countries do not forge a strategic partnership that could be detrimental to U.S. interests. Before the energy boom transformed global markets, China's burgeoning energy demand and Russia's vast energy resources seemed finally poised to bring the two countries closer together. Despite their long history of suspicion and mistrust, energy realities were driving Beijing and Moscow to mend fences and build closer ties. The shift from energy scarcity to energy abundance—although not the only factor that matters in explaining this bilateral relationship—has challenged that trend. Now entertaining numerous options for meeting its energy demands, China is no longer dependent on Russia for its future economic prosperity. Yet Russia is more reliant on China than ever before, as a low price environment (as well as sanctions) calls into question Russia's ability to find new markets. While the two countries are still working closely together, given the disparities, the rapprochement will now occur on China's terms, rather than as an outgrowth of mutual interdependence. Given these realities, the United States could sit back and delight in the irritants emerging in the relationship. Or it could—and should—be more proactive and use this wrinkle in the deepening of relations between Beijing and Moscow to convince China that its broader interests should take it closer to the West.

Finally, the United States should continue to support European efforts to inoculate the union against Russian intimidation and aggression through the political use of its energy trade. America should have modest expectations that it will be able to change the negative trajectory of European shale development. Absent a dramatic technological advance that addresses European environmental considerations, Europe as a whole looks unlikely to embrace its own unconventional resources. Nevertheless, the United States can look for ways to support Europe in its efforts to create a new energy union and further integrate the energy sectors of individual member states. A united Europe—in terms of energy as well as other matters—is a stronger ally and is therefore more in the interests of the United States than a divided one. The energy boom—in conjunction with new security realities in the wake of Russia's annexation of Crimea—has helped create what decades of U.S. diplomacy failed to achieve: a Europe more independent from Russia.

Anticipating New Challenges

Despite the many ways in which the new energy abundance has strengthened sources of American power and created a new, more friendly strategic landscape for the United States, U.S. policymakers cannot be complacent about the future. As is the case with any big strategic shift, some bad will inevitably come with the good. American leaders would be remiss in not anticipating these negative developments and how the United States can manage or resolve them.

Some of the negative geopolitical dimensions of the energy

boom—from a U.S. perspective—are already on display. The most evident is Russia's new petulance, which will continue. The new energy realities have created real hardships for the Russian economy, but have not yet spurred political change or economic reform. While energy prices will stabilize, they will very likely remain well short of what Russia needs to generate real growth, particularly in an economy so rife with corruption. In this situation, Putin will continue to shore up his political standing—and protect his inner circle—by delivering psychological benefits to his people in place of economic ones. Such benefits are most easily generated by reasserting Russia as a global power, challenging the United States, and seeking to divide Europe. As a result, U.S. policymakers should be prepared for continued Russian meddling—and need to be realistic about the extent to which there is a real basis for closer U.S.-Russian cooperation.

Second, policymakers should look at the globe and anticipate the collapse of some states that are heavily reliant on oil and gas revenues. Whereas the countries of the Gulf have taken responsibility for addressing their dire situations, not all countries have either the resources or the leadership to manage this difficult state of affairs. Countries such as Venezuela and Nigeria pose obvious threats, not just to their own people, but to the regions in which they reside given their size and clout.

Finally, U.S. policymakers might do well to keep another, more long-term possibility on their radar—and that is a Middle East that grows in global importance, rather than diminishes, as the new energy abundance continues to unfold. As discussed earlier in this book, a world that uses less oil for any reason is also a world of cheap oil. But it is also a world where high cost producers are squeezed out of the market and the locus of production settles on countries which can produce the most oil for the lowest price: the countries of the Middle East and the Gulf in particular. In these circumstances, Middle Eastern stability will be ever more important, making today's efforts to steady that part of the world good investments in tomorrow.

Policy Prescriptions for All

Even though much of the increased oil and gas production comes from North America, other countries are also affected by the strategic environment generated by the boom and must therefore think strategically about how to react to it. Some needed policy modifications are as obvious as they are unlikely. Russia's economy and energy sector practically scream out for reform if they are to become competitive in the new energy land-scape. Moreover, Russia will continue to struggle to develop expensive LNG projects as long as its foreign policy endeavors lead the United States, Europe, and other countries to keep sanctions in place. While President Putin may have other, more immediate objectives behind his provocative foreign policy behavior, he is engaged in a vicious cycle in which economic duress requires confrontation abroad, which invites more economic pain at home.

For other countries, the steps that would best suit their economies and their pursuit of prosperity are more easily adopted. In Africa and Latin America, resource-wealthy countries should follow the example of Mexico and adapt their investment frameworks to the new reality in which potential foreign investors have a multitude of places to invest— and less capital to make those investments. The days that Steve Coll wrote about in his 2012 book on ExxonMobil, *Private Empire*, are already gone; the phrase "the end of easy oil" is no longer bantered about. Even when prices rise further, investors will have a greater number of options for where to plow their dollars, including the United States. African and Latin American countries should therefore not assume that international companies are hungry to develop their resources regardless of the terms offered and the risks that must be incurred. They will need to offer more competitive terms and more ironclad assurances that those terms will not change. And they will also need to increase transparency within their own systems and—perhaps most important—manage their politics and the inevitable pressures of resources nationalism.

The countries of the Middle East face a different set of challenges. Some might move toward inviting in foreign investment to help develop more difficult-to-access oil and gas. For the Gulf countries in particular, this would be a significant departure from past practice, and over the longer term they have more pressing priorities. First, as discussed, they face the urgent and complicated task of renegotiating their social contracts—a task that many have already begun. Second, as some leaders have started to appreciate, they must navigate a world in which their most precious resources—oil and gas—become less valuable over time.

Eventually, technology and politics will bend the demand curve for oil downward. Preparing a resource-dependent country for this reality, in the Middle East or elsewhere, requires a multi-tiered strategy. Even in countries where the resistance to the idea of peak oil demand (as opposed to peak oil supply) is still strong, leaders should go through the exercise of asking themselves how their short-, medium-, and long-term national strategies for development would differ if they assumed declining demand in the out years. Doing this will, at the very least, sensitize them and their systems to other possible trajectories—and encourage them to take a hard look at the strategies they have currently in place. Ideally, such plans would be resilient in the face of a variety of scenarios for the future, including one of declining demand for oil.

The "risk" that China runs is of a very different variety. As with the United States, China may not take full advantage of the many opportunities the energy boom presents it. To begin with, the new energy abundance is just one reason why China should reevaluate the approaches it has taken toward the development of resources beyond its own borders. While China will continue to have cause to invest in and purchase oil and gas from abroad, there is less need for it to adhere to the model it executed in previous decades in which Chinese entities overpaid for equity investments and, in propping up often nefarious regimes, incurred the condemnation of the international community. The new energy realities give China the space to reevaluate and adjust this going-out strategy in a way that removes many of its downsides, but still allows China to reap the benefits of its upsides.

Even more importantly, the new oil and gas realities provide China huge opportunities at home to tackle the politically and physically toxic environmental issues facing the country. As discussed, even the most sanguine projections for China's future still have the country heavily dependent on coal. Natural gas, which is cleaner—generating approximately half of coal's carbon emissions—could allow China to clean up its environment without sacrificing the economic growth that is so important to the legitimacy of Communist Party rule. Today, China does not appear to be on track to fully exploit the advantages of natural gas. This is the case as China, despite some efforts, has been unable to create adequate demand for natural gas because—even after pricing reforms—it remains nearly three times as expensive as coal when used to generate electricity.

Policy can influence the relative attractiveness of fuel sources, but it is obviously important to do this without jeopardizing economic growth too greatly. Here again the energy abundance offers new possibilities to China. When the price of natural gas was north of $15 per mmbtu in Asia (as it was in 2012), making natural gas competitive with coal would have required radical policy interventions—such as placing an unrealistically high price on carbon. But with the price of natural gas sliced dramatically by the energy boom, policy nudges such as liberalizing end-user prices, encouraging competition in natural gas production and distribution, and allocating more resources to investment in natural gas infrastructure are more practical, more affordable, and politically feasible.

Asia as a whole should also translate this era of energy abundance into more lasting gains. While each country will have individual priorities, the region collectively should use this period of abundance to undertake institutional and market changes that will serve it well in any energy environment. Already, efforts are afoot—in Singapore, Japan, and China—to develop regional natural gas hubs where the purchase and sale of natural gas can be more fluid and more closely tied to the forces of supply and demand, continuing the move away from oil-indexed contracts.

How Could This Book Be Wrong?

When he was chairman of the National Intelligence Council, Joe Nye (now my colleague at Harvard's Kennedy School) always ended meetings by asking: "How could we be wrong?" The future anticipated in this book is not inevitable. The new energy abundance could be, for example, shorter lived or less consequential than this book expects. Here are five possible developments that either independently or collectively could deliver a very different energy future from that presented in these pages.

First, tight oil production could be less responsive to price rises than this book expects. As discussed throughout these pages, no one knows for certain the pace and scale at which tight oil will respond to increased prices. The future portrayed in these pages is one in which tight oil *does* respond with some alacrity to price changes. Production levels change quickly enough both to meet growing demand and to discourage OPEC from trying to manage the markets as it has in the past. There are good reasons to believe this will be the case, but, given the newness of this resource, this is the first opportunity the world has to map out the critical relationship between tight oil production and global prices. As a result, no one knows for certain the pace and scale at which tight oil will respond to increased prices. Another relationship than the one described in these pages could prevail.

Second, demand for oil and gas could grow more robustly than expected over the coming years. The most likely driver of heightened demand would be more vigorous global economic growth. As of April 2017, the IMF foresees that the world economy will expand at 3.5 and 3.6 percent a year in 2017 and 2018. Behind these numbers are weak European growth, positive but anemic U.S. growth, and moderate but slowing growth in China. While any of these prognoses could prove wrong, the variable most likely to confound predictions is growth in China. BP's *Energy Outlook 2035* notes that if China's growth is on average 1.5 percentage points a year lower than the company expects over the course of the next two decades, global energy demand will be nearly half a percentage point lower each year. The reverse also holds true; should China's economic growth pick up, future energy demand will rise well beyond what is anticipated. This would put greater pressure on global energy supplies, perhaps beyond what new resources could satisfy.

Third, technology could surprise to the downside. Thus far, technological advances have been a major driver of the energy boom, as described earlier in this book. They have continued to create new opportunities for resource development, at lower costs. Without these developments, the boom would never have unfolded. The longevity of the energy abundance in part relies on the ability of technology to continue to bring down costs and extract more resources. But it also rests on the expectation that technology will continue to advance in ways that make the usage of oil and gas—and other forms of energy—more efficient over time, thereby tempering demand. For example, falling battery prices could make electric cars much more attractive and cost competitive with those run on combustion engines—and thereby lower demand for oil. Some believe that these advances will come earlier—and with greater impact—than is the norm. But there is also the possibility that technology disappoints, allowing demand to grow faster than expected or causing the costs of developing tight oil and shale gas to level out, rather than continue to decline.

Fourth, a geopolitical calamity or black swan event could suddenly diminish oil or gas supply in a way that has long-term effects. The new energy realities certainly give the world a cushion for geopolitical events of many proportions and for limited durations. But something as monumental as revolution in Saudi Arabia or civil strife in Russia could remove significant quantities of oil and gas from global markets to the point that energy is considered scarce long enough to affect geopolitical calculations. Some will retort that any successor regime in Saudi Arabia or Russia will have the same interests in producing as much oil and gas as possible. History, however, has shown that regimes that emerge following calamitous geopolitical events have struggled to recapture earlier levels of oil and gas production. Libya, for example, never returned to the production levels it had been at before the 1969 coup that brought Qaddafi to power. Nor has Iran come close to what it was producing before its 1979 revolution. Russia did eventually reach production levels similar to those under the Soviet Union, but it took years to do so.

Finally, environmental concerns could either lead to government policies in America that clamp down on fracking or cause companies to lose their social licence to operate. Either scenario could lead to a sharp reversal of the growth in oil and gas supply. Environmental groups in the United States have become more influential in recent years, scoring significant successes in gaining greater protection for local communities as well as action to combat climate change. But a segment of them also has an even more expansive objective: to ensure as few fossil fuels are developed as possible, for any reason and under any circumstances. U.S. federal and local governments face a challenging task of balancing legitimate environmental concerns with the strategic and economic benefits that come with the production and use of fossil fuels. Fortunately, as described earlier in this book, there does seem to be a middle ground, where responsible resource development can be undertaken at reasonable costs.

In 2000, George Mitchell, the crafty and stubborn Texan who helped launch the energy boom, was asked what America's energy landscape would be like in 2020. "Oil in the United States is very hard and expensive to find," he said. "It's necessary in our economy, and a reversal of declining production doesn't seem possible." During the same interview, Mitchell also contended that the United States would be a major natural gas importer by 2020.

Since the time of that interview, the energy realities for the United States—and the world—have taken a turn that even the visionary individual most responsible for the new energy boom could not anticipate. The changes have happened at whiplash speed and, with them, the landscape of foreign affairs and national security has changed. Understanding the connections between the energy boom and geopolitics is a critical prerequisite for anyone, in any country, or nearly any industry or government, hoping to make wise decisions dependent upon how countries, companies, and even individual leaders interact. From the United States to Argentina, Australia to China, Russia to the Middle East, Europe to Venezuela and beyond, policymakers, academics, and investors alike need to size up how the new energy dynamics and geopolitics will transform their futures. This book has sought to illuminate these new dynamics and to articulate some of the fresh approaches that so many actors will need to adopt. Serendipity has carried many quite far, but new strategies are essential to reap further rewards.

Acknowledgments

The first draft of this book, written several years ago, focused on a possible precipitous decline in the price of oil and anticipated the geopolitical fallout that might result. Much has happened since that time, in the world and to this book project. As events rapidly unfolded, I was always seeking to discern the large and lasting trends and separate them from the noise. Writing this book, at this moment in history, has been both exciting and demanding. And, partially as a result, I have incurred many debts of deep gratitude in the process.

My first thanks belongs to Graham Allison, and my many colleagues at the Belfer Center for Science and International Affairs and the Harvard Kennedy School. I arrived at Belfer ten years ago, eager to broaden my areas of expertise beyond the Middle East after a very intense period in government focusing on Iraq and Afghanistan. Graham not only provided a happy home for these intellectual pursuits, but encouraged me as I built out a small team of experts working on the geopolitics of energy and created a project designed to explore the complex connections between energy and foreign policy. My other colleagues—Nick Burns, Bill Clark, Niall Ferguson, Bill Hogan, Henry Lee, Leonardo Maugeri, Joe Nye, and others—all provided me with much needed support and guidance as I dove deeper into the intersection of energy markets and global politics. I am also grateful for the support provided by BP, Loomis and Sayles & Company, and the Middle East Initiative to the Geopolitics of Energy Project at the Harvard Kennedy School, under whose auspices I conducted much of the work for this book. The Environmental and

Natural Resources Program and the International Security Program, both at Harvard's Kennedy School, also provided welcome institutional and other support.

I also owe special thanks to friends and mentors who have played a special role in my intellectual and professional life. In particular, Nick Brady, John Hess, Richard Haass, Steve Hadley, and Bob Zoellick have gone extra distances to create opportunities for me to deepen my energy expertise and have made special efforts to expose me to new ideas and linkages. A wide range of other people also gave of their time and expertise to help me make this book the best it could be. Although they cannot all be acknowledged, I met with hundreds of people in more than twenty countries over the last four years of research for this book. From Latvia to Iraq to Japan to Brazil, I am grateful to every person who sat with me, their words illuminating yet another piece of the complex puzzle I have sought to understand and to explain in the pages of this book.

Closer to home, many others made specific efforts to help me craft the manuscript. Soon after I set my mind to writing this book, I had breakfast with Dan Yergin, the indisputable guru of energy geopolitics, who wrote his wonderful book *The Prize* while at Harvard's Kennedy School. He gave me several good pieces of advice, much of which I followed religiously. I embraced one of Dan's recommendations with particular vigor: hire a fleet of supersmart Harvard graduate students to provide research assistance. And that I did, reaping enormous gains from their intelligence and dedication. To Siddharth Aryan, Joel Bell, Lauren Bloomberg, Rita Chung, Chris Cote, Tobias Cremer, Rani Daher, Cathy Guo, Takuma Iino, Guy Leung, Adam Papa, Scott Quigley, Jaffar al-Rikabi, Hossein Safaei, Razzaq al-Saiedi, Izran Saleh, Scott Siler, Julia Stern, Scott McNally, Blake Meulmester, Ok-Kyoung Song, Sun Ting, Charlie Warren, Alex Yergin, and Aaron Young, I offer my sincere thanks. But above all, I am deeply in debt to two researchers in particular, without whom this book would have been impossible for me to write: Can Soylu and Nikoleta Sremac. They treated this book as if it were their own. Their dedication, goodnature, intelligence, and commitment was extraordinary and for all their work I am very grateful.

This was my first effort to reach a more popular and less academic audience, and to the extent I succeeded, I owe special thanks to Howard and Nathan Means, Richard Todd, and—especially—Michael Carroll. All provided invaluable advice on how to present such a complicated and technical matter to my audience in a way that is (hopefully) easy and pleasurable to digest.

I also leaned on many friends, colleagues, and experts to read the book when it was in its manuscript phase and would like to acknowledge their contributions. Bob Blackwill, Colin Davies, David Gordon, Richard Haass, John Hess, and Bob Zoellick, all provided treasured advice and recommendations on the whole manuscript. John Deutch, Andreas Goldthau, Leslie Palti-Guzman, Bo Kong, Holly Morrow, Morena Skalamera, and Kaho Yu offered me valued expert opinion on different portions of the book. This book is undoubtedly a better piece of work thanks to the help I received from this group.

I would also like to sincerely thank Alice Mayhew and Stuart Roberts—and their colleagues at Simon & Schuster—for all their support and patience throughout the process of writing this book. I am grateful for their strategic guidance and for their tolerance of extended deadlines and heated debates, all of which enabled this book to evolve in a positive direction. I also appreciate the efforts of Andrew Wylie, who has been invaluable in helping me navigate the process of making this book a reality.

I owe a few people for the extraordinary support and care they provided to me during the whole time this book was under way. I have dedicated this book to my parents, Michael and Kathi O'Sullivan, for the encouragement they have given me throughout my whole life, in every realm, but especially in the area of education and intellectual pursuits. My sisters, Kristin and Kate, also provided essential support and reinforcement, as well as much-needed breaks and laughs. And, finally, I want to extend my warmest thanks and appreciation to Arnaud Lacoste. In many ways, he has lived with this book as I have, tolerating piles of papers, dinners propped on books, and all the stresses associated with writing a book. But most important, in his relentless intellectual curiosity and constant efforts to push beyond the obvious, Arnaud was and is a constant source of inspiration for me.

Notes

Preface

ix *He made a forceful and compelling case:* "No country should view its development path on its own," Xi said. "Development is of the people, by the people and for the people." Xi Jinping, "Opening Plenary" (speech, World Economic Forum Annual Meeting 2017, Davos, Switzerland, January 17, 2017), www.weforum.org/agenda/2017/01/full-text-of-xi-jinping-keynote-at-the-world-economic-forum.

ix *"Say no to protectionism":* Ibid.

ix *a newly sworn-in President:* Donald J. Trump, "Inaugural Address" (speech, Washington, D.C., January 20, 2017), www.whitehouse.gov/inaugural-address.

xi *Jervis writes of the tendency:* See Robert Jervis, *Perception and Misperception in International Politics* (Princeton: Princeton University Press, 1976).

Introduction

1 *The United Nations had estimated:* John Heilprin, "UN: Death Toll from Syrian Civil War Tops 191,000," *USA Today*, August 22, 2014,

www.usatoday.com/story/news/world/2014/08/22/united-nations
-syria-death-toll/14429549/. The U.N. special envoy in Syria more
recently estimated that 400,000 people have been killed in the civil
war as of April 2016. John Hudson, "U.N. Envoy Revises Syria Death
Toll to 400,000," *Foreign Policy*, April 22, 2016, http://foreignpolicy
.com/2016/04/22/u-n-envoy-revises-syria-death-toll-to-400000/. The
400,000 estimate is close to more recent reports from the Syrian Center
for Policy Research (470,000 as of February 2016), and the Syria Network
for Human Rights (450,000 as of January 2017).

2 *Since 2010, U.S. crude oil production:* For example, in 2011, the U.S.
Energy Information Administration forecast that in 2012 the United
States would produce 5.4 mnb/d of crude oil; the actual production in
2012 was 6.5 mnb/d. See "U.S. Field Production of Crude Oil," U.S. Energy Information Administration, August 31, 2016, www.eia.gov/dnav
/pet/hist/LeafHandler.ashx?n=pet&s=mcrfpus2&f=a; "Annual Energy
Outlook 2011," U.S. Energy Information Administration, April 26, 2011,
www.eia.gov/forecasts/archive/aeo11/data_side_cases.cfm?filter=oil#
summary; "Annual Energy Outlook 2012," U.S. Energy Information Administration, June 25, 2012, www.eia.gov/forecasts/archive/aeo12/data
_side_cases.cfm?filter=oil#summary; "Annual Energy Outlook 2013,"
U.S. Energy Information Administration, April 15–May 2, 2013, www
.eia.gov/forecasts/archive/aeo13/data_side_cases.cfm?filter=oil#sum
mary.

2 *global oil demand for:* Global oil demand stayed at 92 mnb/d from the
end of 2013 through mid-2014, after having risen continuously throughout 2013. The low oil prices that followed helped bolster demand for oil,
pushing demand growth to highs (1.7 percent or 1.6 mnb/d) not seen in
the previous five years. "Table 1: World Oil Supply and Demand," International Energy Agency—Oil Market Report, February 9, 2016, https://
www.iea.org/oilmarketreport/omrpublic.

3 *In my writings and speeches:* See, for instance, Robert D. Blackwill and
Meghan L. O'Sullivan, "America's Energy Edge: The Geopolitical Consequences of the Shale Revolution," *Foreign Affairs*, March/April 2014,
https://www.foreignaffairs.com/articles/united-states/2014-02-12/
americas-energy-edge, wherein Bob Blackwill and I wrote, "The most
dramatic possible geopolitical consequence of the North American
energy boom is that the increase in U.S. and Canadian oil produc-

tion could disrupt the global price of oil—which could fall by 20 percent or more." Also see Meghan L. O'Sullivan, *North American Energy Remakes the Geopolitical Landscape: Understanding and Advancing the Phenomenon*, New York: Goldman Sachs, June 2014, www .goldmansachs.com/our-thinking/pages/north-american-energy -summit/reports/mos-north-america-energy-remakes-the-geopolitical -landscape.pdf, wherein I wrote, "Although it may not be possible to definitively predict the impact of the unconventional boom on the price of oil, one can claim with conviction that this energy phenomenon has placed— and will continue to place—downward pressure on the oil price by introducing significant new sources of global supply." Also see Meghan L. O'Sullivan, "A Better Energy Weapon to Stop Putin," *Bloomberg View*, March 11, 2014, www.bloombergview.com/articles/2014-03-11/a-better -energy-weapon-to-stop-putin, wherein I wrote, "The U.S., by adding 2.5 million barrels of oil to global markets in the last three years, has prevented the price of oil from edging higher in the face of disruptions in Libya, Iran and elsewhere. Should the U.S. continue to increase its oil production, as is widely assumed, it could create pressure to further lower the price."

3 *Even as the price of oil dipped just below*: "Crude Oil (petroleum); Dated Brent Daily Price," *Index Mundi*, September 6, 2016, www.indexmundi .com/commodities/?commodity=crude-oil-brent&months=60. In June 2012, the Brent price of crude dipped briefly below $100; with that exception, monthly crude prices had exceeded $100 a barrel since February 2011.

3 *Just before I had boarded*: Meghan L. O'Sullivan, "The Geopolitical Implications of Energy Changes" (presentation, U.S.–China Energy Dialogue, New York, NY, September 12, 2014).

3 *My Harvard colleague Leonardo Maugeri*: Leonardo Maugeri, "Oil: The Next Revolution: The Unprecedented Upsurge of Oil Production Capacity and What It Means for the World," discussion paper #2012-10, Geopolitics of Energy Project, Belfer Center for Science and International Affairs, Harvard University, Cambridge, MA, 2012, www.belfercenter.org/pub lication/oil-next-revolution.

3 *Indeed, it predicted that*: Published in May 2014, the International Energy Agency's *World Energy Investment Outlook* anticipated that global oil prices would stabilize "around current levels and [increase] only

moderately to 2035," International Energy Agency, *World Energy Invest-ment Outlook 2014* (Paris: OECD Publishing, 2014), 55, www.iea.org /publications/freepublications/publication/WEIO2014.pdf.

3 *Abdalla Salem El-Badri, the secretary general:* El-Badri stated, "Today, we see a relatively balanced market. And looking at market indicators, we expect this to be the case for the rest of 2014. There is steady demand growth and enough supply to meet demand, with both stocks and spare capacity at comfortable levels." He downplayed the potential for new U.S. production to fundamentally challenge the oil market, emphasizing the "ever-expanding demand from Asia" and calling tight oil "a welcome ad-dition [that] adds depth and diversity to the market." His Excellency Ab-dalla S. El-Badri, "The New Geography of Energy: Business as Usual or a New Era for Energy Supply and Demand?," speech, the 14th International Energy Forum, Moscow, Russia, May 15, 2014, www.opec.org/opec_web /en/2815.htm.

3 *He argued oil markets would:* Prince Abdulaziz bin Salman bin Abdulaziz, the Assistant Minister of Petroleum and Mineral Resources, speech, The Conference of the Arabian Gulf and Regional Challenges at The Institute of Diplomatic Studies and the Gulf Research Centre, Riyadh, September 16, 2014, pp. 3–5; quote on page 5.

3 *Dismissing gyrations in the price of oil:* Ibid.

4 *When I pressed the issue:* Between 2000 and 2010, the countries near the Caspian Sea added 1.5 million of additional production. Over a similar time period, Angola added more than 1 million. See "Oil and Natural Gas Production Is Growing in Caspian Sea Region," U.S. Energy Informa-tion Administration, September 11, 2013, www.eia.gov/todayinenergy /detail.cfm?id=12911; and "Angolan Oil Production Has Doubled Since 2013," U.S. Energy Information Administration, October 14, 2011, www .eia.gov/todayinenergy/detail.cfm?id=3490.

6 *Energy companies worldwide*: The S&P Global Oil Index, which measures the performance of the 120 largest, publicly traded oil and gas companies, was at its lowest levels since 2009 in late 2015. The three worst perform-ers of the S&P 500 in 2015 were all energy companies. Saddled with debt, some U.S. oil and gas companies lost 90 percent of their market value over the course of 2015. See "S&P Global Oil Index," S&P Dow Jones Indices, September 13, 2016, http://us.spindices.com/indices/equity /sp-global-oil-index; Matt Egan, "These are the worst stocks of 2015,"

CNN Money, December 22, 2015, http://money.cnn.com/2015/12/22 /investing/worst-stocks-2015-oil-energy/; Christopher Helman, "Oil Goes Down, Bankruptcies Go Up—These 5 Frackers Could Be Next to Fall," *Forbes*, August 17, 2015, www.forbes.com/sites/christopherhel man/2015/08/17/as-oil-goes-down-bankruptcies-go-up-these-5-frack ers-could-be-the-next-to-fall/#2715e4857a0b56f17add5f33.

6 *The low price of oil itself:* Joe Weisenthal, "BofA: The Oil Crash Is Kicking Off One of the Largest Wealth Transfers in Human History," Bloomberg, January 31, 2016, www.bloomberg.com/news/articles/2016-02-01/bofa -the-oil-crash-is-kicking-off-one-of-the-largest-wealth-transfers-in -human-history.

9 *We have begun to see renewables:* "International Energy Outlook 2016: Executive Summary," U.S. Energy Information Administration, May 11, 2016, www.eia.gov/forecasts/ieo/exec_summ.cfm.

10 *fossil fuels still:* This proportion includes coal; without coal, oil and natu- ral gas account for 55 percent of global energy use.

10 *In a 2014 book,* Game Changers: The authors assessed various technolo- gies in each area, evaluating and categorizing them as commercial, near commercialization, or ones in which commercialization is further out. George P. Shultz and Robert C. Armstrong, eds., *Game Changers: Energy on the Move* (Stanford: Hoover Institution Press, 2014).

One: Behind the Price Plunge

15 *In 2005,* New York Times *columnist John Tierney:* "The $10,000 Ques- tion," *New York Times*, August 23, 2005, www.nytimes.com/2005/08/23 /opinion/the-10000-question.html.

16 *Energy—and increasing competition:* National Intelligence Council, *Global Trends 2025: A Transformed World* (Washington, DC: U.S. Gov- ernment Printing Office, 2008), www.dni.gov/files/documents/News room/Reports%20and%20Pubs/2025_Global_Trends_Final_Report.pdf.

16 *In 2008, Nobuo Tanaka:* International Energy Agency, *World Energy Out- look 2008* (Paris: OECD Publishing, 2008), November 12, 2008, 3, www .worldenergyoutlook.org/media/weowebsite/2008-1994/WEO2008.pdf.

16 *In a similar vein, U.K. defense:* Ben Russell and Nigel Morris, "Armed Forces Are Put on Standby to Tackle Threats of Wars over Water," *The Independ- ent*, February 27, 2006, www.independent.co.uk/environment/armed

-forces-are-put-on-standby-to-tackle-threat-of-wars-over-water
-6108139.html. Also see Michael T. Klare, "There Will Be Blood: Political Violence, Regional Warfare, and the Risk of Great-Power Conflict
over Contested Energy Sources," in *Energy Security Challenges for the
21st Century: A Reference Handbook*, ed. Gal Luft and Anne Korin (Santa
Barbara: Praeger, 2009), 61. Klare envisions "a situation where efforts by
China to secure access to overseas sources of energy would collide with
similar efforts by the United States, producing a direct confrontation between the two."

17 *The authors of the* Global Trends *report:* National Intelligence Council,
Global Trends 2025: A Transformed World, 47–50.

17 *In the first decade:* Between 2000 and 2011, total nonhydro renewable electricity generation quadrupled, growing at 13 percent annually. BP p.l.c.,
BP Statistical Review of World Energy 2016 (BP, June 2016), www.bp.com
/content/dam/bp/excel/energy-economics/statistical-review-2016
/bp-statistical-review-of-world-energy-2016-workbook.xlsx.

17 *This support encouraged expansion:* See Diane Cardwell, "Solar and Wind
Energy Start to Win on Price vs. Conventional Fuels," *New York Times*,
November 23, 2014, www.nytimes.com/2014/11/24/business/energy
-environment-solar-and-wind-start-to-win-on-price-vs-conventional
-fuels.html.

17 *Such growth will continue:* International Energy Agency, "IEA raises its
five-year renewable growth forecast as 2015 marks record year," October 25, 2016, www.iea.org/newsroom/news/2016/october/iea-raises
-its-five-year-renewable-growth-forecast-as-2015-marks-record-year
.html.

17 *After all,* nonhydro *renewable energy:* Non-hydro renewables accounted
for 3.16 percent of global primary energy use in 2016. If one includes
hydro in the mix, this percentage is 10.02 in 2016. See *BP Statistical Review of World Energy 2017*, BP, http://www.bp.com/content/dam/bp/en
/corporate/excel/energy-economics/statistical/review-2017/bp-statisti
cal-review-of-world-energy-2017-underpinning-data-xlsx.

18 *It is the type of extraction process that provides:* For example, as recently
as when Daniel Yergin wrote *The Quest* in 2011, many people considered
deepwater oil and gas unconventional. Today, few do. See Daniel Yergin,
"Unconventional," in *The Quest: Energy, Security, and the Remaking of the
Modern World* (London: Penguin Books, 2011), 244–65.

18 *But oil sands, also:* Oil can also be produced from oil sands through a mining-type process.

19 *In 1952, George Mitchell:* John Kutchin, *How Mitchell Energy & Development Corp. Got Its Start and How It Grew: An Oral History and Narrative Overview* (Boca Raton: Universal Publishers, 2001), 17–20. See also Gregory Zuckerman, *The Frackers: The Outrageous Inside Story of the New Billionaire Wildcatters* (London: Portfolio/Penguin, 2014), 26.

20 *The natural gas Mitchell initially found:* "Before long all the gas above the Barnett Shale was played out," Mitchell said. Brandon Evans, "Mitchell's Gamble Changed an Industry, His Philanthropy Changed the Future," The Cynthia & George Mitchell Foundation (blog), July 31, 2013, http://cgmf.org/blog-entry/64/Mitchell's-gamble-changed-an-industry-his-philanthropy-changed-the-future.html.

20 *The first reference to them:* U.S. Geological Survey National Assessment of Oil and Gas Resources Team, and Laura R. H. Biewick, compiler, "Map of Assessed Shale Gas in the United States" (Reston: U.S. Geological Survey, 2013), 2, http://pubs.usgs.gov/dds/dds-069/dds-069-z/DDS-69-Z_pamphlet.pdf.

20 *Those cracks are then prised open:* See Meagan S. Mauter, Vanessa R. Palmer, Yiqiao Tang, and A. Patrick Behrer, "The Next Frontier in United States Unconventional Shale Gas and Tight Oil Extraction: Strategic Reduction of Environmental Impact," discussion paper 2013-04, Energy Technology Innovation Policy Research Group, Belfer Center for Science and International Affairs, Kennedy School of Government, Harvard University, Cambridge, MA, 2013, http://belfercenter.ksg.harvard.edu/files/mauter-dp-2013-04-final.pdf.

21 *Well aware of the resources trapped:* For more on the U.S. government role, see Alex Trembath et al., "Where the Shale Gas Revolution Came From: Government's Role in the Development of Hydraulic Fracturing in Shale," The Breakthrough Institute, May 23, 2012, http://thebreakthrough.org/blog/Where_the_Shale_Gas_Revolution_Came_From.pdf. For more on Mitchell's own interaction with government programs, see Michael Shellenberger and Ted Nordhaus, "A Boom in Shale Gas? Credit the Feds," *Washington Post*, December 16, 2011, www.washingtonpost.com/opinions/a-boom-in-shale-gas-credit-the-feds/2011/12/07/gIQAecFIzO_story.html.

21 *His heir apparent, Bill Stevens:* See Gregory Zuckerman, *The Frackers:*

The Outrageous Inside Story of the New Billionaire Wildcatters (London: Portfolio/Penguin, 2014), 18.

21 *Mitchell decreased costs:* For a detailed, colorful account of the process of experimentation to find just the right fluid consistency, see ibid., or Gregory Zuckerman, "Breakthrough: The Accidental Discovery That Revolutionized American Energy," *Atlantic,* November 6, 2013, www .theatlantic.com/business/archive/2013/11/breakthrough-the-acci dental-discovery-that-revolutionized-american-energy/281193/.

21 *These advances helped coax:* "Technology Drives Natural Gas Production Growth from Shale Gas Formations," U.S. Energy Information Administration, July 12, 2011, www.eia.gov/todayinenergy /detail.cfm?id=2170. This graph also shows how the number of horizontal wells in the Barnett increased after 2003, when Devon purchased Mitchell Energy, and how natural gas production rose significantly from that point onward.

22 *Devon's knowledge of horizontal drilling:* Mitchell had also experimented with horizontal drilling, but with little positive result. See Zhongmin Wang and Alan Krupnick, "A Retrospective Review of Shale Gas Development in the United States: What Led to the Boom?," RFF discussion paper 13-12, Resources for the Future, Washington, DC, April 2013, 23, www.rff.org/files/sharepoint/WorkImages/Download/RFF-DP-13-12 .pdf.

22 *Within a decade, the number:* The number of wells increased from 2,070 to 17,980 by 2012. Ibid., 27.

22 *production of natural gas:* Natural gas production in the Barnett grew 834 million cubic feet (MMcf) a day in 2003 to 5,752 MMcf a day in 2012. "Texas Barnett Shale Total Natural Gas Production 2000 Through October 2014," Railroad Commission of Texas, Dec. 19, 2014, www.rrc.state .tx.us/media/25828/barnettshale_totalnaturalgas_2000102014.pdf.

22 *Instead, they preferred to:* The EIA posits that the Arctic contains an estimated 13 percent of the world's undiscovered conventional oil resources and 30 percent of its undiscovered conventional natural gas resources. "Arctic Oil and Natural Gas Resources," U.S. Energy Information Administration, January 20, 2012, www.eia.gov/todayinenergy/detail.php ?id-4650.

23 *That same year,* Forbes *ranked them:* Edwin Durgy, "The Forbes 400's Newest Undercover Billionaires: The Wilks Brothers," *Forbes,*

September 26, 2011, www.forbes.com/sites/edwindurgy/2011/09/26/the -forbes-400s-newest-undercover-billionaires-the-wilks-brothers/.

23 *The collective efforts of dozens:* See Leonardo Maugeri, "The Shale Oil Boom: A U.S. Phenomenon," discussion paper 2013-05, Geopolitics of Energy Project, Belfer Center for Science and International Affairs, Kennedy School of Government, Harvard University, Cambridge, MA, 2013, http://belfercenter.ksg.harvard.edu/publication/23191/shale_oil_boom .html.

23 *In 2006, the United States was producing:* "Review of Emerging Resources: U.S. Shale Gas and Shale Oil Plays," U.S. Energy Information Administration, July 8, 2011, https://www.eia.gov/analysis/studies/usshalegas/; "Shale Gas Production," U.S. Energy Information Administration, November 19, 2015, www.eia.gov/dnav/ng/ng_prod_shalegas_s1_a.htm; "Producing Natural Gas from Shale," U.S. Department of Energy, January 26, 2012, http://energy.gov/articles/producing-natural-gas-shale.

23 *By 2015, more than half:* John Staub, "The Growth of U.S. Natural Gas: An Uncertain Outlook for U.S. and World Supply," presentation, U.S. Energy Information Administration Conference, Washington, DC, June 15, 2015, slide 3, www.eia.gov/conference/2015/pdf/presentations/staub.pdf.

23 *Production of tight oil in the Eagle Ford:* The production from Eagle Ford increased from 54,011 barrels a day (b/d) to 1,285,224 b/d during this period. U.S. Energy Information Administration, "Drilling Productivity Report," September 2016.

24 *Tight oil production in the Bakken fields:* Bakken production grew from 132,106 b/d in January 2007 to 1,050,512 b/d in January 2014. Ibid.

24 *In 2014, U.S. tight oil production:* "Table: Oil and Gas Supply," *Annual Energy Outlook 2017*, U.S. Energy Information Administration, https:// www.eia.gov/outlooks/aeo/data/browser/#/?id=14-AEO2017&re gion=0-0&cases=ref2017&start=2015&end=2050&f=A&linechart =~ref2017-d120816a.10-14-AEO2017&ctype=linechart&sid=ref2017 -d120816a.10-14-AEO2017&sourcekey=0. "Production of Crude Oil, NGPL, and Other Liquids 2013," U.S. Energy Information Administration, 2013, https://www.eia.gov/beta/international/data/browser/#?iso =IRQ&c=00000000000000000000001&ct=0&ord=CR&cy=2015&v=H& vo=0&so=0&io=0&start=1980&end=2015&vs=INTL.55-1-IRQ-TBPD .A&pa=00000000000000000000000000000000000bg&f=A&ug=g&tl _type=p&tl_id=5-A.

24 *In the same year:* "Tight Oil Production Pushes U.S. Crude Supply to Over 10% of World Total," U.S. Energy Information Administration, March 26, 2014, www.eia.gov/todayinenergy/detail.cfm?id=15571.

24 *Accounting for nearly half:* Ibid.

24 *An additional 4.9 million barrels:* "Tight oil expected to make up most of U.S. oil production increase through 2040," U.S. Energy Information Administration, February 13, 2017, https://www.eia.gov/todayinenergy /detail.php?id=29932; "Oil Market Report," International Energy Agency, March 11, 2016, www.iea.org/media/omrreports/tables/2016-03-11.pdf.

24 *For instance, in 2005:* "Special Report: Hurricane Katrina's Impact on the U.S. Oil and Natural Gas Markets," U.S. Energy Information Administration, August 31, 2005, www.eia.gov/special/disruptions/hurricane /katrina/eia1_katrina_090205.html; "Cushing, OK WTI Spot Price FOB," U.S. Energy Information Administration, February 1, 2017, https://www .eia.gov/dnav/pet/hist/LeafHandler.ashx?n=PET&s=rwtc&f=D.

27 *Global demand for oil decreased:* Global oil demand decreased from 64 mnb/d in 1979 to 58 mnb/d in 1983. *BP Statistical Review of World Energy* 2017, BP.

27 *In the first half of the 1980s:* Mexico, the U.K., Norway, China, Brazil, and India together increased their production by 63 percent during this time period. This 3.9 million barrels accounted for roughly 6 percent of global crude production in 1980. "International Energy Statistics," U.S. Energy Information Administration, www.eia.gov/cfapps/ipdbproject /iedindex3.cfm?tid=5&pid=55&aid=1&cid=regions&syid=1980&eyid =2013&unit=TBPD.

27 *Saudi Arabia bore the brunt:* In 1985, Saudi Arabia was producing at 37 percent of its 1979 levels. Interestingly, OPEC slashed production by significantly more than non-OPEC country increases, but the price still went down. *BP Statistical Review of World Energy*, BP.

27 *Revenues had plunged:* Revenues declined from $108.2 billion in 1981–82 to $48 billion in 1984–85. See Eliyahu Kanovsky, *Saudi Arabia's Dismal Economic Future: Regional and Global Implications*, Dayan Anter for Middle Eastern and African Studies, The Shiloah Institute (Tel Aviv University, 1984), 34.

27 *Faced with declining prices:* James M. Griffin and Wiewen Xiong, "The Incentive to Cheat: An Empirical Analysis of OPEC," *The Journal of Law & Economics* 40, no. 2 (1997): 306.

27 *Prices, which had tumbled:* By 1986, oil prices had plunged to $13.60 a barrel from $21.60 seven years earlier. OPEC's share of the global market had shrunk from 41 percent of the global market to just 28 percent. *BP Statistical Review* 2016. www.bp.com/content/dam/bp/excel/energy -economics/statistical-review-2016/bp-statistical-review-of-world -energy-2016-workbook.xlsx, and "Real Prices Viewer," U.S. Energy Information Administration, September 7, 2016, www.eia.gov/forecasts /steo/realprices/.

27 *In late 1984, the government raised:* Kanovsky, *Saudi Arabia's Dismal Economic Future: Regional and Global Implications*, 235.

27 *In early 1986, the* Economist *noted:* Ibid., 298; *Economist*, "GCC Survey," February 8, 1986, 18.

28 *After meeting with U.S. energy secretary:* "Shale energy boom helps keep oil markets stable," *Arab News*, January 20, 2014, www.arabnews.com /news/511951.

29 *While global demand for oil:* "Oil Market Report," International Energy Agency, March 11, 2016, https://www.iea.org/media/omrreports/tables /2016-03-11.pdf.

29 *it was initially weakening demand:* In September 2013, Saudi exports were at 7.8 mnb/d. In September 2014, that same number declined to 6.7 mnb/d. For Saudi exports, see "JODI-Oil," Joint Organization Data Initiative, www.jodidb.org/ReportFolders/reportFolders.aspx?sCS_referer=&sCS _ChosenLang=en.

29 *These countries aggressively began:* For details, see Anjli Raval, "The Big Drop: Riyadh's Oil Gamble," *Financial Times*, March 9, 2015, www .ft.com/intl/cms/s/2/25f2d7d6-c3f8-11e4-a02e-00144feab7de.html#ax zz45VgynGqP.

30 *In the eighteen months following January:* Chinese imports of Saudi crude fell from 1.3 mnb/d in January 2013 to 900,000 b/d in August 2014. Ibid.

30 *As a result, even before global prices:* Saudi price cuts began in early 2014 and continued throughout the year. "Saudi Aramco Cuts Price of Crude," *Arabian Oil and Gas*, January 7, 2014, www.arabianoilandgas .com/article-11638-saudi-aramco-cuts-price-of-crude/; Nicole Friedman, Benoit Faucon, and Summer Said, "Saudi Price Cut Upends Oil Market," *Wall Street Journal*, November 3, 2014, www.wsj.com/articles /saudi-oil-price-cut-upends-market-1415063053; Emiko Terazono, "Oil Slides After Saudi Arabia Cuts Prices in US," *Financial Times*, November 4,

2014, www.ft.com/intl/cms/s/0/e15b4e4e-6382-11e4-8a63-00144feabdc0
.html#axzz45VgynGqP.

30 *The world expected OPEC:* "Until about three days ago the absolute and
total consensus in the market was the Saudis would cut," said Robert Mc-
Nally. Ron Bousso and Joshua Schneyer, "Exclusive: Privately, Saudis Tell
Oil Market—Get Used to Lower Prices," Reuters, October 13, 2014, www
.reuters.com/article/us-oil-saudi-policy-idUSKCN0I201Y20141013.

30 *Global supply and demand would subsequently settle:* See "Oil at $50,"
Economist, January 7, 2015, www.economist.com/blogs/graphicdetail
/2015/01/daily-chart-1, which demonstrates a range of fiscal breakeven
prices from the low $70s to $140. The Gulf states and Saudi Arabia would
all have met their fiscal needs at $100 at that time.

30 *But investment into oil:* Francisco Monaldi, "Here's what happens when
oil prices crash—and it is not pretty for producers," *The Guardian,* January, 12, 1985, www.theguardian.com/commentisfree/2015/jan/12/what
-happens-oil-prices-collapse-cheaper-fuel-us-europe.

30 *For instance, OPEC cut millions:* "OPEC Spare Capacity in the First Quarter of
2012 at Lowest Level Since 2008," U.S. Energy Information Administration,
May 24, 2012, https://www.eia.gov/todayinenergy/detail.cfm?id=6410.

31 *Widely referred to as:* "#53 Ali Al-Naimi, The World's Most Powerful Peo-
ple," *Forbes,* November 4, 2014, www.forbes.com/profile/ali-al-naimi/;
Summer Said and Benoit Faucon, "Saudi Arabia's Celebrity Oil Minister
Ali al-Naimi Prepares for Potential OPEC Swan Song," *Wall Street Journal,*
June 4, 2015, www.wsj.com/articles/saudi-arabias-celebrity-oil-minister
-ali-al-naimi-prepares-for-potential-opec-swan-song-1433422421.

31 *Born in 1935 to one of:* For details about al-Naimi, see John Lawton,
"Naimi: 'I Hope to Tell Him Objective Accomplished,'" in *Saudi Aramco
World Magazine* 35, no. 3 (May/June 1984); and David Lamb, *The Arabs*
(New York: Random House, 1987), 276–78.

32 *And Russia was producing:* U.S. Energy Information Administration,
"Russia is world's largest producer of crude oil and lease condensate," August 6, 2015, www.eia.gov/todayinenergy/detail.php?id=22392.

32 *Collectively, non-OPEC producers:* Additional non-U.S., non-OPEC production in this period was 4.7 mnb/d. Derived from data in Table 3.6 of
International Energy Agency, *World Energy Outlook 2015* (Paris: OECD
Publishing, 2015), November 10, 2015, 135, https://www.iea.org/publica
tions/freepublications/publication/WEO2015.pdf.

32 *By some estimates, the field required:* "Kazakhstan's Kashagan Tagged World's Most Expensive Energy Project," *Tengri News*, November 29, 2012, http://en.tengrinews.kz/industry_infrastructure/Kazakhstans-Kashagan -tagged-worlds-most-expensive-energy-project-14913/.

33 *The average cost to drill:* See "Trends in U.S. Oil and Natural Gas Upstream Costs," U.S. Energy Information Administration, March 2016, https://www.eia.gov/analysis/studies/drilling/pdf/upstream.pdf, 4.

34 *Compliance with the 2008 agreement:* "OPEC Cuts Supply, But More Work Needed to Fulfill Deal," *Financial Tribune*, February 5, 2017, https://fi nancial/tribune.com/articles/energy/58885/opec-cuts-supply-but-more -work-needed-to-fulfill-deal.

34 *The meeting, designed to explore:* For more details on this meeting and the proposal on the table, see Jay Solomon and Summer Said, "Why Saudis Decided Not to Prop Up Oil," *Wall Street Journal*, December 21, 2014, www .wsj.com/articles/why-saudis-decided-not-to-prop-up-oil-1419219182.

35 *The Saudis apparently had spent:* Raval, "The Big Drop: Riyadh's Oil Gamble."

35 *Jamie Webster, an oil market analyst:* Jamie Webster, in-person conversation with author, Cambridge, MA, November 13, 2013.

35 *That same month, Ibrahim al-Muhanna:* Solomon and Said, "Why Saudis Decided Not to Prop Up Oil."

35 *Later reports of stress tests done:* Raval, "The Big Drop: Riyadh's Oil Gamble."

35 *Headlines such as "Cancel Thanksgiving":* Miles Udland, "Cancel Thanksgiving: The Most Important OPEC Meeting in Years Is Happening on Thursday," *Business Insider*, November 24, 2014, www.businessinsider .com/opec-november-27-meeting-preview-2014-11.

36 *OPEC's then-twelve members:* Indonesia had rejoined OPEC in 2015, bringing its numbers to thirteen; Indonesia again suspended its membership in November 2016.

36 *Members had agreed to maintain:* Alex Lawler, Amena Bakr, and Dmitry Zhdannikov, "Inside OPEC Room, Naimi Declares Price War on U.S. Shale Oil," Reuters, November 28, 2014, www.reuters.com/article/us -opec-meeting-shale-idUSKCN0JC1GK20141128.

36 *many speaking on a not-for-attribution:* Ibid.

36 *Just as surprising to many:* At a 2014 oil industry event in London, OPEC's secretary-general Abdalla El-Badri told reporters, "If prices stay

at $85, we will see a lot of investment going out of the market. About 65% of the producers, they have high costs. Not OPEC." Alex Lawler and David Sheppard, "OPEC's Badri sees little output change in 2015, says don't panic on oil drop," Reuters, October 29, 2014, www.reuters .com/article/us-opec-oil-idUSK0II0XD20141029.

36 *According to an analyst:* This observation was made in early 2016. Christopher Harder, "Oil Plunge Sparks Bankruptcy Concerns—Energy Journal," *Wall Street Journal*, January 12, 2016, http://blogs.wsj.com /moneybeat/2016/01/12/oil-plunge-sparks-bankruptcy-concerns-energy -journal/.

37 *Russian president Vladimir Putin publicly mused:* Chris Mooney, "Why There Are So Many Kooky Conspiracy Theories About Oil," *Washington Post*, December 23, 2014, www.washingtonpost.com/blogs/wonkblog /wp/2014/12/23/putin-is-trotting-out-conspiracy-theories-about-oil -hes-not-the-first/.

37 *Iranian president Hassan Rouhani was only:* Reuters, "Iranian President Blames Oil Price Fall on Political Economy," CNBC, December 10, 2014, www.cnbc.com/id/102256076#.

37 *Bolivian president Evo Morales:* Mike Whitney, "Did the U.S. and the Saudis Conspire to Push Down Oil Prices?," *Global Research*, December 29, 2014, www.globalresearch.ca/did-the-u-s-and-the-saudis-conspire-to-push -down-oil-prices/5421890.

38 *Writing for the magazine* Foreign Policy: Andrew Scott Cooper, "Why Would the Saudis Deliberately Crash the Oil Markets?," *Foreign Policy*, December 18, 2014, http://foreignpolicy.com/2014/12/18/why-would -the-saudis-crash-oil-markets-iran/. Also see Liam Denning, "OPEC's Weapon of Mass Inaction," *Wall Street Journal*, November 27, 2014, www.wsj.com/articles/opecs-weapon-of-mass-inactionheard-on-the -street-1417105500.

38 *Instead, it would cause:* Izabella Kaminsky, "Why Saudi Arabia's Best Bet May Be to Increase Output," *Financial Times*, October 27, 2014, http:// ftalphaville.ft.com/2014/10/27/2020412/why-saudi-arabias-best-bet -may-be-to-increase-output/.

38 *"If in the process":* Solomon and Said, "Why Saudis Decided Not to Prop Up Oil."

39 *Pointing to a fiscal breakeven price:* Even more dramatic was the more than 50 percent jump in fiscal breakeven prices between 2010 (before

Mubarak's ouster) and 2014. "Breakeven Fiscal Oil Prices, US dollars per barrel," International Monetary Fund, http://data.imf.org/?SK=388D FA60-1D26-4ADE-B505-A05A558D9A42&ss=1479331931186.

39 *Al-Naimi's satisfaction after the November:* Alex Lawler, David Sheppard, and Rania El Gamal, "Saudis Block OPEC Output Cut, Sending Oil Price Plunging," Reuters, November 27, 2014, www.reuters.com/article/us -opec-meeting-idUSKCN0JA0O320141127.

39 *but it was not until later:* Solomon and Said, "Why Saudis Decided Not to Prop Up Oil."

39 *Talking to the* Middle East Economic Survey: "MEES Interview with Ali Naimi: 'OPEC Will Never Plan to Cut,'" *Middle East Economic Survey* 57, no. 51/52 (December 22, 2014), http://archives.mees.com/issues/1562 /articles/52258.

Two: The New Oil Order

41 *For two years:* This would be equivalent to under $20 in prices of the day. "Real Prices Viewer," U.S. Energy Information Administration, April 11, 2017, https://www.eia.gov/outlooks/steo/realprices/.

41 *Despite the dramatic cover, the* Economist: "Drowning in Oil," *Economist,* March 4, 1999, www.economist.com/node/188131.

41 *In 2015 and 2016, for the first time:* See International Energy Agency, *World Energy Outlook 2016* (Paris: OECD Publishing, 2016), 65.

42 *Proponents of this view tend to think:* One such proponent is Tim Gould, head of the oil supply division of the International Energy Agency. See International Energy Agency, *World Energy Outlook 2016* (Paris: OECD Publishing, 2016), 142–43; another is Shell's chief executive Ben van Beurden. Christopher Adams, Anjli Raval, and David Sheppard, "Royal Dutch Shell Warns of Risk of Oil Price Spike," *Financial Times,* October 6, 2015, www.ft.com /intl/cms/s/0/a6226c44-6c1a-11e5-8171-ba1968cf791a.html#axzz4A TfSa7OO. Also see Emily Gosden, "Oil Demand Peak 'Not in Sight' as Stage Set for Boom and Bust, Says IEA," *Telegraph,* November 16, 2016, www.telegraph.co.uk/business/2016/11/16/oil-demand-peak-not -in-sight-as-stage-set-for-boom-and-bust-says/.

43 *A public relations manager:* William M. Alley and Rosemarie Alley, *Too Hot to Touch: The Problem of High-Level Nuclear Waste*

(Cambridge: Cambridge University Press, 2013), 6, http://assets.cambridge .org/97811070/30114/excerpt/9781107030114_excerpt.pdf.

43   *He hung up, returned to the stage:* M. King Hubbert, "Nuclear Energy and the Fossil Fuels," paper, Spring Meeting of the Southern District Division of Petroleum, American Petroleum Institute, Plaza Hotel, San Antonio, Texas, March 7–9, 1956, https://web.archive.org/web/20170525165507 /www.hubbertpeak.com/hubbert/1956/1956.pdf.

44   *it had taken 500 million:* Ibid., 4.

44   *Based on this theory:* Ibid., 22, 24.

44   *Morgan Davis, the head of Humble Oil:* Alley and Alley, *Too Hot to Touch,* 7.

44   *He noted, "The oil business":* Kenneth S. Deffeyes, *Hubbert's Peak: The Impending World Oil Shortage* (Princeton: Princeton University Press, 2001), 3.

44   *Crude oil production in the United States:* "U.S. Field Production of Crude Oil," U.S. Energy Information Administration, October 31, 2016, https:// www.eia.gov/dnav/pet/hist/LeafHandler.ashx?n=pet&s=mcrfpus2&f=m. Crude oil production is one component of "overall oil production" or "total liquids production"—categories that also include natural gas liquids or NGLs.

44   *In 2005, U.S. EIA:* Guy Caruso, "When Will World Oil Production Peak?," presentation, 10th Annual Asia Oil and Gas Conference, Kuala Lumpur, Malaysia, June 13, 2005, www.industrializedcyclist.com/EIA speakoil4cast.pdf.

44   *In fact, the notion that the world:* Robert L. Hirsch, Roger Bezdek, and Robert Wendling, *Peaking of World Oil Production: Impacts, Mitigation, and Risk Management* (Washington, DC: U.S. Department of Energy, February 2005).

45   *In 2010, a think tank associated:* See Stefan Schultz, "Military Study Warns of a Potentially Drastic Oil Crisis," *Der Spiegel,* September 1, 2010, www .spiegel.de/international/germany/peak-oil-and-the-german-govern ment-military-study-warns-of-a-potentially-drastic-oil-crisis-a-715138 .html.

45   *it has been the inspiration:* See Paolo Bacigalupi, *The Windup Girl* (San Francisco: Night Shade Books, 2012); "A Picky List of Peak Oil Resources," Transition Voice, accessed November 5, 2016, http://transition voice.com/a-snarky-guide-to-peak-oil/peak-oil-resources/.

45   *Many declared the end:* Nathan Bomey and Roger Yu, "Low Oil Prices End

21st Century Gold Rush," *USA Today*, March 17, 2016, www.usatoday
.com/story/money/2016/03/17/crude-oil-prices-us-economy/81318012/.

46 *But eventually, after U.S. crude oil production:* Overall U.S. production
includes crude oil plus natural gas liquids—or "NGLs" as they are frequently known. See "U.S. Field Production of Crude Oil," U.S. Energy
Information Administration, https://www.eia.gov/dnav/pet/hist/Leaf
Handler.ashx?n=pet&s=mcrfpus2&f=m.

46 *Despite lower prices, in their:* In 1970, the overall U.S. oil production
(crude plus natural gas liquids) was approximately 11.4 mnb/d. (Crude
oil production alone was 9.6 mnb/d). According to the IEA's New Policy Scenario (its "most likely" scenario), overall U.S. oil production
could reach 14.1 mnb/d by 2020. According to the EIA reference scenario (its "most likely" scenario), overall U.S. oil production could reach
14.5 mnb/d by 2020. These numbers were calculated from the following sources: "IEA Oil Information Statistics: World oil statistics," International Energy Agency, http://stats.oecd.org.ezp-prod1.hul.harvard
.edu/deliverdotstat?cid=id10763&institution_name=Harvard+Universi
ty+Library&baseurl=http%3a%2f%2fstats.oecd.org%2fwbos%2fbrand
edview.aspx&doi=data-00474en&return_url1=http%3a%2f%2fwww
.oecd-ilibrary-org%2fcontent%2fahahtoc%3ffmt%3dahah&lang=en&o
ecdstat=oil-data-en&itemId=%2fcontent%2fdata%2fdata-00474-en;
"U.S. Gas Plant Production of Natural Gas Liquids and Liquid Refinery
Gases,"U.S.EnergyInformationAdministration,https://www.eia.gov/dnav
/pet/hist/LeafHandler.ashx?n=PET&s=MNGFPUS2&f=A; "U.S. Field
Production of Crude Oil," U.S. Energy Information Administration,
https://www.eia.gov/dnav/pet/hist/LeafHandler.ashx?n=pet&s=mcrf
pus2&f=a; "Table: Petroleum and Other Liquids Supply and Disposition,"
Annual Energy Outlook 2017, U.S. Energy Information Administration,
www.eia.gov/outlooks/aeo/data/browser/#/?id=11-AEO2017&region
=0-0&cases=ref2017~lowprice&start=2015&end=2050&f=A&linechar
t=lowprice-d120816a.3-11-AEO2017~lowprice-d120816a.19-11-AEO
2017&ctype=linechart&sid=ref2017-d120816a.19-11-AEO
2017~ref2017-d120816a.3-11-AEO2017~lowprice-d120816a.3-11
-AEO2017~lowprice-d120816a.19-11-AEO2017&sourcekey=0; International Energy Agency, *World Energy Outlook 2016* (Paris: OECD Publishing, 2016), 136.

46 *There are shale fields:* Dan Murtaugh, "Texas Isn't Scared of $30 Oil,"

Bloomberg, February 3, 2016, www.bloomberg.com/news/articles/2016
-02-03/texas-toughness-in-oil-patch-shows-why-u-s-still-strong-at-30.

46 *Even in this low-price scenario:* In the EIA's low-price scenario, overall
U.S. oil production (crude plus natural gas liquids) reaches 12.4 mnb/d
in 2020, roughly one million barrels higher than the 1970s high of 11.4
mnb/d. In 2050, this amount falls to 11.4 mnb/d, the same as the 1970
peak. "Table: Petroleum and Other Liquids Supply and Disposition,"
Annual Energy Outlook 2017, U.S. Energy Information Administration,
https://www.eia.gov/outlooks/aeo/data/browser/#/?id=11-AEO2017&re
gion=0-0&cases=ref2017~lowprice&start=2015&end=2050&f=A&line
chart=lowprice-d120816a.3-11-AEO2017~lowprice-d120816a.19-11
-AEO2017&ctype=linechart&sid=ref2017-d120816a.19-11-A
EO2017~ref2017-d120816a.3-11-AEO2017~lowprice-d120816a.3-11
-AEO2017~lowprice-d120816a.19-11.

47 *in 2015, the United States and Canada:* Rosalie Starling, "EIA Expands
Global Shale Oil and Natural Gas Resource Assessment," Hydrocarbon
*Engineering,*www.energyglobal.com/downstream/gas-processing/15122015
/EIA-expands-global-shale-oil-and-natural-gas-resource-assessment
-1979/.

47 *The fact is that:* See International Energy Agency, *World Energy Outlook
2016* (Paris: OECD Publishing, 2016), 128.

47 *As of the end of 2015:* Ibid.

48 *Argentina, Russia, Mexico, Colombia:* "World Tight Oil Production to
More Than Double from 2015 to 2040," U.S. Energy Information Administration, August 12, 2016, www.eia.gov/todayinenergy/detail.cfm?id
=27492.

48 *At that point, tight oil production:* See ExxonMobil, *The Outlook for Energy:
A View to 2040* (Irving, TX: ExxonMobil, 2016), 59, http://cdn.exxonmobil
.com/~/media/global/files/outlook-for-energy/2016/2016-outlook
-for-energy.pdf.

48 *"For the last forty years":* Bob Belfer, in-person conversation with author,
Cambridge, MA, May 5, 2016.

49 *The decade from 1997 to 2007:* From 1997 to 2007, overall global energy consumption grew at 2.4 percent a year on average. This number
was 0.89 percent for OECD countries and 4.93 percent for non-OECD
countries. BP p.l.c., *BP Statistical Review of World Energy 2016* (BP, June
2016), www.bp.com/content/dam/bp/excel/energy-economics/statisti

cal-review-2016/bp-statistical-review-of-world-energy-2016-workbook
.xlsx.

49 *China, whose focus on heavy industry:* "GDP Growth (Annual %)," The
World Bank, accessed November 5, 2016, http://data.worldbank.org
/indicator/NY.GDP.MKTP.KD.ZG?locations=CN.

49 *In just the few years:* Fredrich Kahrl and David Roland-Holst, "Growth
and Structural Change in China's Energy Economy," *Energy* 34 (2009):
894. BP p.l.c., *BP Statistical Review of World Energy 2016* (BP, June
2016), www.bp.com/content/dam/bp/excel/energy-economics/statistical
-review-2016/bp-statistical-review-of-world-energy-2016-workbook
.xlsx.

49 *By 2007, China's needs:* In 2007, China's energy consumption (including
that of Hong Kong) accounted for 18.6 percent of what the whole world
consumed that year. If we consider energy demand *growth*, as opposed to
looking at energy demand in the absolute, we find that China, including
Hong Kong, accounted for 45.2 percent of global demand growth. BP p.l.c.,
BP Statistical Review of World Energy 2016 (BP, June 2016), www.bp.com
/content/dam/bp/excel/energy-economics/statistical-review-2016/bp
-statistical-review-of-world-energy-2016-workbook.xlsx.

49 *Global energy demand growth:* BP p.l.c., *BP Energy Outlook 2035: January 2017* (London: BP, 2017), http://www.bp.com/content/dam/bp/excel
/energy-economics/energy-outlook-2017/bp-energy-outlook-2017
-summary-tables.xlsx.

49 *Looking forward, various companies:* For instance, see the data tables of
BP Energy Outlook 2017 and EIA IEO 2016: BP p.l.c., *BP Energy Outlook 2035: January 2017* (London: BP, 2017), http://www.bp.com/content
/dam/bp/excel/energy-economics/energy-outlook-2017/bp-energy-out
look-2017-summary-tables.xlsx.; "Table: World total primary energy
consumption by region," *International Energy Outlook 2016*, U.S. Energy
Information Administration, https://www.eia.gov/outlooks/aeo/data
/browser/#/?id=1-IEO2016&region=0-0&cases=Reference&start=201
0&end=2040&f=A&linechart=Reference-d021916a.28-1-IEO2016&c
type=linechart&sid=Reference-d021916a.28-1-IEO2016&sourcekey=0.

49 *In part, these trends are due to:* With the exception of the expansion of the
middle class—the rate of growth of these parameters will slow either in the
immediate years ahead out to 2020 and beyond, or in the case of economic
growth, in the decade beyond 2030. For example, people will continue to

move to urban areas, boosting city dwellers from just over half of the world's
population in 2014 to two-thirds in the midpoint of this century. But the
quintupling of urban residents from 1950 to 2014 will neither be matched
in absolute numbers nor in rates of growth from 2014 to 2050. Similarly,
over the years to 2025, the global urban population is expected to grow at
1 percent a year, but this is still significantly less that the 1.3 percent annual
growth rates realized from 1992 to 2012. See United Nations, Department
of Economic and Social Affairs, Population Division, *World Urbanization
Prospects: The 2014 Revision, Highlights* (United Nations, 2014), www.esa
.un.org/unpd/wup/Publications/Files/WUP2014-Highlights.pdf.

49 *For example, according to one projection:* BP p.l.c., *BP Energy Outlook: 2016
    edition* (London: BP, 2016), slide 91, www.bp.com/content/dam/bp/pdf
    /energy-economics/energy-outlook-2016/bp-energy-outlook-2016.pdf.

49 *This phenomenon will not be limited:* Ibid., slide 45.

49 *India and the economies of Africa:* "Country insights: India," BP Global,
    www.bp.com/en/global/corporate/energy-economics/energy-outlook
    /country-and-regional-insights/india-insights.html; "Regional insights:
    Africa," BP Global, www.bp.com/en/global/corporate/energy-economics
    -energy-outlook/country-and-regional-insights/africa-insights.html.

50 *When it comes to oil—as opposed to:* ExxonMobil, *The Outlook for Energy:
    A View to 2040*, 72.

50 *One reason to expect positive growth is that:* More than 90 percent of the
    world's transportation runs on petroleum-based fuels. "IEO2016: Energy
    use: Liquids," U.S. Energy Information Administration, https://www.eia
    .gov/outlooks/aeo/data/browser/#/?id=15-IEO2016&region=4-0&cas
    es=Reference&start=2010&end=2040&f=A&linechart=Reference
    -d021916a.2-15-IEO2016.4-0~Reference-d021916a.26-15-IEO2016.4
    -0~Reference-d021916a.34-15-IEO2016.4-0~Reference-d021916a.18
    -15-IEO2016.4-0~Reference-d021916a.10-15-IEO2016.4-0&map=&c
    type=linechart&sid=Reference-d021916a.26-15-IEO2016.4-0~Ref
    erence-d021916a.34-15-IEO2016.4-0~Reference-d021916a.18-15
    -IEO2016.4-0~Reference-d021916a.10-15-IEO2016.4-0~Reference
    -d021916a.2-15-IEO2016.4-0&sourcekey=0. In contrast, 56 percent of
    global petroleum consumption is used for transportation.

50 *By one account, the need for fuel:* BP p.l.c., *BP Energy Outlook: 2016 edi-
    tion* (London: BP, 2016), slide 23, www.bp.com/content/dam/bp/pdf
    /energy-economics/energy-outlook-2016/bp-energy-outlook-2016.pdf.

50 *With 57 cars per 100 people:* ExxonMobil, *The Outlook for Energy: A View to 2040*, 17.

50 *This bump-up in global ownership:* ExxonMobil anticipates that fuel efficiency will improve from 25 miles per gallon in 2014 to 45 mpg in 2040; BP expects movement from 30 mpg to 50 mpg in 2035. ExxonMobil, *The Outlook for Energy: A View to 2040*, 18; BP p.l.c., *BP Energy Outlook: 2016 edition*, slide 25.

50 *Others—such as Fatih Birol:* See Fatih Birol, "Special Briefing: World Energy Outlook," (keynote speech, The Atlantic Council Global Energy Summit, Abu Dhabi, United Arab Emirates, January 13, 2017), www.atlanticcouncil.org/news/transcripts/special-briefing-world-energy-outlook.

51 *As Lincoln Moses, the first head:* See Frederic Murphy, "Ideas for the new NEMS Liquid Fuel Market Model" (white paper, Fox School of Business, Temple University, Philadelphia, PA, September, 2009), 4, www.eia.gov/outlooks/documentation/workshops/pdf/fred%20murphy%20lfmm%20white%20paper.pdf.

51 *BP, in fact, identifies:* BP p.l.c., *BP Energy Outlook: 2016 edition* (London: BP, 2016), 74–76, www.bp.com/content/dam/bp/pdf/energy-economics/energy-outlook-2016/bp-energy-outlook-2016.pdf.

51 *Growth in world energy demand:* Oil demand would decline by 5 percent; natural gas demand by 7 percent; and coal by 8 percent. Ibid.

51 *Certainly, there is sufficient uncertainty:* Even apart from overall rates of growth, the pace of the restructuring of the Chinese economy will also have a significant impact on energy demand. See ibid., slides 58–59.

51 *Nearly two-thirds of them:* See Andrew Browne, "China Bulls Become an Extinct Species," *Wall Street Journal*, April 5, 2016, www.wsj.com/articles/china-bulls-become-an-extinct-species-1459829989.

51 *The second scenario BP identifies:* BP p.l.c., *BP Energy Outlook: 2016 edition*, slides 78–81.

51 *Policies such as putting a significant:* Ibid., 79.

52 *By convention, the reference case scenarios:* Some improvements in existing technologies, however, are factored in over time. The AEO2015 Reference case projection is a "business-as-usual trend estimate, given known technology and technological and demographic trends." U.S. Energy Information Administration, *Annual Energy Outlook 2015; With Projections to 2040* (Washington, DC: U.S. Department of Energy), iii, https://www.eia.gov/forecasts/archive/aeo15/pdf/0383(2015).pdf.

52 *For example, ExxonMobil:* ExxonMobil assesses that in 2014, 94 percent of world's transport ran on oil. It expects this amount to decrease to 89 percent in 2040. The company forecasts that 40 percent of new vehicles purchased in 2040 will be conventional hybrids. But the company sees only marginal growth in electric vehicles, given current constraints; plug-in hybrids and electric cars will, in ExxonMobil's view, account for less than 10 percent of all car sales in 2040. ExxonMobil, *The Outlook for Energy: A View to 2040*, 19, 20, 23, 72.

52 *However, BP's 2017:* See "The impact of electric cars on oil demand," BP Global, 2017, www.bp.com/en/global/corporate/energy-economics/en ergy-outlook/electric-cars-and-oil-demand.html.

52 *The company also acknowledges:* Spencer Dale and Thomas D. Smith, "Back to the future: electric vehicles and oil demand," (Speech, Bloomberg New Energy Finance: The Future of Energy, EMEA Summit, London, UK, October 2016), http://www.bp.com/en/global/corporate/media /speeches/back-to-the-future-electric-vehicles-and-oil-demand.html.

52 *Rather than focusing on the low:* Specifically, Dr. Ghouri's work predicts a decline of 13.8 mnb/d of oil in comparison to what would have otherwise been consumed in 2040. Dr. Salman Saif Khan Ghouri, in-person conversation with the author, Qatar Petroleum Headquarters, Doha, Qatar, August 22, 2016.

52 *When I asked about the reaction:* Ibid.

53 *She posits a number:* Belova's "Status quo" scenario sees global oil demand leveling off by 2020 due to fuel economy standards and decreases in demand from the power and industry sector. In her "most likely" scenario (which she calls the base case), the world sees a 7 percent drop in oil demand by 2035; her "innovative" scenario sees an absolute drop by 17 percent in the same time period. See Grigory Vygon et al., "Technological Progress in Motor Transport: How Close is Peak Oil Demand?" (executive summary, VYGON Consulting, Moscow, Russia, October 2016), https://vygon.consulting/upload/iblock/e46/vygon_consulting_oil_de mand_2016_executive_summary_en.pdf.

53 *With this advance in mind:* Tom Randall, "Here's How Electric Cars Will Cause the Next Oil Crisis," Bloomberg, February 25, 2016, www .bloomberg.com/features/2016-ev-oil-crisis/.

53 *According to BNEF, this would displace:* "Electric Vehicles to Be 35% of Global New Car Sales by 2040," Bloomberg, February 25, 2016, http://

about.bnef.com/press-releases/electric-vehicles-to-be-35-of-global-new
-car-sales-by-2040/.

53 *BNEF made various calculations:* Randall, "Here's How Electric Cars Will
Cause the Next Oil Crisis."

53 *Anthony Yuen of Citigroup suggested:* Anthony Yuen, in-person conversation with author, Aspen, CO, July 7, 2016.

53 *perhaps between 300 and 400:* It is also useful to note that, in the United
States, there are about 800 vehicles per 1000 people, while many other
developed countries, such as Canada, Japan, and those in Europe, have
a range of 500 to 700 vehicles per 1000. "All countries compared for
transport; motor vehicles per 1000 people," NationMaster, 2017, www
.nationmaster.com/country-info/stats/Transport/Road/Motor-Vehicles
-per-1000-people. To learn more about a number of studies predicting
future Chinese vehicle use, see Huo et al., "Projection of Chinese Motor
Vehicle Growth, Oil Demand, and CO_2 Emissions Through 2050," *Transportation Research Record: Journal of the Transportation Research Board*,
no. 2038 (2007): 69–77, www.ntl.bts.gov/lib/32000/32100/32149/fulltext
.pdf.

53 *Others still attribute slower:* BP p.l.c., *BP Energy Outlook: 2017 edition*
(London: BP, 2017), 73, www.bp.com/content/dam/bp/pdf/energy-eco
nomics/energy-outlook-2017/bp-energy-outlook-2017.pdf.

54 *Speaking to a large audience in Houston:* Steven Poruban, "IHS CER-
AWeek: Al-Naimi 'Optimistic' That the World Oil Market Will Rebalance,"
Oil & Gas Journal, February 23, 2016, www.ogj.com/articles/2016/02/ihs
-ceraweek-al-naimi-optimistic-that-world-oil-market-will-rebalance
.html.

54 *There is no need for "meddling":* Ibid.

54 *He told a small gaggle of reporters:* Stanley Reed, "Saudi Oil Chief Khalid al-Falih Tells OPEC Changes Are Coming," *New York Times*, June 2,
2016, www.nytimes.com/2016/06/03/business/energy-environment/opec
-meeting-oil-production-saudi-arabia.html?ref=world.

54 *In 1881, readers gobbled up:* Henry Demarest Lloyd, "Monopoly on the
March," *Atlantic Monthly*, March 1881.

55 *In fact, OPEC later modeled:* David F. Prindle, "Railroad Commission,"
Texas State Historical Association, June 15, 2010, www.tshaonline.org
/handbook/online/articles/mdr01.

55 *When, near the start of the Great Depression:* See Frederic M. Scherer,

Industry Structure, Strategy, and Public Policy (New York: HarperCollins, 1996), 67.

55 *Rampant cheating was so persistent:* Lucile Silvey Beard, "The History of the East Texas Oil Field" (master's thesis, Graduate School of Hardin-Simmons University, Overton, Texas, 1938), 51, www.texasranger.org /E-Books/History_of_the_East_Texas_Oil_Field_(Silvey).pdf.

55 *The idea of a cartel among them:* The leaders of Anglo-Persian Oil (later British Petroleum), Royal Dutch Shell, and Standard Oil (later Exxon) were there that night. Four other large companies later joined their plans. For more, see Anthony Sampson, *The Seven Sisters: The Great Oil Companies and the World They Shaped* (London: Hodder & Stoughton, 1988).

55 *From the 1940s until OPEC:* Ibid.

55 *In 1953, they controlled almost 90 percent:* The Seven Sisters controlled 87.1 percent of the world's oil production in 1953. A. F. Alhajji and David Huettner, "OPEC and Other Commodity Cartels: A Comparison," *Energy Policy* 28 (2001): 1158.

56 *When this cohesion has been at its highest:* For more on OPEC's ability to influence the market, see Charles F. Doran, "OPEC Structure and Cohesion: Exploring the Determinants of Cartel Policy," *The Journal of Politics* 42, no. 1 (February 1980): 82–101; James Richard, "New Cohesion in OPEC's Cartel? Pricing and Politics," *Middle East Review of International Affairs* 3, no. 2 (June 1999): 18–23, www.rubincenter.org/meria/1999/06/richard.pdf; Nazli Choucri, "OPEC: Calming a Nervous World Oil Market," *Technology Review* 83, no. 1 (October 1980), http://web.mit.edu/polisci/nchoucri/pub lications/articles/D-9_Choucri_OPEC_Calming_Nervous_World_Oil _Market.pdf; Bassam Fattouh, "OPEC Pricing Power: The Need for a New Perspective," WPM 31, Oxford Institute for Energy Studies, March 2007, https://www.oxfordenergy.org/wpcms/wp-content/uploads/2010/11 /WPM31-OPECPricingPowerTheNeedForANewPerspective-BassamFat touh-2007.pdf.

56 *Ten of OPEC's members:* Benoit Faucon, Nathan Hodge, and Summer Said, "Oil-producing Countries Agree to Cut Output Along with OPEC," *Wall Street Journal*, December 10, 2016, www.wsj.com/articles/opec-russia -upbeat-about-securing-oil-output-deal-as-meeting-begins-1481362580.

56 *in 2016, OPEC countries had only earned:* Nayla Razzouk, Angelina Rascouet, and Golnar Motevalli, "OPEC Confounds Skeptics, Agrees to First Oil Cuts in 8 Years," *Bloomberg Markets*, November 30, 2016, www

.bloomberg.com/news/articles/2016-11-30/opec-said-to-agree-oil-pro
duction-cuts-as-saudis-soften-on-iran.

57 *Oil ministers from Saudi Arabia*: See Vladimir Soldatkin, Rania El Gamal,
and Alex Lawler, "OPEC, Non-Opec, Agree First Global Oil Pact Since
2001," Reuters, December 10, 2016, www.reuters.com/article/us-opec
-meeting-idUSKBN13Z0J8; Rachael Boothroyd Rojas, "Venezuela Celebrates 'Historic' OPEC Deal," Venezuelanalysis.com, December 1, 2016,
https://venezuelanalysis.com/news/12813.

57 *In the first month of the agreement*: In January 2017, the month the agreement came into force, Iran, Libya, and Nigeria brought 270,000 barrels
of new oil to market, while other OPEC members cut 840,000. Angelina Rascouet and Julian Lee, "OPEC Cuts Oil Output, But More Work
Needed to Fulfill Deal," *Bloomberg Markets*, February 2, 2017, www
.bloomberg.com/news/articles/2017-02-02/opec-cuts-oil-production
-but-more-work-needed-to-fulfill-deal.

57 *Put succinctly by Abdalla El-Badri*: Dan Murtaugh and Javier Blas, "OPEC
Unsure How It Can 'Live Together' with Shale Oil," Bloomberg, February 22, 2016, www.bloomberg.com/news/articles/2016-02-22/opec-s-el
-badri-doesn-t-know-how-to-live-together-with-shale-oil.

58 *Rather than quickly curtailing production*: In fact, despite the price drop
that started in mid-2014, U.S. crude oil production rose, not declined, until
April 2015 when it peaked; flush credit markets and effective cost-cutting
measures buoyed production despite the falling price. After April 2015,
tight oil production began to drop, hitting a bottom in September 2016
after shedding 11 percent of production compared to its 2015 peak. See
"U.S. Gas Plant Production of Natural Gas Liquids and Liquid Refinery
Gases," U.S. Energy Information Administration, March 31, 2017, www
.eia.gov/dnav/pet/hist/LeafHandler.ashx?n=pet&s=mngfpus2&f=m.

58 *From September 2016—when rumors*: "Drilling Productivity Report," U.S.
Energy Information Administration, June 12, 2017, https://www.eia.gov
/petroleum/drilling/#tabs-summary-2; "Drilling Productivity Report," U.S.
Energy Information Administration, September 12, 2016, https://www.eia
.gov/petroleum/drilling/archive/2016/09/#tabs-summary-2.

58 *Technological advances made in the past*: For more innovations bringing down the cost of tight oil production, see Clifford Krauss, "Drillers Answer Low Oil Prices with Cost-Saving Innovations," *New York
Times*, May 11, 2015, https://www.nytimes.com/2015/05/12/business

/energy-environment/drillers-answer-low-oil-prices-with-cost-saving -innovations.html?_r=0; Alison Sider and Erin Ailworth, "Oil Companies Tap New Technologies to Lower Production Costs," *Wall Street Journal*, September 13, 2015, https://www.wsj.com/articles/oil-companies -tap-new-technologies-to-lower-production-costs-1442197712.

58 *John B. Hess, the CEO of Hess Corporation:* John B. Hess, in-person conversation with author, Washington, DC, December 13, 2016.

58 *Roger Diwan, an energy market expert:* Roger Diwan, in-person conversation with author, Houston, Texas, March 9, 2017.

58 *In the closing days of 2014, Nick Butler:* Nick Butler, "Forget Opec—Sex and Technology Shape the Oil Market Now," *Financial Times*, December 21, 2014, http://blogs.ft.com/nick-butler/2014/12/21/forget-opec-sex -and-technology-shape-the-oil-market-now/.

59 *Experts from former Federal Reserve:* In April 2015, ConocoPhillips CEO Ryan Lance discussed the emergence of U.S. drillers as the world's swing suppliers. Bradley Olson, "Conoco Phillips CEO Bets Farm on Shale, Sees Oil Rebound," www.bloomberg.com/news/articles/2015-04-08/con ocophillips-ceo-bets-farm-on-shale-sees-oil-rebound; Alan Greenspan, "OPEC has Ceded to the US Its Power Over Oil Price," *Financial Times*, February 19, 2015, www.ft.com/intl/cms/s/0/92ab80e4-b827-11e4-b6a5 -00144feab7de.html#axzz4ATfSa7OO.

59 *American tight oil has no ability:* If the United States wanted to act more like a swing producer, it could consider making dramatic changes to how it uses its Strategic Petroleum Reserve. This is discussed in Chapter Six.

60 *Those licking their chops:* See Gal Luft, "Fifty Years to OPEC: Time to Break the Oil Cartel," *Journal of Energy Security*, September 29, 2010, www.ensec.org/index.php?option=com_content&view=article&id =263:fifty-years-to-opec-time-to-break-the-oil-cartel&catid=110:en ergysecuritycontent&Itemid=366; Nancy Brune, "50 Years Later: OPEC's Continuing Threat to American Security," *Journal of Energy Security*, September 29, 2010, http://ensec.org/index.php?option =com_content&view=article&id=267:50-years-later-opecs-continu ing-threat-to-american-security&catid=110:energysecuritycontent &Itemid=366.

60 *three researchers from Oxford University:* Zoheir Ebrahim, Oliver R. Inderwildi, and David A. King, "Macroeconomic Impacts of Oil Price Volatility: Mitigation and Resilience," *Frontiers in Energy* 8, no.

1 (2014): 11, www.smithschool.ox.ac.uk/news/FEP-14003-EZ-proof
-checked.pdf.

60 *Surveys of hundreds of senior:* John R. Graham, Campbell R. Harvey, and
Shiva Rajgopal, "Value Destruction and Financial Reporting Decisions,"
Columbia Business School, September 6, 2006, 12, https://www0.gsb.co
lumbia.edu/mygsb/faculty/research/pubfiles/12924/Rajgopal_value.pdf.

61 *"The big dogs, the Saudis":* Lynn Doan and Dan Murtaugh, "Shale as
World's Swing Producer Signals 'Jagged' Oil Future," Bloomberg, April 20,
2015, www.bloomberg.com/news/articles/2015-04-20/shale-as-world-s
-swing-producer-signals-jagged-future-for-oil.

61 *At a minimum, companies working:* It is, however, worth noting that there
are many drilled-but-uncompleted (DUC) wells in the United States.
After drilling the well, companies decided not to produce the oil in the
price environment of the time. As a result, when prices reach above a
certain threshold, these wells can begin to produce oil in very short order.

Three: Natural Gas Becomes More Like Oil

64 *An old wildcatters' joke:* This quip is also reproduced in *Economist*, "Vapour
Trails," *Economist*, July 1, 2010, www.economist.com/node/16488892.

64 *Put another way, a barrel of:* Calculations were made using: Derek Supple,
"Units & Conversions Fact Sheet," MIT Energy Club, accessed November 5, 2016, http://cngcenter.com/wp-content/uploads/2013/09/UnitsAnd
Conversions.pdf; "US Barrels (Oil) to Liters," Metric Conversions, November 3, 2016, www.metric-conversions.org/volume/us-oil-barrels-to-liters
.htm; "Energy Units and Calculators Explained," U.S. Energy Information
Administration, August 9, 2016, www.eia.gov/energyexplained/index
.cfm/index.cfm?page=about_energy_units.

65 *Compared to oil, the relatively complicated:* While cross-border trade in
natural gas is clearly the trade relevant for geopolitics, it is important to
note that the majority of natural gas consumed is produced domestically.

65 *Until roughly the year 2000:* As can be seen from Figure 3.1, up until
1999, fewer than ten countries were importing LNG. See International
Gas Union, *IGU World LNG Report: 2017 Edition* (Barcelona: International Gas Union, April 2017), 7, http://www.igu.org/sites/default/files
/103419-World_IGU_Report_no%20crops.pdf.

66 *In 1967, given the fluidity:* In contrast, in 1973, the Arab members of

OPEC employed a much more effective tactic; rather than simply embargoing certain countries, they instituted a series of mounting production cuts that drove up prices.

66   *Although the disruptions were minor:* For more details on both the 2006 and 2009 crises, see Simon Pirani, Jonathan Stern, and Katja Yafimava, "The Russo-Ukrainian Gas Dispute of January 2009: A Comprehensive Assessment," (NG 27, Oxford Institute for Energy Studies, Oxford University, Oxford, U.K., February 2009), https://www.oxfordenergy.org /wpcms/wp-content/uploads/2010/11/NG27-TheRussoUkrainianGas DisputeofJanuary2009AComprehensiveAssessment-JonathanStern SimonPiraniKatjaYafimava-2009.pdf.

66   *Although Russia's decision to cut off gas:* During this crisis, at least eleven people froze to death. Dan Bilefsky and Andrew E. Kramer, "Deal to End Russia's Cutoff of Gas Remains Uncertain," *New York Times*, January 09, 2009, www.nytimes.com/2009/01/10/world/europe/10gazprom .html.

66   *At the time, 80 percent:* Andrew E. Kramer, "Russia Cuts Gas, and Europe Shivers," *New York Times*, January 6, 2009, www.nytimes.com/2009/01/07 /world/europe/07gazprom.html?pagewanted=all&_r=0.

66   *So not only did Ukraine:* See Aleksander Kovacevic, "The Impact of the Russia-Ukraine Gas Crisis in South Eastern Europe," (NG 29, Oxford Institute for Energy Studies, Oxford University, Oxford, U.K., March 2009), https://www.oxfordenergy.org/wpcms/wp-content/uploads/2010/11 /NG29-TheImpactoftheRussiaUkrainianCrisisinSouthEastern Europe-AleksandarKovacevic-2009.pdf.

66   *Recognizing that the transit of its gas:* See Morena Skalamera and Andreas Goldthau, "Russia: Playing Hardball or Bidding Farewell to Europe?," discussion paper 2016-03, Geopolitics of Energy Project, Belfer Center for Science and International Affairs, Harvard University, Cambridge, MA, June 2016, http://belfercenter.hks.harvard.edu/files/ Russia%20Hardball %20-%20Web%20Final.pdf.

66   *Since deregulation of natural gas:* For more on the deregulation of U.S. natural gas markets, see Andreij Juris, "Development of Competitive Natural Gas Markets in the United States," (Note No. 141, Public Policy for the Private Sector, The World Bank Group, April 1998), http://siteresources .worldbank.org/EXTFINANCIALSECTOR/Resources/282884-1303327 122200/141juris.pdf.

67 *The fact that, even in 2016:* According to the U.S. EIA, transportation consumed 56 percent of all global oil consumption in 2016. "IEO2016: Energy use: Liquids," U.S. Energy Information Administration, https://www.eia .gov/outlooks/aeo/data/browser/#/?id=15-IEO2016&region=4-0&cases =Reference&start=2010&end=2040&f=A&linechart=Reference -d021916a.2-15-IEO2016.4-0~Reference-d021916a.26-15-IEO2016.4 -0~Reference-d021916a.34-15-IEO2016.4-0~Reference-d021916a.18 -15-IEO2016.4-0~Reference-d021916a.10-15-IEO2016.4-0&map=&c type=linechart&sid=Reference-d021916a.26-15-IEO2016.4-0~Reference -d021916a.34-15-IEO2016.4-0~Reference-d021916a.18-15-IEO2016.4 -0~Reference-d021916a.10-15-IEO2016.4-0~Reference-d021916a.2-15 -IEO2016.4-0&sourcekey=0.

67 *Anita George, former senior director:* Tim Ward, "5 Questions for the World Bank on Ending Routine Gas Flaring and Climate Change," *Huffington Post,* December 7, 2015, www.huffingtonpost.com/tim-ward/5 -questions-for-the-world_b_8733362.html.

67 *Even as late as 2016:* Fred Julander, in-person conversation with the author, Aspen, CO, July 6, 2016.

67 *The agenda of the two leaders:* Richard C. Bush, "Shinzo Abe's Visit to Washington," Brookings, February 22, 2013, https://www.brookings.edu /blog/up-front/2013/02/22/shinzo-abes-visit-to-washington/.

68 *But amidst a program:* See "Abe to ask Obama to boost shale gas exports to Japan," *Japan Times,* April 20, 2014, www.japantimes.co.jp /news/2014/04/20/national/politics-diplomacy/abe-to-ask-obama-to -boost-shale-gas-exports-to-japan/#.WVpsQo-cHrg.

68 *In addition, natural gas:* "How Much Carbon Dioxide Is Produced When Different Fuels Are Burned?" U.S. Energy Information Administration, June 14, 2016, https://www.eia.gov/tools/faqs/faq.cfm?id=73&t=11.

68 *at least in certain parts of the world:* Europe is the notable exception where advocates of natural gas have been unable to persuade policymakers and citizens that natural gas holds unique benefits. See, for instance, Jonathan Stern, "The Future of Gas in Decarbonising European Energy Markets: The need for a new approach" (OIES PAPER: NG 116, The Oxford Institute for Energy Studies, Oxford University, UK, January, 2017), https://www .oxfordenergy.org/wpcms/wp-content/uploads/2017/01/The-Future -of-Gas-in-Decarbonising-European-Energy-Markets-the-need-for-a -new-approach-NG-116.pdf.

68 *President Obama, in his 2014:* President Barack Obama, "State of the Union Address," The White House Office of the Press Secretary, January 28, 2014, https://www.whitehouse.gov/the-press-office/2014/01/28/president-barack-obamas-state-union-address.

68 *Scholars Maximilian Kuhn:* Maximilian Kuhn and Frank Umbach, "The Triple 'A' Argument for Natural Gas," *International Association for Energy Economics* (First Quarter 2012), 34–38, https://www.google.com/url?sa=t&rct=j&q=&esrc=s&source=web&cd=1&cad=rja&uact=8&ved=0ahUKEwiqjKKNrpXQAhVK2oMKHa5dAUUQFggdMAA&url=https%3A%2F%2Fiaee.org%2Fen%2Fpublications%2Fnewsletterdl.aspx%3Fid%3D160&usg=AFQjCNGa_DV9Yj3wmKkCgK5Bz6nKDX29tg&sig2=YVqXxnMwYJwlWqL-XlCUQQ.

68 *Some industry experts, such as Oklahoma:* See Robert Hefner III, "The Age of Energy Gases: The Importance of Natural Gas in Energy Policy," paper, Aspen Institute's Aspen Strategy Group conference "The Global Politics of Energy," Aspen, Colorado, August 3–8, 2007, www.ghkco.com/downloads/ASG-ImportanceofNaturalGasinEnergyPolicy08.07.doc.

69 *In this scenario, the use of natural gas:* In this vision of the world in 2035, the IEA anticipates China's demand for gas rising from roughly equivalent to Germany's demand in 2010 to be equal to the demand of the whole European Union. Demand for natural gas from the Middle East almost doubles in this time period and India's quadruples. See International Energy Agency, *World Energy Outlook 2011 Special Report: Are We Entering a Golden Age of Gas?* (Paris: OECD Publishing, 2011), 7–13, www.worldenergyoutlook.org/media/weowebsite/2011/WEO2011_GoldenAgeofGasReport.pdf.

69 *In the United States, as well as in China:* IEA, *World Energy Outlook 2011 Special Report*, 8.

69 *According to this vision of the future:* The scenario perceived as the one most likely to materialize is usually called a "reference case" scenario. For the IEA this most likely scenario is termed the New Policies Scenario. The IEA notes that, in the golden age of gas, "global energy-related CO_2 emissions in 2035 are only slightly lower than those in the New Policies Scenario." Ibid.

69 *In 2012, in its Shale Gas Five-Year Plan:* The 2012 Shale Gas Five-Year Plan included targets of producing between 2.1 trillion cubic feet (tcf)

to 3.5 tcf of shale gas per year by 2020. David Sandalow et al., "Meeting China's Shale Gas Goals," working draft, Center on Global Energy Policy, Columbia School of International and Public Affairs, September 2014, http://energypolicy.columbia.edu/sites/default/files/energy/China%20 Shale%20Gas_WORKING%20DRAFT_Sept%2011_0.pdf. This amount would equal roughly 2 percent to 3 percent of the world's overall natural gas production, or 40–70% of Chinese demand in 2012. "IEO2016: Total natural gas production: Total World," U.S. Energy Information Administration, November, 2015, https://www.eia.gov/forecasts/aeo/data /browser/#/?id=41-IEO2016&region=0-0&cases=Reference&start=201 0&end=2040&f=A&linechart=Reference-d021916a.47-41-IEO2016&c type=linechart&sid=Reference-d021916a.47-41-IEO2016&sourcekey=0.

69 *In 2011, Prime Minister:* Tusk also anticipated that Poland could be "dependent mainly on our own gas" by 2035. "Commercial Extraction of Shale Gas Possible Already in 2014," Prime Minister's Office of Poland, September 18, 2011, https://www.premier.gov.pl/en/news/news/commercial-ex traction-of-shale-gas-possible-already-in-2014.html.

70 *Shale gas production boomed:* The 2011 IEA report anticipated that, in 2035, the United States would be producing 779 billion cubic meters (27.5 tcf); in fact, the EIA reports that in 2015, U.S. natural gas production had already reached 27.2 tcf. International Energy Agency, *World Energy Outlook 2011 Special Report: Are We Entering a Golden Age of Gas?*, 27; "Table: Natural Gas Supply, Disposition, and Prices," U.S. Energy Information Administration, 2017, www.eia.gov/outlooks/aeo/data/browser /#/?id=13-AEO2017&cases=ref2017&sourcekey=0.

70 *Natural gas seemed nearly ubiquitous:* Timothy Puko, "Natural Gas Falls to All-Time Inflation-Adjusted Low," *Wall Street Journal*, December 16, 2015, www.wsj.com/articles/natural-gas-dips-below-all-time-inflation-adjust ed-low-1450280571.

70 *In mid-2016, the spot price:* In October 2005, the Henry Hub spot price was $13.42 per mmbtu; in May 2016, it was $1.92 per mmbtu. "Henry Hub Natural Gas Spot Price," U.S. Energy Information Administration, November 2, 2016, www.eia.gov/dnav/ng/hist/rngwhhdM.htm.

70 *In 2016, largely as a result:* "Electricity Data Browser: Net Generation for All Sectors, Annual," U.S. Energy Information Administration, https://www.eia.gov/electricity/data/browser/#/topic/0?agg=2,0,1& fuel=vvg&geo=g&sec=g&linechart=ELEC.GEN.ALL-US-99.A

~ELEC.GEN.COW-US-99.A~ELEC.GEN.NG-US-99.A&column
chart=ELEC.GEN.ALL-US-99.A&map=ELEC.GEN.ALL-US-99.A&
freq=A&start=2008&end=2016&ctype=linechart<ype=pin&rtype=s&pin
=&rse=0&maptype=0

70 *Additionally, as a resurgence of cheap coal:* See Guy Chazan and Gerrit
    Wiesmann, "Shale Gas Boom Sparks EU Coal Revival," *Financial Times*,
    February 3, 2013, https://www.ft.com/content/d41c2e8a-6c8d-11e2-953f
    -00144feab49a.

70 *In the first four years:* China and Argentina did reach modest levels of
    commercial production in 2015. Faouzi Aloulou, "Shale Gas and Tight
    Oil Are Commercially Produced in Just Four Countries," U.S. Energy
    Information Administration, February 13, 2015, www.eia.gov/todayin
    energy/detail.cfm?id=19991. According to the EIA, China drilled approx-
    imately six hundred shale gas wells and produced 0.5 bcf/d a day of shale
    gas from 2010 to 2015. Faouzi Aloulou and Victoria Zaretskaya, "Shale
    Gas Production Drives World Natural Gas Production Growth," U.S. En-
    ergy Information Administration, August 15, 2016, www.eia.gov/todayin
    energy/detail.php?id=27512.

70 *Just two years after rolling out:* Chen Aizhu, Judy Hua, and Charlie
    Zhu, "China Finds Shale Gas Challenging, Halves 2020 Output Target,"
    Reuters, August 7, 2014, www.reuters.com/article/2014/08/07/us-china
    -shale-target-idUSKBN0G71FX20140807.

71 *Instead, in 2016, talk of an enduring:* The IEA's June 2016 Medium Term
    Gas Outlook predicted oversupplied natural gas markets for the next
    several years. International Energy Agency, *Medium-Term Gas Market
    Report 2016* (Paris: OECD Publishing, 2016), http://www.iea.org/book
    shop/721-Medium-Term_Gas_Market_Report_2016; the International
    Gas Union, in 2017, wrote of a "looming LNG supply glut." International
    Gas Union, *IGU World LNG Report: 2017 Edition* (Barcelona: Interna-
    tional Gas Union, 2017), 32 and 46, http://www.igu.org/sites/default
    /files/103419-World_IGU_Report_no%20crops.pdf.

71 *Looking ahead, natural gas:* See International Energy Agency, *World
    Energy Outlook 2016* (Paris: OECD Publishing, 2016), 64; "Table: De-
    livered energy consumption by end-use sector and fuel," U.S. Energy
    Information Administration, May, 2015, https://www.eia.gov/outlooks
    /aeo/data/browser/#/?id=15-IEO2016&region=4-0&cases=Reference
    &start=2010&end=2040&f=A&linechart=~Reference-d021916a.60-1

5-IEO2016.4-0~Reference-d021916a.57-15-IEO2016.4-0~Reference
-d021916a.56-15-IEO2016.4-0~Reference-d021916a.55-15-IEO2016.4
-0&map=&ctype=linechart&chartindexed=0&sid=Reference
-d021916a.55-15-IEO2016.4-0~Reference-d021916a.56-15-IEO2016.4
-0~Reference-d021916a.57-15-IEO2016.4-0~Reference-d021916a.60-15
-IEO2016.4-0&sourcekey=0.

71 *Moreover, even in scenarios:* The IEA's 450 Scenario sees the world keeping emissions below the amount required to be consistent with limiting a rise in global temperatures to 2 degrees Celsius—and increasing
its consumption of natural gas out to 2040. In contrast, a new joint scenario by the IEA and the International Renewable Energy Agency has
greater climate ambition; it sees increases in natural gas consumption
rise until 2030, and then begin to decline thereafter. International Energy
Agency, *World Energy Outlook 2016* (Paris: OECD Publishing, 2016),
64; International Energy Agency and International Renewable Energy
Agency, *Perspectives for the Energy Transition: Investment Needs for a
Low-Carbon Energy System* (Paris: OECD/IEA and IRENA Publishing,
2017), 57, http://www.irena.org/DocumentDownloads/Publications/Per
spectives_for_the_Energy_Transition_2017.pdf.

71 *Either of two policy extremes:* See, for instance, BP p.l.c., *BP Energy Outlook: 2017 edition* (London: BP, 2017), 82-85, www.bp.com/content
/dam/bp/pdf/energy-economics/energy-outlook-2017/bp-energy-out
look-2017.pdf.

72 *On a hazy Sunday:* "First LNG Shipment Imminent as Tanker Docks
at US Cheniere's Sabine Pass," Reuters, February 21, 2016, www.reuters
.com/article/cheniere-energy-lng-idUSL3N1611GE.

72 *Natural gas, which had been cooled:* "Watch Cheniere Energy Ship First
Ever Lower 48 LNG Cargo," YouTube, 1:09, posted by "Nikkiso Cryo,"
April 25, 2016, https://www.youtube.com/watch?v=uXWFxhCgkGI.

72 *A few days later:* Jacob Gronholt-Pedersen, "U.S. Exports First Shale Gas
as LNG Tanker Sails from Sabine Pass Terminal," Reuters, February 24,
2016, www.reuters.com/article/us-shale-export-idUSKCN0VY08B.

72 *For Cheniere, the American company:* The celebration was real, if qualified, given the narrowing window between regional natural gas prices
(which limits the potential for arbitrage) and with global enthusiasm for
long-term contracts ebbing. See "Cheniere Aims to Export Disruption to
Global LNG Market," Houston Business Journal video, 04:01, April 25,

2016, www.bizjournals.com/houston/video/Axd2ozMzE6tKbjUU6SDilS -jCZZ2WWis.

73 *By 2016, shale accounted:* See "Oil and Gas: Crude Oil: Lower 48 Average Wellhead Price," *Annual Energy Outlook 2017*, U.S. EIA, https:// www.eia.gov/outlooks/aeo/data/browser/#/?id=14-AEO2017&region=0-0&cases=ref2017&start=2015&end=2050&f=A&linechart =ref2017-d120816a.34-14-AEO2017~ref2017-d120816a.37-14-AEO 2017&ctype=linechart&sid=ref2017-d120816a.37-14-AEO2017 ~ref2017-d120816a.34-14-AEO2017&sourcekey=0.

73 *The U.S. EIA has already assessed:* "World Shale Resource Assessments," U.S. Energy Information Administration, September 24, 2015, https:// www.eia.gov/analysis/studies/worldshalegas/.

73 *According to those initial estimates:* See Table 4.2, International Energy Agency, *World Energy Outlook 2016* (Paris: OECD Publishing, 2016), 176.

73 *If we include other types:* International Energy Agency, *World Energy Outlook 2015*, 233; U.S. Energy Information Administration, *Annual Energy Outlook 2016*, www.eia.gov/forecasts.aeo/data/browser/#/?id=14-AEO2016&region=0-0&cases.

73 *Given the difficulties of replicating:* Ibid.

73 *By 2040, the United States will still:* Ibid.

73 *By some assessments, by 2035:* BP p.l.c., *BP Energy Outlook: 2016 edition* (London: BP, 2016), 32–33, www.bp.com/content/dam/bp/pdf/energy -economics/energy-outlook-2016/bp-energy-outlook-2016.pdf.

73 *However, at the age of fourteen:* Dennis Allen Jacobs and Karen Anita Branden, *From McEnergy to EcoEnergy: America's Transition to Sustainable Energy* (Pittsburgh: Whitmore Publishing, 2008), 64.

73 *Yet it was not until nearly a century:* The significant lag between the first commercial plant and the *Methane Pioneer* was in part due to the failure of the Cleveland project and an explosion costing 130 lives. Malcolm Abbott, *The Economics of the Gas Supply Industry* (London: Routledge, 2016), 141.

74 *Despite the success of this shipment:* "Strategy for LNG Market Development: Challenges and Countermeasures Toward the Creation of Flexible LNG Market and LNG Trading Hub in Japan," presentation, Ministry of Economy, Trade and Industry, Tokyo, Japan, May 2, 2016, 2, http://www .meti.go.jp/english/press/2016/pdf/0502_01a.pdf.

74 *Now, according to Leslie Palti-Guzman:* Leslie Palti-Guzman, in-person conversation with the author, Washington, DC, July 7, 2016.

74 *Countries such as Egypt, Pakistan, and Turkey:* International Gas Union, *IGU World Gas LNG Report: 2017 edition* (Barcelona: International Gas Union, April 2017), 7, www.igu.org/sites/default/files/103419-World_ IGU_Report_no%20crops.pdf. Turkey's FSRU came into operation in 2016. *Daily Sabah,* "Turkey's First FSRU Opened, Ready to Boost Energy Supply Security by 20M cbm Daily," *Daily Sabah Energy,* December 23, 2016, https://www.dailysabah.com/energy/2016/12/24/turkeys-first-fsru -opened-ready-to-boost-energy-supply-security-by-20m-cbm-daily.

75 *In an absolute sense:* "World LNG Trade More Than Doubles, from About 12 Tcf in 2012 to 29 Tcf in 2040," Chapter 3, "Natural Gas," U.S. Energy Information Administration, May 11, 2016, www.eia.gov/forecasts/ieo /nat_gas.cfm.

75 *Remarkably, nearly half this growth:* BP p.l.c., *BP Energy Outlook: 2016 edition* (London: BP, 2016), 35, www.bp.com/content/dam/bp/pdf/energy -economics/energy-outlook-2016/bp-energy-outlook-2016.pdf.

75 *According to BP, LNG volumes:* See BP p.l.c., *BP Energy Outlook: 2017 edition* (London: BP, 2017), 57, www.bp.com/content/dam/bp/pdf/energy -economics/energy-outlook-2017/bp-energy-outlook-2017.pdf.

75 *The advent of U.S. LNG:* Holly Morrow writes that "Australia's CBM industry has experienced such robust growth that CBM-based LNG projects are slated to make up nearly 30 percent of Australia's LNG exports by 2020, even amid strong growth in conventional gas exports." Holly Morrow, "Unconventional Gas: Lessons Learned from Around the World," Geopolitics of Energy Project, Belfer Center for Science and International Affairs, Harvard University, Cambridge, MA, October 2014, http://belfercenter.ksg.harvard.edu/files/Unconventional%20Gas-%20 Lessons%20Learned%20from%20Around%20the%20World.pdf.

76 *Assuming these costs are between:* Jurgen Weiss et al., *LNG and Renewable Power: Risk and Opportunity in a Changing World* (Cambridge, MA: The Brattle Group, 2016), 8, http://www.brattle.com/system/pub lications/pdfs/000/005/249/original/LNG_and_Renewable_Power _-_Risk_and_Opportunity_in_a_Changing_World.pdf?1452804455.

76 *If all the applications:* "Long Term Applications Received by DOE/FE to Export Domestically Produced LNG from the Lower-48 States (as of July 2016)," U.S. Department of Energy, http://energy.gov/sites/prod

/files/2016/07/F33/Summary%20of%20LNG%20Export%20Applica
tions.pdf.

76 *that is nearly twice:* In 2016, global LNG trade was 258 million tons. International Gas Union, *IGU World LNG Report: 2017 edition*, 7. Japan LNG
imports equaled 87.5 million tons in 2013. "UPDATE2-Japan's 2013 LNG
imports hit record highs on nuclear woes," Reuters, January 27, 2014,
www.renters.com/article/energy-japan-mof-idUSL3N0L103N20140127.

76 *Yet the volumes are set:* According to the U.S. EIA, U.S. LNG exports could
be 7.8 bcf a day in 2020 and 12 bcf a day in 2035. See "Table: Natural Gas Imports and Exports," AEO2017, U.S. Energy Information Administration,
https://www.eia.gov/outlooks/aeo/data/browser/#/?id=76-AEO2017&re
gion-0-0&cases=ref2017&start=2015&end=2050&f=Q&linechart
=ref2017-d120816a.3-76-AEO2017~~ref2017-d120816a.16-76
-AEO2017&ctype=linechart&sourcekey=0.

76 *According to the U.S. EIA:* "Chapter 3: Natural Gas," *International Energy
Outlook 2016*, U.S. Energy Information Administration, https://www.eia
.gov/outlooks/ieo/nat_gas.php.

77 *The price to import natural gas:* "European Union Natural Gas Import
Price," YCharts, November 3, 2016, https://ycharts.com/indicators/europe
_natural_gas_price.

77 *In Asia, where roughly 70 percent:* "WoodMac: 2014 Asian LNG Demand
Much Lower Than Expected," *LNG World News*, January 5, 2015, www
.lngworldnews.com/woodmac-2014-asian-lng-demand-much-lower
-than-expected/.

77 *In Japan, the price of importing LNG:* "Japan Liquefied Natural Gas Import Price," YCharts, November 4, 2016, https://ycharts.com/indicators
/japan_liquefied_natural_gas_import_price.

78 *Their arrival gave European utilities:* Norway and Russia collectively
supplied nearly 70 percent of Europe's natural gas imports. BP p.l.c., *BP
Statistical Review of World Energy 2012* (BP, June 2012), https://www
.laohamutuk.org/DVD/docs/BPWER2012report.pdf.

78 *Tying the price of natural gas:* De Pous held this ministerial position from
1959 to 1963. The architect of the first European cross-border trade in
natural gas, de Pous supposedly laid out the framework for natural gas
trade in a ten-page policy document and with only half a day's debate in
the Dutch parliament. Part of his structure involved indexing the price
of natural gas to the price of oil, a practice that was originally known as

"Nota de Pous." For more on the history of oil indexation, see Ludovico Grandi, "European Gas Markets: From Oil Indexation Prices to Spot Prices?" *Energy Brains—Energy Analysis,* June 2014, www.energybrains .org/docs/EA/EnergyBrains_EA_NatGasPricing_LG_2014.pdf.

78 *The simplicity of this linkage:* The simplicity of de Pous's proposal was also appealing given that this pricing mechanism was expected to be short-lived, given anticipations that the gas trade would soon be made obsolete by the rise of nuclear power. Nuclear power plants were just beginning to spring up and were considered to be safe and clean alternatives. For further information, see Henk Kamp, "The Bright Past and Challenging Future of Natural Gas," speech, symposium, "The Bright Past and Challenging Future of Natural Gas," October 3, 2013, Government of the Nether-lands, https://www.government.nl/documents/speeches/2013/10/03/the -bright-past-and-challenging-future-of-natural-gas.

78 *From 2005 to 2012, the amount of natural gas:* According to the Inter-national Gas Union, gas-on-gas pricing in Europe increased around 7 percent in 2005 to more than 60 percent in 2015 while oil indexation decreased from more than 80 to below 40 percent during the same pe-riod. International Gas Union, *IGU World LNG Report: 2017 edition,* 17.

78 *In Asia, the shift was less dramatic:* According to the IEA, in 2005, in 95 percent of gas trade in the Asia-Pacific, the price was indexed to oil; in 2010, this number was 88 percent. Warner ten Kate, László Varró, and Anne-Sophie Corbeau, *Developing a Natural Gas Trading Hub in Asia: Obstacles and Opportunities* (Paris: OECD Publishing, 2013), www.iea .org/media/freepublications/AsianGasHub_WEB.pdf. Also see "Figure 3-26, Global LNG trade by contract type, 2010 and 2014," *International Energy Outlook 2016,* U.S. Energy Information Administration, www.eia .gov/outlooks/ieo/excel/figure3-26_date.xls.

79 *The percentage of LNG traded:* International Gas Union, *IGU World LNG Report, 2017 edition,* 15–20.

79 *Regional trading hubs are slowly beginning:* Victoria Zaretskaya and Scott Bradley, "Natural Gas Prices in Asia Mainly Linked to Crude Oil, but Use of Spot Indexes Increases," U.S. Energy Information Administration, September 29, 2015, www.eia.gov/todayinenergy/detail.cfm?id=23132. Also see Jonathan Stern and Howard Rogers, "Challenges to JCC Pricing in Asian LNG Markets," *The Oxford Institute for Energy Studies,* 2014; Warner Ten Kate, Lászlo Varró, and Anne-Sophie Corbeau, *Developing*

a Natural Gas Trading Hub in Asia: Obstacles and Opportunities (Paris: International Energy Agency, 2013).

79 *First, the price of the overwhelming majority:* Eighty percent of U.S. LNG volumes for projects under construction as of 2015 were linked to Henry Hub or under a hybrid pricing mechanism linked to Henry Hub. Victoria Zaretskaya and Scott Bradley, "Natural Gas Prices in Asia Mainly Linked to Crude Oil, but Use of Spot Indexes Increases," U.S. Energy Information Administration, September 29, 2015, www.eia.gov/todayinenergy/detail .cfm?id=23132.

79 *Jonathan Stern, an Oxford University scholar:* James Nicoli, "The Emergence of an Asia-Pacific LNG Trading Hub," *Petroleum Economist*, April 12, 2016, www.petroleum-economist.com/articles/midstream-downstream /lng/2016/the-emergence-of-an-asia-pacific-lng-trading-hub.

79 *Moreover, increased LNG trade:* See Kenneth B. Medlock III, "Global Natural Gas Markets: Recent Trends and Emerging Fundamentals," presentation, Harvard Energy Conference, Harvard Business School, November 14, 2013, 9, http://bakerinstitute.org/research/global-natural -gas-markets-recent-trends-and-emerging-fundamentals/.

Four: America's Unrequited Love

84 *Speaking as a candidate in May 2016:* Andrew Restuccia, "Trump Calls for 'Complete American Energy Independence,'" *Politico*, May 26, 2016, http://www.politico.com/story/2016/05/donald-trump-energy-drilling -fossil-fuels-223628.

84 *only two months into his presidency:* Donald J. Trump, "Presidential Executive Order on Promoting Energy Independence and Economic Growth," The White House Office of the Press Secretary, March 28, 2017, https:// www.whitehouse.gov/the-press-office/2017/03/28/presidential-execu tive-order-promoting-energy-independence-and-economi-1.

84 *In 2010, President Barack Obama exhorted:* "President Obama's Oval Office Address on the BP Oil Spill: 'A Faith in the Future that Sustains us as a People,'" Barack Obama White House Archives, June 16, 2010, https:// obamawhitehouse.archives.gov/blog/2010/06/16/president-obamas -oval-office-address-bp-oil-spill-a-faith-future-sustains-us-a-peopl.

84 *his predecessor, George W. Bush, urged:* President George W. Bush, "State

of the Union Address by the President" (speech, United States Capitol, Washington, D.C., January 31, 2006), https://georgewbush-whitehouse .archives.gov/stateoftheunion/2006/.

85 *In 1988, Vice President George H. W. Bush:* George Bush, "Address Accepting the Presidential Nomination" (speech, Republican National Convention, New Orleans, LA, August 18, 1988), http://www.presidency .ucsb.edu/ws/?pid=25955.

85 *while President Jimmy Carter:* Jimmy Carter, "Address to the Nation on Energy and National Goals: 'The Malaise Speech'" (speech, Oval Office, The White House, Washington, D.C., July 15, 1979), The American Presidency Project, http://www.presidency.ucsb.edu/ws/?pid=32596.

85 *Just four years earlier:* Gerald Ford, "State of the Union Address by the President," (speech, United States Capitol, Washington, DC, January 15, 1957), www.fordlibrarymuseum.gov/LIBRARY/speeches/750028.htm.

85 *In 1973, his predecessor, Richard Nixon:* Charles Homans, "The Best-Laid Plans," *Foreign Policy*, January 3, 2012, http://foreignpolicy .com/2012/01/03/the-best-laid-plans/?wp_login_redirect=0.

85 *In the 2000s, concerns arose:* In 2003 testimony before Congress, Federal Reserve chairman Alan Greenspan cautioned that "we are not apt to return to earlier periods of relative abundance and low prices anytime soon" and therefore urged Congress to prepare for more LNG imports. *Natural Gas Supply, Before the Committee on Energy and Natural Resources*, 108th Cong. (July 10, 2003) (testimony of Alan Greenspan, Chairman of the Federal Reserve), www.federalreserve.gov/boarddocs/testimony /2003/20030710/default.htm.

85 *Of the ten U.S. recessions:* James D. Hamilton, "Oil and the Macroeconomy," working paper, Department of Economics, University of California, San Diego, La Jolla, August 25, 2005, 1, www.econweb.ucsd.edu /~jhamilto/JDH_palgrave_oil.pdf.

85 *Although economists disagree:* According to the more conservative linear VAR model of a U.S. Federal Reserve study examining seven oil price shocks since 1979, the average cumulative effect on GDP over a horizon of eight quarters following the price shock (after controlling for the effects of other macroeconomic indicators) was a 0.91 percent decrease in 1979 and a 0.74 percent decrease in 2007. Lutz Kilian and Robert J. Vigfusson, "The Role of Oil Price Shocks in Causing U.S. Recessions," International Finance

discussion paper 1114, Board of Governors of the Federal Reserve System, August 2014, www.federalreserve.gov/pubs/ifdp/2014/1114/ifdp1114.pdf.

85 *Secretary of State Condoleezza:* Rice is quoted in Gal Luft, "Dependence on Middle East Energy and Its Impact on Global Security," in *Energy and Environmental Challenges to Security*, eds. Stephen Stec and Besnik Baraj (Springer Science & Business Media, 2006), 197.

86 *That same year,* New York Times *columnist:* Thomas L. Friedman, "Fill 'Er Up with Dictators," *New York Times*, September 27, 2006, www.ny times.com/2006/09/27/opinion/27friedman.html.

86 *The Energy Policy and Conservation Act:* "The Energy Policy and Conservation Act (P.L. 94-163, 42 U.S.C. 6201)," *William & Mary Environmental Law and Policy Review* 1, no. 2 (1976), http://scholarship.law.wm.edu/cgi /viewcontent.cgi?article=1488&context=wmelpr.

86 *It also sought to increase:* One of the goals of the Energy Policy and Conservation Act of 1975 was "increasing domestic energy production and supply." Ibid.

86 *Tax incentives for the increased production:* Molly F. Sherlock and Jeffrey M. Stupak, *Energy Tax Incentives: Measuring Value Across Different Types of Energy Resources* (CRS Report No. R41953) (Washington, DC: Congressional Research Service, March 19, 2015), 12, https://www.fas.org /sgp/crs/misc/R41953.pdf.

86 *These measures paid off temporarily:* Net imports decreased from 8.6 mnb/d in 1977 to 4.3 mnb/d in 1985. "U.S. Net Imports of Crude Oil and Petroleum Products," U.S. Energy Information Administration, October 31, 2016, www.eia.gov/dnav/pet/hist/LeafHandler.ashx?n=pet&s=mttntus2&f=a; "U.S. Product Supplied of Crude Oil and Petroleum Products," U.S. Energy Information Administration, www.eia.gov/dnav/pet/hist/LeafHandler.ashx ?n=PET&s=MTTUPUS2&f=A.

86 *But such policies could not keep up:* "U.S. Field Production of Crude Oil," U.S. Energy Information Administration, April 28, 2017, https://www .eia.gov/dnav/pet/hist/LeafHandler.ashx?n=pet&s=mcrfpus2&f=m.

86 *By 2007, American imports:* "U.S. Net Imports of Crude Oil and Petroleum Products," U.S. Energy Information Administration, October 31, 2016, www.eia.gov/dnav/pet/hist/LeafHandler.ashx?n=pet&s=mttntus2&f=a.

86 *Nearly two-thirds of America's:* To be precise, 59 percent of U.S. consumption of oil was met by imports in 2007. "Annual Energy Outlook 2010 Table:

Liquid Fuels Supply and Disposition," U.S. Energy Information Administration, https://www.eia.gov/outlooks/aeo/data/browser/#/?id=11-AEO 2010&cases=aeo2010r&sourcekey=0.

87 *Co-led by FedEx chairman:* Matthew L. Wald, "Executives Urge Action to Cut Dependence on Foreign Oil," *New York Times,* December 13, 2006, www.nytimes.com/2006/12/13/business/worldbusiness/13energy.html.

87 *Carville, a Democratic strategist:* See Thomas L. Friedman, "The Energy Mandate," *New York Times,* October 13, 2006, www.nytimes.com/2006 /10/13/opinion/13friedman.html.

87 *Enter serendipity:* "Serendipity," *English Oxford Living Dictionaries,* 2017, https://en.oxforddictionaries.com/definition/serendipity.

87 *As noted in earlier chapters:* U.S. crude oil production was 5 mnb/d in 2008; it reached 9.6 mnb/d in 2015. "U.S. Field Production of Crude Oil," U.S. Energy Information Administration, October 31, 2016, www.eia.gov/ dnav/pet/hist/LeafHandler.ashx?n=pet&s=mcrfpus2&f=a.

88 *Natural gas production increased:* The increase in U.S. natural gas production of 40 percent refers to all forms of natural gas. "U.S. Natural Gas Gross Withdrawals," U.S. Energy Information Administration, October 31, 2016, www.eia.gov/dnav/ng/hist/n9010us2m.htm.

88 *The United States became:* When measured as petroleum and other liquids, U.S. production surpassed that of Saudi Arabia in 2013, although in terms of just crude oil, Saudi Arabia was still a bigger producer. "Total Petroleum and Other Liquids Production 2016," U.S. Energy Information Administration: Beta, 2016, https://www.eia.gov/beta/international /data/browser/#/?pa=00000000000000000000000000000000g&c= ruvvvvvfvtvnvv1urvvvvfvvvvvvfvvvou20evvvvvvvvvnvvuvo&ct=0&tl _id=5-A&vs=INTL.53-1-AFG-TBPD.A&vo=0&v=H&start=1980 &end=2016; "Production of Crude Oil Including Lease Condensate 2016," U.S. Energy Information Administration: Beta, 2016, https://www .eia.gov/cfapps/ipdbproject/iedindex3.cfm?tid=5&pid=57&aid=1&cid =SA,US,&syid=2005&eyid=2014&unit=TBPD. For numbers of U.S. and Russian natural gas production, see "Gross Natural Gas Production 2014," U.S. Energy Information Administration: Beta, https://www.eia.gov /beta/international/data/browser/#/?pa=g0q&c=0000000000000000000 0000000000000000000000400000002&ct=0&tl_id=3002-A&vs=INTL.3-1 -RUS-BCF.A&vo=0&v=H&start=1980&end=2014.

88 *Given a further leg-up:* U.S. total petroleum consumption came down 8.9 percent from 2005 to 2013. Consumption did however begin to rise in response to lower prices in 2014 and 2015. "Total Petroleum Consumption 2015," U.S. Energy Information Administration, 2015, www.eia.gov /cfapps/ipdbproject/iedindex3.cfm?tid=5&pid=5&aid=2&cid=r1,&sy id=2005&eyid=2013&unit=TBPD.

88 *these production increases reduced:* "Table 3.1 Petroleum Overview," U.S. Energy Information Administration, https://www.eia.gov/totalenergy /data/browser/?tbl=T03.01#/?f=A&start=1949&end=2016&charted =12-15; "Table 3.3a Petroleum Trade Overview," U.S. Energy Information Administration, https://www.eia.gov/totalenergy/data/browser/?tbl=T03 .03A#/?f=A; "Table 3.3b Petroleum Trade: Imports and Exports," U.S. Energy Information Administration, https://www.eia.gov/totalenergy/data /browser/?tbl=T03.03B#/?f=A.

88 *The specter of a growing:* See, for instance, Gary J. Schmitt, "Energy Conundrums: Natural Gas: The Next Energy Crisis?" *Issues in Science and Technology* 22, no. 4 (Summer 2006), http://issues.org/22-4/schmitt-3/.

88 *Demand for natural gas continued:* "U.S. Liquefied Natural Gas Imports," U.S. Energy Information Administration, October 31, 2016, www.eia .gov/dnav/ng/hist/n9103us2m.htm.

88 *There is even less uncertainty:* The U.S. Energy Information Administration forecasts a modest but steady upward curve in U.S. natural gas production out to 2040. "Table: World Total Natural Gas Production by Region," U.S. Energy Information Administration, www.eia.gov/outlooks/aeo/data /browser/#/?id=45-IEO2013&region=0-0&cases=Reference&start=200 9&end=2039&f=A&linechart=Reference-d041117.1-45_IEO2013~~Re ference-d041117.3-45-IEO2013&ctype=linechart&sid=&sourcekey=0.

90 *The scale and swiftness with which:* One might draw a comparison with Pyongyang's acquisition of a nuclear bomb, which propelled North Korea from a mere menace to a country absorbing the attention of China, Russia, and the United States. Or one could point to the discovery of oil or the development of desalination technology without which the establishment of large cities in the Arabian Peninsula would be impossible and today's Saudi Arabia unimaginable. Or perhaps one could highlight the construction of the Suez and Panama Canals, which in the span of a decade solidified Egypt and Panama as strategic pieces of real estate, if not great powers.

90   *In 2005, Andre Hoth:* "Singing for Change: A New Anthem for Energy Independence," PR Newswire, April 5, 2006, www.prnewswire.com /news-releases/singing-for-change-a-new-anthem-for-energy-indepen dence-55953472.html.

90   *But a world in which:* Energy independence has also inspired poetry. Sidi J. Mahtrow, "Energy Independence," PoemHunter.com, July 30, 2008, www.poemhunter.com/poem/energy-independence-2/.

> Then now is the time to act
> And put the Nation back on track
> To be energy independent from all
> Those that seek our downfall!

91   *It is true that virtually:* "Total Biofuels Consumption 2012," U.S. Energy Information Administration, 2012, www.eia.gov/cfapps/ipdbproject /IEDIndex3.cfm?tid=79&pid=79&aid=2#; "Total Primary Energy Pro-duction 2013," U.S. Energy Information Administration, 2013, www.eia .gov/cfapps/ipdbproject/IEDIndex3.cfm?tid=1&pid=1&aid=24.

91   *But oil self-sufficiency:* Recent forecasts suggest that, under some circum-stances, the United States could in fact become a net oil exporter in the 2030s. According to the IEA's *World Energy Outlook of 2016*, if remaining recoverable reserves are larger than currently estimated, the United States could become a net exporter of oil by the early 2030s. International En-ergy Agency, *World Energy Outlook 2016*, 134.

91   *While U.S. oil imports:* The United States is already a net exporter of refined petroleum products. Data can be found at "Liquid Fuels: Net Product Im-ports," U.S. Energy Information Administration, 2016, www.eia.gov/fore casts/aeo; "Table: Petroleum and Other Liquids Supply and Disposition," U.S. Energy Information Administration, *Annual Energy Outlook, 2017*, https://www.eia.gov/outlooks/aeo; International Energy Agency, *World Energy Outlook, 2016*, 134; BP, *BP Energy Outlook, 2017*, http://www .bp.com/content/dam/bp/pdf/energy-economics/energy-outlook-2017 /bp-energy-outlook-2017.pdf.

91   *In the face of cheap oil:* John M. Rothgeb Jr., *U.S. Trade Policy: Balanc-ing Economic Dreams and Political Realities* (Washington, DC: CQ Press, 2001), 146, footnote 3.

91   *Ultimately, Eisenhower was swayed:* Ibid., 146–47.

91 *After failed attempts to induce:* Dwight D. Eisenhower, "Proclamation 3279—Adjusting Imports of Petroleum and Petroleum Products into the United States" (March 10, 1959), The American Presidency Project, www .presidency.ucsb.edu/ws/?pid=107378; Dwight D. Eisenhower, "Statement by the President Upon Signing Proclamation Governing Petroleum Imports," The American Presidency Project, March 10, 1959, http://www .presidency.ucsb.edu/ws/?pid=11676.

91 *He stressed that it would:* See Richard A. Melanson and David Mayers, eds., *Reevaluating Eisenhower: American Foreign Policy in the Fifties* (Champaign: University of Illinois Press, November 1, 1988), 157–58; and Vito Stagliano, *A Policy of Discontent: The Making of a National Energy Strategy* (Nashua, NH: PennWell Publishing, July 15, 2011), 12.

92 *A task force initiated by:* Stagliano, *A Policy of Discontent: The Making of a National Energy Strategy*, 13.

92 *To be sure, the quota:* "U.S. Field Production of Crude Oil," U.S. Energy Information Administration, October 31, 2016, www.eia.gov/dnav/pet /hist/LeafHandler.ashx?n=PET&s=MCRFPUS2&f=A.

92 *While their stated objective at the time:* For more, see Daniel Yergin, "OPEC and the Surge Pot," in *The Prize: The Epic Quest for Oil, Money and Power* (New York: Free Press, 2008), 502–5.

93 *Instead, because there were:* Tim Worstall, "The Economic Effects of Lifting the Crude Export Ban: WTI/Brent Spread Disappears," *Forbes*, December 28, 2015, https://www.forbes.com/sites/timworstall/2015/12/28 /the-economic-effects-of-lifting-the-crude-export-ban-wtibrent-spread -disappears/#5733db8946a3; "WTI-Brent Crude Oil Price Spread has Reached Unseen Levels," U.S. Energy Information Administration, February 28, 2011, https://www.eia.gov/todayinenergy/detail.php?id=290.

93 *John Hess, CEO of:* John Hess, "The Oil Export Ban: A Relic of the 1970s," *Wall Street Journal*, April 24, 2015, www.wsj.com/articles/the-oil-export -ban-a-relic-of-the-1970s-1429913717.

93 *Scholars and institutes from across:* See, for example, Charles K. Ebinger and Heather L. Greenley, "Lifting the U.S. ban on crude oil exports: Let's use data over ideology," Brookings, September 16, 2015, https://www .brookings.edu/blog/planetpolicy/2015/09/16/lifting-the-u-s-ban-on -crude-oil-exports-lets-use-data-over-ideology/; "Effects of Lifting the Crude Oil Export Ban," Bipartisan Policy Center, https://bipartisanpolicy .org/wp-content/uploads/2015/09/BPC-Energy-Crude-Oil-Export-Ban

-Gas-Prices.pdf; Nicolas Loris, "Time to Lift the Ban on Crude Oil Exports," The Heritage Foundation, May 15, 2014, http://www.heritage.org /environment/report/time-lift-the-ban-crude-oil-exports; Jason Bordoff and Trevor Houser, "Navigating the U.S. Oil Export Debate," Center on Global Energy Policy, School of International and Public Affairs, Columbia University, New York, NY, January 2015, http://energypolicy.colum bia.edu/sites/default/files/energy/Navigating%20the%20US%20Oil%20 Export%20Debate_January%202015.pdf.

93 *If one standardizes all energy sources:* The U.S. EIA could see the United States becoming a net exporter of energy as early as 2026. BP's Energy Outlook forecasts that the United States could reach this status in 2023. See "Table: Total Energy Supply, Disposition, and Price Summary," *Annual Energy Outlook 2017*, U.S. Energy Information Administration, https://www.eia.gov/outlooks/aeo/data/browser/#/?id=1-AEO2017 &region=0-0&cases=ref2017&start=2015&end=2050&f=A&linecha rt=ref2017-d120816a.19-1-AEO2017~ref2017-d120816a.25-1-AEO 2017&ctype=linechart&sourcekey=0; "Country Insights: US," BP Global, 2017, http://www.bp.com/en/global/corporate/energy-economics/ener gy-outlook/country-and-regional-insights/us-insights.html.

94 *The United States would likely export:* "U.S. Energy Imports and Exports to Come into Balance for the First Time Since 1950s," U.S. Energy Information Administration, April 15, 2015, https://www.eia.gov/todayin energy/detail.cfm?id=20812.

94 *Various companies and agencies differ:* According to the IEA, if remaining recoverable reserves are larger than currently estimated in its default scenario, the United States could become a net exporter of oil by the early 2030s. BP's Energy Outlook's main case forecasts that the United States will be a net exporter of oil by 2035. International Energy Agency, *World Energy Outlook 2016* (Paris: OECD Publishing, 2016), 134; BP p.l.c., *BP Energy Outlook 2035: January 2017* (London: BP, 2016), http://www.bp .com/content/dam/bp/excel/energy-economics/energy-outlook-2017 /bp-energy-outlook-2017-summary-tables.xlsx.

95 *In the words of a March 14:* "When Is a Bargain Not a Bargain," *New York Times*, March 14, 1909, http://query.nytimes.com/mem/archive-free/pdf ?res=9A03E4D8173EE033A25757C1A9659C946897D6CF.

95 *In its heyday as:* In 2004, Mexico was producing 3.8 mnb/d of oil per day and 1.5 tcf of gas annually. "Gross Natural Gas Production—Mexico,"

U.S. Energy Information Administration, https://www.eia.gov/beta/in
ternational/data/browser/#?iso=MEX&c=000000000000000000000000
0001&ct=0&cy=2013&start=1980&end=2013&ord=SA&v=T&vo=0&s
o=0&io=0&vs=INTL.3-1-MEX-BCF.A&pin=p&pa=g1q&f=A&ug=g&tl
_type=a&tl_id=1-A. "Total Petroleum and Other Liquids Production
—Mexico," U.S. Energy Information Administration: Beta, https://www
.eia.gov/beta/international/data/browser/#/?pa=000000000000000000
0000000000000000vg&c=000000000000000000000000000001&ct=0&tl
_id=5-A&vs=INTL.53-1-MEX-TBPD.A&cy=2014&vo=0&v=T&start=1
980&end=2016.

96 *Despite Mexico's large oil reserves:* "Unfixable Pemex," *Economist*, August 10,
    2013, www.economist.com/news/business/21583253-even-if-government
    -plucks-up-courage-reform-it-pemex-will-be-hard-fix-unfixable.
    From 2005 to 2013, crude oil production declined by 25 percent. "Pro-
    duction of Crude Oil Including Lease Condensate 2015," U.S. En-
    ergy Information Administration, 2015, www.eia.gov/cfapps/ipdb
    project/iedindex3.cfm?tid=5&pid=57&aid=1&cid=r1,&syid=2005&ey
    id=2013&unit=TBPD.

96 *Petróleos Mexicanos, or PEMEX:* "Unfixable Pemex," *Economist*, Au-
    gust 15, 2013.

96 *Over the course of:* Production dropped from 3.8 mnb/d of oil in
    2004 to 3 mnb/d of oil in 2009. "Total Petroleum and Other Liquids
    Production—Mexico," U.S. Energy Information Administration: Beta,
    https://www.eia.gov/beta/international/data/browser/#/?pa=000000000
    0000000000000000000000vg&c=000000000000000000000000000001
    &ct=0&tl_id=5-A&vs=INTL.53-1-MEX-TBPD.A&cy=2014&vo=0&v
    =T&start=1980&end=2016.

96 *A year before President:* Kenneth B. Medlock III and Ronald Soligo,
    "Scenarios for Oil Supply, Demand and Net Exports for Mexico," paper
    prepared for the study, "The Future of Oil in Mexico/El Futuro del Sec-
    tor Petrolero en México," James A. Baker III Institute for Public Policy,
    Rice University, and The Mexican Studies Programme, Nuffield Col-
    lege, Oxford University, April 29, 2011, http://bakerinstitute.org/media
    /files/Research/f1a1ca0e/EF-pub-MedlockSoligoScenarios-04292011
    .pdf.

96 *Galloping domestic use of oil:* Mexican demand for oil more than quadru-
    pled between 1971 and 2010. Rice University, "Mexico Could Become

Oil Importer by 2020 Without New Investment," *Science Daily*, April 29, 2011, www.sciencedaily.com/releases/2011/04/110429095117.htm.

96 *Notwithstanding its significant:* The composition of Mexican LNG imports has changed over time, with Egypt fading as a supplier and Peru and others entering the fray. "Mexico Week: Record Mexican Natural Gas Imports Include Higher Flows from U.S.," U.S. Energy Information Administration, May 16, 2013, www.eia.gov/todayinenergy/detail .cfm?id=11291.

97 *initial surveys, Mexico:* Mexico is believed to hold the tenth largest unconventional oil reserves, amounting to 3 percent of global shale oil and the sixth largest shale gas reserves amounting to 7 percent of shale gas in the world. "World Shale Resource Assessments," U.S. Energy Information Administration, September 24, 2015, www.eia.gov/analysis/studies /worldshalegas/.

97 *Instead of the vague language:* Mexican constitutional reforms require a two-thirds majority vote in Congress and the approval of the majority of the legislatures in Mexican states.

97 *The electricity sector was also reformed:* Jude Webber, "Electricity Reform: What Mexico Deserves," *Financial Times*, February 24, 2015, http://blogs.ft .com/beyond-brics/2015/02/24/electricity-reform-what-mexico-deserves/.

97 *Neither the discovery of a snake:* "Snake on the Loose Fails to Keep Deputies from Energy Laws," *Mexico News Daily*, August 2, 2014, http:// mexiconewsdaily.com/news/snake-loose-fails-keep-deputies-approving -energy-laws/; "Mexican Congress Approves Controversial Oil and Gas Bill," *BBC News*, December 13, 2013, www.bbc.com/news/world-latin -america-25350993.

98 *ExxonMobil, Statoil, Chevron:* See Adrián Lajous, "Mexico's Deepwater Auctions," Center on Global Energy Policy, School of International and Public Affairs, Columbia University, New York, NY, January 9, 2017, http:// energypolicy.columbia.edu/sites/default/files/energy/CGEP_Mexico %E2%80%99s%20Deepwater%20Auctions_Lajous.pdf.

98 *The U.S. EIA suggested that:* Shortly after the reforms were enacted, projections were very optimistic. For example, in 2014, the EIA anticipated that Mexican oil production could increase by as much as 75 percent by 2040. Since that time, the optimism has been tempered, although agencies still anticipate considerable production increases by 2040. See "Energy Reform Could Increase Mexico's Long-Term Oil Production by

75%," U.S. Energy Information Administration, August 25, 2014, https://www.eia.gov/todayinenergy/detail.cfm?id=17691. Also see "AEO2017: International Liquids: Production: Non-OPEC OECD: Mexico and Chile," U.S. Energy Information Administration, 2017, https://www.eia.gov/outlooks/aeo/data/browser/#/?id=19-AEO2017&region=0-0&cases=ref2017&start=2015&end=2050&f=A&linechart=ref2017-d120816a.46-19-AEO2017&ctype=linechart&sid=ref2017-d120816a.46-19-AEO2017&sourcekey=0; International Energy Agency, *Mexico Energy Outlook: World Energy Outlook Special Report* (Paris: OECD Publishing, 2016), 66, https://www.iea.org/publications/freepublications/publication/MexicoEnergyOutlook.pdf.

98 *Mexican natural gas production:* International Energy Agency, *Mexico Energy Outlook: World Energy Outlook Special Report* (Paris: OECD Publishing, 2016), 69, https://www.iea.org/publications/freepublications/publication/MexicoEnergyOutlook.pdf.

98 *Yet, even with such uncertainties:* The IEA projects that Mexican production will hit its nadir below 2 mnb/d around 2020 and then begin to increase as reforms "bear fruit." By 2040, Mexico could be producing 2.4 mnb/d of crude oil, or 3.4 mnb/d of crude and natural gas liquids. International Energy Agency, *Mexico Energy Outlook: World Energy Outlook Special Report* (Paris: OECD Publishing, 2016), 11, www.iea.org/publications/freepublications/publication/MexicoEnergyOutlook.pdf.

98 *A Lemonade Stand With One Customer:* Elizabeth McSheffrey, "Canadian Oil Lobbyists Fear Running a Lemonade Stand with Only One Customer," *National Observer*, June 7, 2016, http://www.nationalobserver.com/2016/06/07/news/canadian-oil-lobbyists-fear-running-lemonade-stand-only-one-customer.

98 *Between 1980 and 2014:* "Total Petroleum and Other Liquids Production 2014," U.S. Energy Information Administration, https://www.eia.gov/beta/international/data/browser/#/?pa=000000000000000000000000000000000vg&c=0000001&ct=0&tl_id=5-A&vs=INTL.53-1-CAN-TBPD.A&cy=2014&vo=0&v=H&start=1980&end=2016; International Energy Agency, *World Energy Outlook 2016*, 136.

99 *Dating advice columns urged:* Tenille Bonoguore, "Still Single? Time to Move West," *Globe and Mail*, April 3, 2009, www.theglobeandmail.com/life/still-single-time-to-move-west/article1078717/.

99 *Demand for housing skyrocketed:* Paul Haavardsrud, "Calgary House Prices: Nothing Stays Up Forever," CBC News, September 15, 2015, http://www.cbc.ca/news/business/calgary-house-prices-nothing-stays-up-forever-1.3226602.

99 *Calgary's annual Stampede festival:* Bill Kaufmann, "Calgary Stampede Ends on a High with Second-Highest Attendance on Record," *Calgary Sun,* July 14, 2014, www.calgarysun.com/2014/07/14/calgary-stampede-ends-on-a-high-with-second-highest-attendance-on-record.

99 *A Canadian trade group:* "Facts and Statistics," Alberta Energy, 2017, www.energy.alberta.ca/oilsands/791.asp; Jeff Gerth, "Canada Builds a Large Oil Estimate on Sand," *New York Times,* June 18, 2003, www.nytimes.com/2003/06/18/business/canada-builds-a-large-oil-estimate-on-sand.html; "Crude Oil Proved Reserves 2016," U.S. Energy Information Administration, 2016, www.eia.gov/cfapps/ipdbproject/iedindex3.cfm?tid=5&pid=57&aid=6&cid=regions&syid=2000&eyid=2015&unit=BB.

99 *Several years later, after:* These new definitions became effective on January 1, 2010. Charles Kraus and Kristi Kasper, "Top Five Things Canadian Issuers Need to Know About the SEC's New Oil and Gas Reporting Requirements," Stikeman Elliott, February 11, 2009, www.stikeman.com/cps/rde/xchg/se-en/hs.xsl/12187.htm.

99 *As long as the price:* For a wider range of possible breakeven prices for Canadian oil sands, see J. Peter Findlay, "The Future of the Canadian Oil Sands" (OIES Paper: WPM 64, The Oxford Institute for Energy Studies, University of Oxford, Oxford, U.K., February 2016), 35–36, https://www.oxfordenergy.org/wpcms/wp-content/uploads/2016/02/The-Future-of-the-Canadian-Oil-Sands-WPM-64.pdf.

99 *As of June 2016:* Canadian Association of Petroleum Producers (CAPP), 2016 *Crude Oil Forecast, Markets & Transportation* (Calgary, Alberta: CAPP, June 2016), 3–4, www.documentcloud.org/documents/3679754-2016-CAPP-Crude-Oil-Forecast-Markets.html. The U.S EIA and IEA also anticipate similar modest growth. See International Energy Agency, *World Energy Outlook* 2016, 136; "Petroleum and Other Liquids Production," *International Energy Outlook 2016,* U.S. Energy Information Administration, https://www.eia.gov/outlooks/aeo/data/browser/#/?id=38-IEO2016&region=0-0&cases=Reference&start=2010&end=2

040&f=A&linechart=Reference-d021916a.1-38-IEO2016~Refer
ence-d021916a.10-38-IEO2016&ctype=linechart&sourcekey=0.

100 *America's shale gas bonanza:* See Robert Tuttle and Dan Murtaugh, "Shale
Oil Growth in U.S. Hurts Canadians as Well as OPEC," Bloomberg, January 20, 2015, www.bloomberg.com/news/articles/2015-01-21/shale-oil
-growth-in-u-s-hurts-canadians-as-well-as-opec.

100 *Nearly three-quarters of Canada's:* To be exact, 73 percent of Canada's oil
production and 47 percent of its gas production was imported by the United
States in 2016. Calculations derived from "2016 Oil Exports Statistics
Summary," National Energy Board, www.neb-one.gc.ca/nrg/sttstc/crdl
ndptr/mpdct/stt/crd/smmr/2016/smmry2016-eng.html; "2016 Natural
Gas Exports and Imports Summary," National Energy Board, www.neb
-one.gc.ca/nrg/sttstc/ntrlgs/rprt/ntrlgssmmr/2016/smmry2016-eng
.html. Joint Organizations Data Initiative, www.jodidb.org/ReportFold
ers/reportFolders.aspx?sCS_referer=&sCS_ChosenLang-en.

100 *Having no other outlet:* From 2006 to 2014 Canada's natural gas production declined by approximately 20 percent. "Gross Natural Gas Production 2014," U.S. Energy Information Administration, 2014, https://
www.eia.gov/beta/international/data/browser/index.cfm#/?pa=g0q
&c=0000001&ct=0&tl_id=3002-A&vs=INTL.3-1-CAN-BCF.A&vo=0&
v=H&start=2007&end=2014.

100 *As of 2016, Canada:* Christopher Adams, "Here are the Major Canadian
Pipelines the Oil Patch Wants Built," *National Observer*, September 22,
2016, http://www.nationalobserver.com/2016/09/22/analysis/here-are
-major-canadian-pipelines-oil-patch-wants-built; Tracy Johnson, "LNG
Exports Begin from U.S. as Canada Sits on Sidelines," *CBC News*, February
25, 2016, www.cbc.ca/news/canada/calgary/will-lng-in-canada-happen
-1.3463008.

100 *North America as a whole:* "Geopolitical Implications of North American
Energy Independence," Global Horizons Service—Risks & Uncertainties Insight, Wood Mackenzie, September 2013, www.woodmacresearch
.com/content/portal/energy/highlights/wk4__13/Wood_Mackenzie_Re
port_Geopolitical_implications_of_North_American_energy_indepen
dence.pdf.

100 *Indeed, the United States:* This assessment is derived from data found at
U.S. Energy Information Administration, *Annual Energy Outlook 2017*,
https://www.eia.gov/outlooks/aeo;"Natural Gas Consumption: OECD:

OECD Americas," *International Energy Outlook 2016*, U.S. Energy Information Administration, https://www.eia.gov/outlooks/ieo.

101 *One recent indication:* The capacity of the Keystone XL Pipeline was 830,000 barrels per day, although 100,000 barrels per day was allocated to move crude from the Bakken in North Dakota southward.

101 *The argument was* not *that Keystone XL:* See Juliet Eilperin and Steven Mufson, "State Department Releases Keystone XL Final Environmental Impact Statement," *Washington Post*, January 31, 2014, www .washingtonpost.com/business/economy/state-to-release-keystones -final-environmental-impact-statement-friday/2014/01/31/3a9bb25c-8a 83-11e3-a5bd-844629433ba3_story.html.

102 *multiple State Department assessments:* For the final State Department environmental impact statement, see United States Department of State Bureau of Oceans and International Environmental and Scientific Affairs, "Final Supplemental Environmental Impact Statement for the Keystone XL Project: Executive Summary," Applicant for Presidential Permit: Trans Canada Keystone Pipeline, LLC, January, 2014, https://keystonepipe line-xl.state.gov/documents/organization/221135.pdf.

102 *President Obama might have:* See Scott Brown, "Prime Minister Justin Trudeau Says Kinder Morgan Pipeline Part of Canada's Climate Plan," *Vancouver Sun*, December 20, 2016, http://vancouversun.com/news/local -news/trudeau-says-kinder-morgan-pipeline-part-of-governments-cli mate-plan.

102 *Yet, President Obama's words:* President Barack Obama, "Statement by the President on the Keystone XL Pipeline," The White House Office of the Press Secretary, November 6, 2015, https://www.whitehouse.gov/the -press-office/2015/11/06/statement-president-keystone-xl-pipeline.

103 *Previous to the experience:* Edward Greenspon et al., "How Obama Shocked Harper as Keystone Frustrator-in-Chief," Bloomberg, April 26, 2014, www.bloomberg.com/news/articles/2014-04-24/how-obama -shocked-harper-as-keystone-frustrator-in-chief.

103 *But after a tense 2011:* Ibid.

103 *Neither Harper nor his cabinet:* Harper expressed his "profound disappointment" to President Obama when he heard of the application's rejection. He told President Obama that Canada will "continue to work to diversify its energy exports." "Harper Builds Oil Link with China After Obama Keystone 'Slap,'" *Bloomberg*, January 25, 2012, www.bloomberg

.com/news/articles/2012-01-25/harper-builds-oil-links-with-china
-after-obama-slap-on-keystone-pipeline. Similarly, Canadian natural
resource minister Joe Oliver told reporters that the "decision by the
Obama administration underlines the importance of diversifying and
expanding our markets, including the growing Asian market." Theoph-
ilis Arigitis and Jeremy Loon, "Obama's Keystone Denial Prompts Canada to Focus on China," *Bloomberg*, January 19, 2012, www.bloomberg.
com/news/articles/2012-01-19/canada-pledges-to-sell-oil-to-asia-after
-obama-rejects-keystone-pipeline.

103 *As former U.S. ambassador:* Gordon Giffin, in-person conversation with
author, Montreal, Canada, November 8, 2014.

103 *Numerous pipeline proposals have:* Alistair MacDonald and Paul
Vieira, "Canada's Own Oil Pipeline Problem," *Wall Street Journal*,
April 19, 2015, www.wsj.com/articles/canadas-own-oil-pipeline-prob
lem-1429479110; Kai Nagata, "Is Northern Gateway Dead?," *The Tyee*,
April 13, 2015, http://thetyee.ca/Opinion/2015/04/13/Northern-Gate
way-Dead/.

104 *Meanwhile, without the full participation:* Given the natural gas wealth in
the United States, the continent could easily still be self-sufficient should
Canada find a way to move more of its natural gas (as opposed to its oil)
to Asian markets.

104 *In 2013, shortly before:* Dolia Estevez, "Mexican Leftist Leader López
Obrador Warns ExxonMobil: Investing In Mexico's Oil 'Tantamount
To Piracy,'" *Forbes Magazine*, November 12, 2013, https://www.forbes
.com/sites/doliaestevez/2013/11/12/mexican-leftist-leader-lopez-obra
dor-warns-exxonmobil-investing-in-mexicos-oil-tantamount-to-piracy
/#d620c424af71; David Alire Garcia, "In third bid to lead Mexico, fiery
leftist puts oil reform in crosshairs," Reuters, August 12, 2016, www.reuters
.com/article/uk-Mexico-politics-energy-idUKKCN10N1ZQ.

105 *As the IEA assessed:* International Energy Agency, *Mexico Energy Outlook*, 14, http://www.iea.org/publications/freepublications/publication
/MexicoEnergyOutlook.pdf.

105 *In contrast, the "no reform case":* Ibid., 107–13.

105 *Lourdes Melgar, a soft spoken:* Lourdes Melgar, in-person conversation
with author, Cambridge, MA, April 14, 2017.

105 *With almost two-thirds:* "Mexico: Analysis," U.S. Energy Information

Administration: Beta, December 9, 2016, https://www.eia.gov/beta/international/analysis.cfm?iso=MEX.

105 *Throughout the 2016 U.S.*: Maggie Severns, "Trump pins NAFTA, 'worst trade deal ever,' on Clinton," Politico, September 26, 2016, http://www.politico.com/story/2016/09/trump-clinton-come-out-swinging-over-nafta-228712.

106 *A senior industry official*: Senior industry official, telephone conversation with author, April 21, 2017.

106 *The mechanisms established*: "Overview of the Dispute Settlement Provisions," NAFTA Secretariat, 2014, https://www.nafta-sec-alena.org/Home/Dispute-Settlement/Overview-of-the-Dispute-Settlement-Provisions.

Five: Hard Power Accelerator

108 *The real median family*: "Real Median Family Income in the United States [MEFAINUSA672N], U.S. Bureau of the Census, retrieved from FRED, Federal Reserve Bank of St. Louis, www.fred.stlouisfed.org/series/MEFAINUSA672N.

108 *Meanwhile, the poverty*: Statistics in the next two sentences are from Peter Ferrera, "Reaganomics vs. Obamanaomics: Facts and Figures," *Forbes*, May 5, 2011, www.forbes.com/sites/peterferrara/2011/05/05/reaganomics-vs-obamanomics-facts-and-figures/.

108 *As Reagan recalled in his memoir*: Ronald Reagan, *An American Life* (New York: Simon & Schuster, 2011), 333.

109 *Bernard Brodie, a military*: See Bernard Brodie, *Strategy in the Missile Age* (Santa Monica: RAND, 1959), 358.

109 *Battling the worst recession*: See U.S. Bureau of Labor Statistics, United States Department of Labor, "The Recession of 2007–2009," Spotlight on Statistics, February 2012, 2, www.bls.gov/spotlight/2012/recession/pdf/recession_bls_spotlight.pdf.

109 *On average, household consumer spending*: See ibid., 15.

109 *The situation was so dire*: "Warren Buffett: This Is an 'Economic Pearl Harbor,'" YouTube, 3:17, posted by "GettingtotheTruth2," October 2, 2008, www.youtube.com/watch?v=LkMS3Oz1xkc.

109 *Hank Paulson, secretary of the treasury*: See Henry M. Paulson, *On the*

Brink: Inside the Race to Stop the Collapse of the Global Financial System (New York: Business Plus, 2010), 241.

109 *In 2011, with the economic:* Barack Obama, "Remarks by the President on the Way Forward in Afghanistan," The White House Office of the Press Secretary, June 22, 2011, www.whitehouse.gov/the-press-office /2011/06/22/remarks-president-way-forward-afghanistan.

110 *In a 2013 public opinion poll:* "Public Sees U.S. Power Declining as Support for Global Engagement Slips," Pew Research Center, December 3, 2013, www.people-press.org/2013/12/03/public-sees-u-s-power-declining -as-support-for-global-engagement-slips/.

110 *According to one study:* IHS Energy and IHS Economics, "Unleashing the Supply Chain: Assessing the Economic Impact of a US Crude Oil Free Trade Policy," IHS, March 2015, 4, www.energy.senate.gov/public/index .cfm/files/serve?File_id=6fcbe64a-7a34-4ac8-ba14-b7be1cfe1bee.

110 *A 2015 Harvard Business School/Boston Consulting Group:* Michael E. Porter, David S. Gee, and Gregory J. Pope, "America's Unconventional Energy Opportunity: A Win-Win Plan for the Economy, the Environ- ment, and a Lower-Carbon, Cleaner-Energy Future," Harvard Business School and Boston Consulting Group, June 2015, 19, www.hbs.edu/com petitiveness/Documents/america-unconventional-energy-opportunity .pdf.

110 *This amount translates into:* The $831 billion American Recovery and Re- investment Act of 2009 sought to boost the flagging economy through investment in infrastructure, education, renewable energy, and health. Congressional Budget Office, "Estimated Impact of the American Recov- ery and Reinvestment Act on Employment and Economic Output from October 2011 Through December 2011," Congressional Budget Office, Washington, DC, February, 2012, www.cbo.gov/sites/default/files/cbofiles /attachments/02-22-ARRA.pdf.

110 *The burgeoning energy sector:* Unemployment remained higher than during any other recovery from a recession since the 1970s. Natalia A. Kolesnikova and Yang Liu, "Jobless Recoveries: Causes and Conse- quences," Federal Reserve Bank of St. Louis, April 2011, www.stlouis fed.org/Publications/Regional-Economist/April-2011/Jobless-Recover ies-Causes-and-Consequences.

110 *Estimates of the impact:* "Net" jobs would also take into account whether such activities resulted in the loss of jobs in other sectors.

110 *Moody's Analytics, a consultancy:* Chris Lafakis, Moody's Analytics, presentation to Atlantic Council Task Force on the Energy Boom and National Security, April 2015, quoted in Atlantic Council Task Force on the U.S. Energy Boom and National Security, *Empowering America: How Energy Abundance Can Strengthen U.S. Global Leadership* (Washington, DC: Atlantic Council Global Energy Center), 5, www.atlanticcouncil.org /images/publications/Task_Force_Report_PDF.pdf.

110 *Each of these directly:* Ibid.

111 *These new jobs were roughly equivalent:* December 2007 to June 2009 is the official duration of the recession according to the National Bureau of Economic Research, the official arbiter of recessions. "Employment, Hours, and Earnings from the Current Employment Statistics Survey (National)," Bureau of Labor Statistics, United States Department of Labor, November 10, 2016, http://data.bls.gov/timeseries/CES3000000001.

111 *In contrast, the 2015 Harvard Business School:* According to this study, the total number of American jobs added between 2005 and mid-2015 was 4.9 million. Porter, Gee, and Pope, "America's Unconventional Energy Opportunity: A Win-Win Plan for the Economy, the Environment, and a Lower-Carbon, Cleaner-Energy Future," 6. The study cites "Current Employment Statistics," Bureau of Labor Statistics, http://www.bls.gov /data/.

111 *The two companies describe:* "About Us," Gulf Coast Growth Ventures, www.gulfcoastgrowthventures.com/about-us; "Gulf Coast Growth Ventures Announces Site Decision," Gulf Coast Growth Ventures, April 19, 2017, www.gulfcoastgrowthventures.com/press_releases.php?action=sub mit&story_id=51.

111 *The new $1 billion steel mill:* Bret Schulte, "Can Natural Gas Bring Back U.S. Factory Jobs?" *National Geographic,* February 1, 2014, http:// news.nationalgeographic.com/news/energy/2014/01/140131-natural -gas-manufacturing-jobs/; "Groundbreaking for a New Generation of Steelworks to Be Initially Built in Arkansas, USA," *MPT International* 1 (2015): 24–29, http://bigriversteel.com/wp-content/uploads/2015/03 /BRS-Metallurgical-Plant-and-Tech-Article-3_18_15.pdf.

111 *In early 2016, the mill:* Ted Evanoff, "Big River Steel Begins Hiring at Osceola," *Commercial Appeal,* March 18, 2016, www.commercialap peal.com/business/jobs-wages/Big-River-Steel-begins-hiring-at-Osce ola-372566682.html.

111 *Several years later, Dow:* Jack Kaskey, "Chemical Companies Rush to the U.S. Thanks to Cheap Natural Gas," Bloomberg, July 25, 2013, www .bloomberg.com/bw/articles/2013-07-25/chemical-companies-rush-to -the-u-dot-s-dot-thanks-to-cheap-natural-gas.

111 *Such stories are indicative:* Barack Obama, "State of the Union Address," January, 28, 2014, The White House Office of the Press Secretary, www .whitehouse.gov/the-press-office/2014/01/28/president-barack-obamas -state-union-address. In a subsequent testimony, Daniel Yergin references a report of $117 billion of new investment in the petrochemicals sector alone. Prepared testimony of Daniel Yergin, "America's Natural Gas Revolution: What It Means for Jobs and Economic Growth," Joint Economic Committee of the United States Congress, 103rd Cong. (June 24, 2014), 7, www.jec.senate.gov/public/_cache/files/e4afd2b5-0353-4070-8a15-37f8 eec794e9/yergin-testimony.pdf.

111 *Of course, low energy prices:* Some analysts rightly point out limits to how sweeping this U.S. "manufacturing renaissance" could be. The expansion of manufacturing frequently takes place in the context of a weak currency, which tends to boost demand for a country's products overseas. The energy boom, by taming the current account deficit (as a result of fewer energy imports) and strengthening the dollar, would therefore seem to some to run contrary to a growth in U.S. manufacturing beyond 2015. See Trevor Houser and Shashank Mohan, "Economic Impact," Chapter Four in *Fueling Up: The Economic Implications of America's Oil and Gas Boom* (Washington, D.C.: Peterson Institute for International Economics, 2014).

111 *But these sectors, which include:* See "Value Added by Industry as a Percentage of Gross Domestic Product," Bureau of Economic Analysis, April 21, 2017, https://www.bea.gov/iTable/iTable.cfm?ReqID=51&step =1#reqid=51&step=51&isuri=1&5101=1&5114=a&5113=gdpva, 31gva,331va,332va,3361mvva,3364otva,322va,325va,441va &5112=2016&5111=2005&5102=5.

112 *Increased oil and gas production:* Porter, Gee, and Pope, "America's Unconventional Energy Opportunity: A Win-Win Plan for the Economy, the Environment, and a Lower-Carbon, Cleaner-Energy Future," 55. An earlier (2013) IHS report included similar findings, stating that "The full value chain of industrial activity and employment associated with unconventional oil and natural gas contributed more than $74 billion

in federal and state government revenues in 2012. Tax receipts will rise to more than $125 billion annually by 2020 and reach $138 billion by 2025." "U.S. Unconventional Oil and Gas Revolution to Increase Disposable Income by More than $2,700 per Household and Boost U.S. Trade Position by More than $164 billion in 2020, New IHS Study Says," IHS Markit, September 4, 2013, http://press.ihs.com/press-release/economics /us-unconventional-oil-and-gas-revolution-increase-disposable -income-more-270.

112 *For context, this increase:* The 2014 Defense Budget was $520 billion, once the cap was raised by Congress. Lawrence J. Korb, Max Hoffman, and Kate Blakeley, "A User's Guide to the Fiscal Year 2015 Defense Budget," Center for American Progress, April 24, 2014, 3, https://cdn.american progress.org/wp-content/uploads/2014/04/DoDbudget-brief.pdf.

112 *The Brookings Institution found:* While the report notes that all consumers benefited, producers on the whole experienced a loss, as their gains from the expansion of supply were outweighed by the fall in the price of gas. The gains to consumers are however greater than the loss to producers. See Catherine Hausman and Ryan Kellogg, "Welfare and Distributional Implications of Shale Gas," *Brookings Papers on Economic Activity*, Spring 2015, www.brookings.edu/about/projects/bpea/papers /2015/welfare-distributional-implications-shale-gas. An American Petroleum Institute–ICF International study found that the "reduction in petroleum product prices have saved U.S. consumers an estimated $63 to $248 billion in 2013 and estimated cumulative savings of between $165 and $624 billion from 2008 to 2013." ICF International, "U.S. Oil Impacts: The Impacts of Horizontal Multi-stage Hydraulic Fracturing Technologies on Historical Oil Production, International Oil Costs, and Consumer Petroleum Product Costs," presentation, The American Petroleum Institute, Washington, DC, October 30, 2014, www.api.org/~/media/Files /Policy/Hydraulic_Fracturing/ICF-Hydraulic-Fracturing-Oil-Impacts .pdf.

112 *Homeowners around the country:* "Natural Gas Prices," U.S. Energy Information Administration, October 31, 2016, www.eia.gov/dnav/ng/ng _pri_sum_dcu_nus_a.htm.

112 *In part by purchasing its supply:* "PSE&G Proposes to Cut Residential Gas Bills by 5.7 Percent This Coming Winter," Public Service

Enterprise Group, June 1, 2015, https://www.pseg.com/info/media /newsreleases/2015/2015-06-01.jsp#.VZrc9YteHvM.

112 *One firm estimated that:* Harold L. Sirkin, Michael Zinser, and Justin Rose, "How Cheap Natural Gas Benefits the Budgets of U.S. Households," BCG Perspectives, February 3, 2014, https://www.bcgperspectives.com/con tent/articles/lean_manufacturing_energy_environment_how_cheap _natural_gas_benefits_budgets_us_households/; *The Ongoing Rise of Shale Gas: The Largest Revolution the Energy Landscape Has Seen in Two Decades*, BCG e-book, October 9, 2014, https://www.bcgperspectives .com/content/articles/energy_environment_ongoing_rise_shale_gas/.

112 *Had American innovators and entrepreneurs:* Even Saudi Arabia would have not been able to compensate for the millions of barrels lost to the market, given limits on their spare capacity at the time.

113 *By one estimate, oil prices:* An American Petroleum Institute study conducted in October 2014 in collaboration with ICF International found that "that international Brent crude oil prices would have averaged $122 to $150 per barrel in 2013 without U.S. HMSHF crude oil and condensate production increases." (HMSHF stands for "horizontal multistage hydraulic fracturing.") ICF International, "U.S. Oil Impacts: The Impacts of Horizontal Multi-stage Hydraulic Fracturing Technologies on Historical Oil Production, International Oil Costs, and Consumer Petroleum Product Costs," presentation, The American Petroleum Institute, Washington, DC, October 30, 2014, www.api.org/~/media/Files/Policy/Hy draulic_Fracturing/ICF-Hydraulic-Fracturing-Oil-Impacts.pdf. "Europe Brent Spot Price FOB," U.S. Energy Information Administration, November 9, 2016, www.eia.gov/dnav/pet/hist/LeafHandler.ashx?n=pet&s =rbrte&f=a.

113 *Similarly, another study suggested:* A Brookings Institution study ran a counterfactual scenario where the authors matched the 2007 supply curve with 2013 to conclude that had the shale boom not occurred, the price of natural gas would be $7.33 per thousand cubic feet, with the reduction achieved by the shale boom amounting to $3.45 (a 47 percent reduction). The study uses thousand cubic feet (mcf) rather than mmbtu, so a conversion is done to determine the price in mmbtu ($7.07). Hausman and Kellogg, "Welfare and Distributional Implications of Shale Gas," 3, 15–16.

113 *The International Monetary Fund estimates:* In 2002, the IMF estimated that an increase in the price of oil by $5 a barrel, sustained over time,

would adversely affect global growth by 0.3 percent after one year. A World Bank paper written in 2013 anticipates a similar impact, when it estimates that a 50% increase in oil prices could dampen global economic growth in 2020 by 1.29%. See Hiromi Kato, "Effects of the Oil Price Upsurge on the World Economy," The Institute of Energy Economics, December 2005, 14, http://eneken.ieej.or.jp/en/data/pdf/313.pdf; "The Price of Fear," *Economist*, March 3, 2011, www.economist.com/node/18285768; Govinda R. Timilsina, "How Much Does an Increase in Oil Prices Affect the Global Economy? Some Insights from a General Equilibrium Analysis" (policy research working paper 6515, Environment and Energy Team, Development Research Group, The World Bank, Washington, D.C., June 2013), 11, http://documents.worldbank.org/curated/en/405231468331249876/pdf/WPS6515.pdf.

113 *When the world is growing:* "World Economic Outlook," The International Monetary Fund, April 2017, http://www.imf.org/external/pubs/ft/weo/2017/01/weodata/download.aspx.

113 *One calculation suggests that:* Atlantic Council Task Force on the U.S. Energy Boom and National Security, *Empowering America: How Energy Abundance Can Strengthen U.S. Global Leadership*, 5–7.

113 *Estimates of the savings incurred:* "Daniel Yergin Congressional Testimony—Joint Economic Committee of the United States," IHS Markit, June 24, 2014, http://press.ihs.com/press-release/energy-power-media/daniel-yergin-congressional-testimony-joint-economic-committee-unit.

113 *The Harvard Business School/Boston Consulting Group:* Porter, Gee, and Pope, "America's Unconventional Energy Opportunity: A Win-Win Plan for the Economy, the Environment, and a Lower-Carbon, Cleaner-Energy Future," 55.

114 *U.S. government revenues:* Ibid.

114 *More than Wine for Cloth:* This is a reference to David Ricardo's 1817 discussion of trade in *On the Principles of Political Economy and Taxation*.

114 *Looking only at the volume:* Wilson, *Growing Together: Economic Ties between the United States and Mexico*, 7.

114 *In 2016, Mexico was the first:* These statistics are gleaned from data compiled by the U.S. Census Bureau. "Foreign Trade," United States Census Bureau, https://www.census.gov/foreign-trade/statistics/state/data/index.html.

114 *Take this example from:* Wilson, *Growing Together: Economic Ties between the United States and Mexico*, 7.

114 *Such an example clarifies:* This 40 percent stands in contrast to 25 percent for Canada and 4 percent for China. See Robert Koopman et al., "Give Credit Where Credit Is Due: Tracing Value Added in Global Production Chains," (Working Paper 16426, National Bureau of Economic Research, Cambridge, MA, September 2010), 7–8, http://www.nber.org/papers/w16426.pdf.

114 *There is a risk that U.S. policymakers:* On January 26, 2017, just a few days after taking office, President Trump tweeted, "The U.S. has a 60 billion dollar trade deficit with Mexico. It has been a one-sided deal from the beginning of NAFTA with massive numbers." Donald J. Trump, Tweet, January 26, 2017, 5:51 a.m., https://twitter.com/realDonaldTrump/status/824615820391305216; For information about jobs reliant on U.S.-Mexico trade, see Wilson, *Growing Together: Economic Ties between the United States and Mexico.*

115 *By some assessments, Mexican manufacturing labor costs:* Harold L. Sirkin, Michael Zinser, and Justin R. Rose, "The Shifting Economics of Global Manufacturing," BCG, August 2014, 8, https://www.bcgperspectives.com/Images/The_Shifting_Economics_of_Global_Manufacturing_Aug_2014.pdf; Harold L. Sirkin, "China vs. the U.S.: It's Just as Cheap to Make Goods in the USA," Bloomberg, www.bloomberg.com/bw/articles/2014-04-25/china-vs-dot-the-u-dot-s-dot-its-just-as-cheap-to-make-goods-in-the-u-dot-s-dot-a.; Pan Kwan Yuk, "Want Cheap Labour? Head to Mexico, Not China," *Financial Times*, January 14, 2016, https://www.ft.com/content/bddc8121-a7a0-3788-a74c-cd2b49cd3230; "International Comparisons of Hourly Compensation Costs in Manufacturing, 2015—Summary Tables," The Conference Board, April 12, 2016, https://www.conference-board.org/ilcprogram/index.cfm?id=38269.

115 *Since NAFTA was signed:* "U.S. Natural Gas Exports and Re-Exports by Country," U.S. Energy Information Administration, April 28, 2017, www.eia.gov/dnav/ng/ng_move_expc_s1_a.htm. This volume of exports only accounts for roughly 4 percent of overall U.S. natural gas production, but 60 percent of all U.S. natural gas exports in 2015; BP p.l.c., *BP Statistical Review of World Energy 2016* (BP, June 2016), 22, 28, www.bp.com/content/dam/bp/pdf/energy-economics/statistical-review-2016/bp-statistical-review-of-world-energy-2016-full-report.pdf.

115 *The volume of exports nearly doubled:* Ibid.

115 *Unless politicians take steps:* International Energy Agency, *Energy Policies*

Beyond IEA Countries—Mexico 2017, 2017, 119, https://www.eia.org
/publications/freepublications/publication/EnergyPoliciesBeyondIEA
CountriesMexico2017.pdf.

115 *this will help Mexico:* "New U.S. border-crossing pipelines bring shale
gas to more regions in Mexico," U.S. Energy Information Administration, December 1, 2016, https://www.eia.gov/todayinenergy/detail
.php?id=28972.

115 *In 2014, Mexican industries paid:* See "The Power and the Glory," *Economist*, July 5, 2014, http://www.economist.com/news/americas/2160
6269-foreigners-enthuse-over-enrique-pe-nietos-reforms-mexicans-are
-warier-power-and. Since this time, electricity prices in Mexico have
abated somewhat.

115 *As a result, high-energy-intensive industries:* Ibid.

116 *These scholars envision that:* Jorge Alvarez and Fabian Valencia, "Made
in Mexico: Energy Reform and Manufacturing Growth," WP/15/45, International Monetary Fund, Western Hemisphere Department, February 2015, 12, https://www.imf.org/external/pubs/ft/wp/2015/wp1545
.pdf.

116 *This gathering, boisterous as it was:* "Millions Join Anti-war Protests
Worldwide," *BBC News*, February 17, 2003, http://news.bbc.co.uk/2/hi
/europe/2765215.stm.

116 *The Rome protest reportedly involved:* "Largest Anti-War Rally," *Guinness
World Records*, archived from the original on September 4, 2004, http://
web.archive.org/web/20040904214302/http:/www.guinnessworldre
cords.com/content_pages/record.asp?recordid=54365.

116 *Speaking to London's largest:* "Millions Join Anti-war Protests Worldwide," *BBC News*, February 17, 2003, http://news.bbc.co.uk/1/hi/world
/europe/2765215.stm.

116 *Academics writing about the date:* Stefaan Walgrave and Dieter Rucht,
"Introduction," in *The World Says No to War: Demonstrations Against the
War in Iraq*, ed. Stefaan Walgrave and Dieter Rucht (Minneapolis: Uni-
versity of Minnesota Press, 2010), xiii.

116 *To understand why, it is important:* See Meghan L. O'Sullivan, "The En-
tanglement of Energy, Grand Strategy, and International Security," in
The Handbook of Global Energy Policy, ed. Andreas Goldthau (New York:
John Wiley & Sons, 2013).

117 *Hilter's generals once presented him:* James Marriott and Mika Minio-

Paluello, *The Oil Road: Journeys from the Caspian Sea to the City of London* (London: Verso, 2013), 129.

117 *Foreign investment in the development:* The ownership of Kuwaiti natural resources by foreign companies is prohibited under the Kuwaiti constitution. This has been interpreted to prevent common product-sharing agreements that international oil firms prefer when making major investments. Jareer Elass, "Domestic Politics Slow Kuwait's Oil Production Expansion Plans," *Arab Weekly*, June 19, 2016, www.thearabweekly.com/Economy/5512/Domestic-politics-slow-Kuwait%E2%80%99s-oil-production-expansion-plans.

118 *A large part of the rationale:* Iraq and Kuwait comprised one-fifth (19.7 percent) of the world's proved crude oil reserves at the time. "Crude Oil Proved Reserves 2014," U.S. Energy Information Administration, 2014, www.eia.gov/cfapps/ipdbproject/iedindex3.cfm?tid=5&pid=57&aid=6&cid=regions&syid=1991&eyid=1991&unit=BB.

118 *If it had, such an invasion:* The reserve of Iraq, Kuwait, and Saudi Arabia were 45.7 percent of global reserves at the time. Derived from ibid.

118 *In 2015, U.S. tight oil:* "Crude Oil and Natural Gas Proved Reserves, Year-end 2015," U.S. Energy Information Administration, December 14, 2016, https://www.eia.gov/naturalgas/crudeoilreserves/#4; "International Energy statistics," U.S. Energy Administration, https://www.eia.gov/beta/international.

118 *In 2016, U.S. crude production:* "International Liquids: Crude Oil Prices: Brent," Annual Energy Outlook 2017, U.S. Energy Information Administration, https://www.eia.gov/outlooks/aeo/data/browser/#/?id=19-AEO2017&region=0-0&cases=ref2017&start=2015&end=2017&f=A&linechart=ref2017-d120816a.4-19-AEO2017&sourcekey=0; "Oil and Gas: Crude Oil: Lower 48 Average Wellhead Price," Annual Energy Outlook 2017, U.S. Energy Information Administration, https://www.eia.gov/outlooks/aeo/data/browser/#/?id=14-AEO2017&cases=ref2017&sourcekey=0.

119 *More immediately, the new energy abundance:* "U.S. Forces Afghanistan Memorandum from General David Petraeus: Supporting the Mission with Operational Energy," Office of the Under Secretary of Defense for Acquisition, Technology and Logistics," June 7, 2011, http://www.acq.osd.mil/eie/Downloads/OE/U.S.%20Forces%20Afghanistan%20Memo%20Gen%20David%20Petraeus_06-07-11.pdf.

119	*The Pentagon is the largest:* Jeremy Scahill, "Fueling War: Pentagon Still Buying Most of Its Oil and Gas from BP," *Nation*, June 9, 2010, www .thenation.com/article/fueling-war-pentagon-still-buying-most-its-oil -and-gas-bp; Gregory J. Lengyel, Colonel, USAF, "Department of Defense Energy Strategy: Teaching an Old Dog New Tricks" (21st Century Defense Initiative, Foreign Policy Studies, The Brookings Institution, Washington, D.C., August 2007), 7, https://web.archive.org/web/20140529175335 /http://www.brookings.edu/~/media/research/files/papers/2007/8/de fense%20lengyel/lengyel20070815.pdf.

119	*as of 2013:* "International Energy Statistics," U.S. Energy Information Administration; Defense Logistics Agency Energy, *Fiscal Year 2015 Fact Book* (Washington, DC: Department of Defense, 2015), 25–26.

119	*The Pentagon's oil intake:* Paul Dimotakis et al., "Reducing DoD Fossil-Fuel Dependence" (McLean, VA: The MITRE Corporation, September 2006), 13, https://fas.org/irp/agency/dod/jason/fossil.pdf.

119	*In 2007, when American forces:* "Factbox—US Military Fuel Spending," Reuters, March 20, 2008, www.reuters.com/article/2008/03/20 /idUSN20416568.

119	*In 2013, the Pentagon consumed virtually:* "International Energy Statistics," U.S. Energy Information Administration. In 2014, the Nigerian economy overtook the South African one as the continent's largest economy.

119	*For instance, the Arleigh Burke:* Peter Hoy, "The World's Biggest Fuel Consumer," *Forbes*, June 5, 2008, https://www.forbes.com /2008/06/05/mileage-military-vehicles-tech-logistics08-cz_ph_0605fuel .html; Daniel Engber, "FYI: What Kind of Gas Mileage Can You Get From a Naval Warship?" *Popular Science*, October 17, 2012, www.popsci.com /technology/article/2012-09/fyi-what-kind-of-mileage-can-you-get-na val-warship.

120	*During World War II:* "U.S. military in Iraq fuels the gouge of fuel costs," NBC News, April 2, 2008, www.nbcnews.com/id/23922063/ns/world _news_mideast_n_africa/t/us-military-iraq-fuels-gouge-fuel-costs/.

120	*In Iraq, where heavy equipment:* Ibid.

120	*In the late 2000s, the Pentagon changed:* "Seven Step FBCF Methodology and the JLTV," The DoD Energy Blog: Rethinking Military Power, July 9, 2009, www.dodenergy.blogspot.com/2009/07/seven-step-fbcf-method ology-and-jltv.html.

120 *In some of the most remote:* Defense Science Board, Report of the Defense
Science Board Task Force on DoD Energy Strategy: 'More Fight—Less
Fuel' (Washington, D.C.: Office of the Under Secretary of Defense for
Acquisition, Technology, and Logistics, February 2008), 30, http://www
.acq.osd.mil/dsb/reports/2000s/ADA477619.pdf.

120 *In the fiscal year of 2013:* Defense Logistics Agency Energy, *Fiscal Year 2015
Fact Book* (Washington, D.C.: Department of Defense, 2015), 25–26, http://
www.dla.mil/Portals/104/Documents/Energy/Publications/E_Fiscal
2015FactBookLowResolution_160707.pdf?ver=2016-07-08-124636-630.

120 *Two years later, the costs:* Ibid.

120 *The $6 billion saved:* "Military Expenditure by Country, in Constant
(2015) US$ m., 1988–1996," Stockholm International Peace Research
Institute, 2017, https://www.sipri.org/sites/default/files/Milex-constant
-2015-USD.pdf.

120 *As former Deputy Secretary:* Bill Lynn, "Energy for the War Fighter: The
Department of Defense Operational Energy Strategy," The White House
Barack Obama, June 14, 2011, https://obamawhitehouse.archives.gov
/blog/2011/06/14/energy-war-fighter-department-defense-operational
-energy-strategy.

120 *As Senator Jack Reed:* Martin Matishak, "DOD Eyes Savings in Sink-
ing Oil Prices," *The Hill,* January 8, 2015, http://thehill.com/policy/de
fense/228849-dod-eyes-savings-in-sinking-oil-prices.

121 *2011, General David Petraeus:* "U.S. Forces Afghanistan Memorandum
from General David Petraeus: Supporting the Mission with Operational
Energy," Office of the Under Secretary of Defense for Acquisition, Technology and Logistics," June 7, 2011, http://www.acq.osd.mil/eie/Down
loads/OE/U.S.%20Forces%20Afghanistan%20Memo%20Gen%20
David%20Petraeus_06-07-11.pdf.

121 *In calling on his commanders:* Ibid.

121 *With nearly half of all convoys:* Gordon Feller, "Casualty Costs of Fuel and
Water Resupply Convoys in Afghanistan and Iraq," February 26, 2010,
Army Technology.com, www.army-technology.com/features/feature77200.

121 *One study by the army:* David S. Eady, Steven B. Siegel, R. Steven Bell,
and Scott H. Dicke, "Sustain the Mission Project: Casualty Factors for
Fuel and Water Resupply Convoys: Final Technical Report," Army Environmental Policy Institute, September 2009, www.aepi.army.mil/docs
/whatsnew/SMP_Casualty_Cost_Factors_Final1-09.pdf.

121 *Numbers in the same report:* Gordon Feller, "Casualty Costs of Fuel and Water Resupply Convoys in Afghanistan and Iraq."

121 *A second study concluded:* Paul Skalny, director of the Army's National Automotive Center, from William Matthews, "A Different Kind of Hybrid," *Defense News*, November 2, 2009; Schuyler Null, "Defense Sustainability: Energy Efficiency and the Battlefield," *Global Green USA*, February 2010, 13, https://static1.squarespace.com/static/5548ed90e4b 0b0a763d0e704/t/55548eb1e4b0df117522923c/1431604913196/publica tion-112-1.pdf.

121 *Such statistics put in context:* Chief Warrant Officer 2 Kenneth Hudak, "Lengthening the Tether of Fuel in Afghanistan," U.S. Army, March 6, 2013, https://www.army.mil/article/97879/Lengthening_the_Tether_of _Fuel_in_Afghanistan/.

121 *Early one morning in January:* The source of this anecdote and a fuller examination of Levey's efforts can be found at: Robin Wright, "Stuart Levey's War," *New York Times Magazine*, October 31, 2008, www.nytimes .com/2008/11/02/magazine/02IRAN-t.html?pagewanted=all.

122 *"That could spark the right":* Ibid.

122 *With the backing of the highest echelons:* Ibid.

122 *Iran's oil exports fell 60 percent:* Kenneth Katzman, *Iran Sanctions* (CRS Report No. 7-5700) (Washington, DC: Congressional Research Service, 2016), 55, https://www.fas.org/sgp/crs/mideast/RS20871.pdf.

122 *Tehran's financial difficulties were compounded:* In December 2011, as part of a defense authorization act, Congress passed a provision that essentially forced banks to make a choice between doing business with Iran's central bank, or maintaining or opening accounts in the United States. This provision was critical, as oil importers used the central bank to pay Iran hard currency for its oil exports. In March 2012, Iranian banks were disconnected from the Society for Worldwide Interbank Financial Telecommunication (SWIFT), further frustrating efforts to conduct transactions in hard currency. Ibid., 20, 41, 50.

122 *The rial, Iran's currency:* Facts from ibid., 56.

123 *Before sanctions proved so critical:* Dick Kirschten, "Chicken Soup Diplomacy," *National Journal*, January 4, 1997, 13–17, http://archives.us aengage.org/archives/news/970104nj.html.

123 *Efforts to corral others:* These secondary sanctions were devised by Congress, not the executive branch. Europe threatened to bring any

imposition of such sanctions to the World Trade Organization. See "EU Regrets Extension of US Sanctions Law Against Iran and Libya: Statement by Commissioner for External Relations, Chris Patten," European Commission Press Release Database, July 31, 2001, http://europa.eu /rapid/press-release_IP-01-1162_en.htm.

123 *Efforts to marginalize Iran:* International Energy Agency, *Key World Energy Statistics 2007* (Paris: OECD Publishing, 2007), 11, 23, www.copro cem.com/documents/key_stats_2007.pdf.

123 *Some analysts even suggested:* "Broad economic sanctions, comparable to the isolation of Iraq in the 1990s, are no longer feasible. Unlike the cheap oil of the 1990s, oil prices today are at or near record levels. Given tight global supplies, Iran has greater leverage to counter sanction [*sic*] major oil consuming nations by cutting back its oil exports. Few producing nations have the spare capacity to increase shipments to offset potential Iranian cutbacks, so prices would likely rise sharply. Iran would sell less . . . and earn more." Statement by Jeffrey J. Schott, "Economic Sanctions, Oil, and Iran," Hearing on "Energy and the Iranian Economy" before the Joint Economic Committee, 109th Cong. (July 25, 2006), www.iranwatch.org /sites/default/files/us-congress-jec-schott-iran-energy-072506.pdf. Also see Thijs Van de Graaf, "The 'Oil Weapon' Reversed? Sanctions Against Iran and U.S.-EU Structural Power," *The Middle East Policy Council* 20, no. 3 (Fall 2013), www.mepc.org/oil-weapon-reversed-sanctions-against -iran-and-us-eu-structural-power?print.

124 *As the perception of the threat:* UN Resolutions 1737, 1747, and 1803 imposed sanctions on Iran's nuclear program and weapons of mass destruction infrastructure.

124 *As Tom Donilon, former national:* See "A Review of the 'Asia Rebalance' and a Preview of the President's Trip to the Region: A Conversation with Thomas E. Donilon," March 6, 2014, transcript, The Brookings Institution, Washington, D.C., https://www.brookings.edu/wp-content/up loads/2014/02/20140306_donilon_asia_transcript.pdf.

124 *The late 2009 exposure of Fordo:* Julian Borger and Patrick Wintour, "Why Iran Confessed to Secret Nuclear Site Built Inside Mountain," *Guardian,* September 25, 2009, www.theguardian.com/world/2009/sep/25/iran-nu clear-uranium-enrichment-intelligence; William J. Broad, "Nuclear Plant in Iranian Desert Emerges as Flash Point in Talks," *New York Times,*

April 3, 2015, www.nytimes.com/2015/04/04/world/middleeast/nuclear
-plant-in-iranian-desert-emerges-as-flash-point-in-talks.html?_r=0;
Karen DeYoung and Michael D. Shear, "US, Allies Say Iran Has Secret
Nuclear Facility," *Washington Post*, September 26, 2009, www.washington
post.com/wp-dyn/content/article/2009/09/25/AR2009092500289_2
.html?sid=ST2009092601752.

124 *To make matters worse:* "Growth of the Real Gross Domestic Product
(GDP) in the European Union and the Euro Area from 2010 to 2020
(Compared to the Previous Year)," Statista, www.statista.com/statistics
/267898/gross-domestic-product-gdp-growth-in-eu-and-euro-area/.

124 *While Donilon and his national:* Donilon recounted this tension in a pub-
lic event at the Brookings Institution on March 6, 2014, "A Review of the
'Asia Rebalance' and a Preview of the President's Trip to the Region: A
Conversation with Thomas E. Donilon."

124 *Burgeoning U.S. tight oil production:* See Daniel Yergin, "America's Unconventional Revolution, Energy Security and Innovation," *Manufacturing
Engineering*, July 10, 2013, www.sme.org/MEMagazine/Article.aspx?id
=74235&taxid=1476#sthash.mFCVB2ql.dpuf.

124 *In Beijing, New Delhi, and Tokyo:* See, for example, "Saudi Arabia Issues Oil-Supply Assurance," *Daily Star Lebanon*, January 31, 2012, www
.dailystar.com.lb/Business/Middle-East/2012/Jan-31/161641-saudi-ara
bia-issues-oil-supply-assurance.ashx; CNN Wire Staff, "Oil Minister: Saudi
Arabia Can Make Up for Iranian Crude," CNN, January 17, 2012, www
.cnn.com/2012/01/16/world/meast/saudi-oil-production/; Ayesha Daya,
"Saudi Arabia Can Raise Output 25% if Needed, Naimi Says," Bloomberg,
March 20, 2012, www.bloomberg.com/news/articles/2012-03-20/saudi
-arabia-can-increase-oil-output-25-if-needed-naimi-says.

124 *U.S. officials were also armed:* See James Hamilton, "US Tight Oil Production Surging," *Econbrowser*, December 22, 2013, http://econbrowser.com
/archives/2013/12/us_tight_oil_pr.

125 *A provision in U.S. legislation:* See National Defense Authorization Act
for Fiscal Year 2012, Pub. L. No. 112-81, 125 Stat. 1298 (2011), https://
www.treasury.gov/resource-center/sanctions/Programs/Documents
/ndaa_publaw.pdf.

125 *Europe instituted its own ban:* Justyna Pawlak and Parisa Hafezi, "Exclusive:
EU Agrees to Embargo on Iranian Crude," Reuters, January 4, 2012, www

.reuters.com/article/2012/01/04/us-iran-idUSTRE8031DI20120104. The ban came into effect on July 1, 2012.

125 *Tokyo was initially unenthusiastic:* See Irina Mironova, "Japan Weighs the Pros and Cons of Imposing Sanctions on Russia," *Russia Direct*, September 19, 2014, www.russia-direct.org/analysis/japan-weighs-pros -and-cons-imposing-sanctions-russia; Aaron Sheldrick, "Japan Is Worried That Western Sanctions on Russia Would Hurt Tokyo," *Business Insider*, March 5, 2014, www.businessinsider.com/r-japans -embrace-of-russia-under-threat-with-ukraine-crisis-2014-05.

125 *At the time, Russia was the fourth:* Alexander Martin, "Japan Announces Fresh Russia Sanctions," *Wall Street Journal*, September 24, 2014, www .wsj.com/articles/japan-announces-new-russia-sanctions-1411553420.

125 *The amount of natural gas:* Japan increased its LNG imports from Russia by 40 percent, from 6.0 million tons in 2010 to 8.4 million tons in 2014.

125 *The two had met five times:* Michael Lipin, "Why is Japan's Abe Seeking Better Ties with Russia's Putin?" VOA News, February 12, 2014, http:// www.voanews.com/a/why-is-abe-seeking-better-ties-with-russia-putin /1850393.html.

125 *Nevertheless, despite Tokyo's wariness:* Alexander Martin, "Japan Announces Fresh Russia Sanctions," *Wall Street Journal*, September 24,2014, www.wsj.com/articles/japan-announces-new-russia-sanc tions-1411553420; "Japan Steps Up Sanctions as Tensions Rise with Russia," *BBC News*, September 24, 2014, www.bbc.com/news/world -asia-29345451.

126 *In their book,* Economic Sanctions Reconsidered: Gary Clyde Hufbauer, Jeffrey J. Schott, Kimberly Ann Elliott, and Barbara Oegg, *Economic Sanctions Reconsidered*, 2nd rev. ed., 2 vols. (Washington, DC: Peterson Institute for International Economics, 1990).

126 *University of Chicago professor Robert Pape:* Robert A. Pape, "Why Economic Sanctions Do Not Work," *International Security* 22, no. 2 (Fall 1997): 90–136, https://www.jstor.org/stable/2539368?seq=1#page_scan _tab_contents; Robert A. Pape, "Why Economic Sanctions Still Do Not Work," *International Security* 23, no. 1 (Summer 1998): 66–77, https:// www.jstor.org/stable/2539263?seq=1#page_scan_tab_contents.

126 *While its rejuvenated use may pertain:* See Meghan L. O'Sullivan, *Shrewd Sanctions: Statecraft and State Sponsors of Terrorism* (Washington, D.C.: Brookings, 2003).

Six: Soft Powering Up

128 *Writing in 1990, Joe Nye:* Joseph S. Nye, Jr., "Soft Power and American For-
     eign Policy," *Political Science Quarterly* 119, no. 2 (Summer 2004): 255–70.

128 *According to Nye, the soft power:* Joseph S. Nye, "What China and Russia
     Don't Get About Soft Power," *Foreign Policy*, April 29, 2013, http://for
     eignpolicy.com/2013/04/29/what-china-and-russia-dont-get-about-soft
     -power/.

128 *To the extent that these three factors:* Since Nye developed this con-
     cept, it has become common parlance in academic settings, as well as
     in the practice of foreign policy; it has also been embraced in the busi-
     ness world, where it has been used to refer to intangible factors that in-
     fluence negotiations or specific qualities that individuals bring to the
     table. See "Soft Power: How to Open Doors and Influence People in
     France," *Economist*, June 7, 2014, www.economist.com/news/business
     /21603466-how-open-doors-and-influence-people-france-soft-power;
     Diane Coutu, "Smart Power," *Harvard Business Review*, November 2008,
     https://hbr.org/2008/11/smart-power.

128 *America has always relied:* The United States has never been the sole
     purveyor of soft power. The Soviets also wielded significant soft power,
     at least until their use of military force in Hungary and Czechoslovakia
     undermined it. More recently, the Chinese have embraced the concept.
     In 2007, then Chinese president Hu Jintao had declared at the 17th Com-
     munist Party Congress that China must increase its soft power. Seven
     years later, his successor, Xi Jinping, emphasized the same, affirming the
     need to "increase China's soft power, give a good Chinese narrative, and
     better communicate China's message to the world." Jane Perlez, "Leader
     Asserts China's Growing Importance on Global Stage," *New York Times*,
     November 30, 2014, www.nytimes.com/2014/12/01/world/asia/leader
     -asserts-chinas-growing-role-on-global-stage.html.

128 *During the Cold War, the attractiveness:* For instance, Washington-funded
     radio broadcasts reached 70–80 percent of Eastern Europeans during the
     height of the Cold War. Joseph S. Nye, Jr., "The Decline of America's Soft
     Power," *Foreign Affairs* 83, no. 3 (May/June 2004): 16–20, https://www
     .foreignaffairs.com/articles/2004-05-01/decline-americas-soft-power.

129 *The United States unquestionably still claims:* According to the U.S. De-
     partment of Homeland Security, there were 1.23 million international

students studying in the United States in 2016, up from 1.19 million the previous year. *SEVIS by the Numbers: General Summary Quarterly Review* (Washington, D.C.: U.S. Immigration and Customs Enforcement, November 2016), 2, https://www.ice.gov/doclib/sevis/pdf/byTheNumbers Dec2016.pdf.

129 *For much of the 2000s:* See, for instance, "A Superpower in Decline: Is the American Dream Over?" *Der Spiegel*, November 1, 2010, http://www.spiegel.de/international/world/a-superpower-in-decline-is-the-american-dream-over-a-726447.html.

129 *In 2009, China's government-run:* Li Hongmei, "The U.S. Hegemony Ends, the Era of Global Multipolarity Enters," *People's Daily Online*, February 24, 2009, http://en.people.cn/90002/96417/6599374.html.

129 *In 2011, while speaking:* Al Lewis, "Blood-Sucking Nation," *Wall Street Journal*, August 7, 2011, www.wsj.com/articles/SB1000142405311190345450457649254118958176.

130 *Gregory Zuckerman, author of* The Frackers: Bret Stephens, "The Marvel of American Resilience," *Wall Street Journal*, December 22, 2014, www.wsj.com/articles/bret-stephens-the-marvel-of-american-resilience-1419292945.

130 *In his far-reaching book:* Henry Kissinger, *World Order* (London: Penguin, 2015), 362–63.

130 *The liberal economic order has been:* A 2013 study done by Martin Ravallion assesses that in the absence of big changes in political and economic thinking that occurred around 1950, 1.5 billion more people would be in poverty today than there are today. See Martin Ravallion, "The Idea of Antipoverty Policy" (Working Paper 19210, National Bureau of Economic Research, Cambridge, MA, July 2013), 30, http://www.nber.org/papers/w19210.pdf.

131 *Wars have not ceased:* See Steven Pinker, *The Better Angels of Our Nature: Why Violence Has Declined* (New York: Viking, 2011).

131 *This liberal international order, however:* See, for instance, Kissinger, *World Order*, 365–67; also see Charles P. Kindleberger, *The World in Depression: 1929–1939* (Oakland, CA: University of California Press, 1986).

131 *First is self-doubt:* Kissinger, *World Order*, 365.

131 *The second challenge—also clear in today's world:* Ibid., 366–67.

131 *At a minimum, China desires:* For two views on the changes China desires, see Thomas Fingar, "China's Vision of World Order," in *Strategic Asia*

2012–13: China's Military Challenge, ed. Ashley J. Tellis and Travis Tanner (Washington, D.C.: October 2012), http://nbr.org/publications/element.as px?id=624 and John G. Ikenberry, "The Future of the Liberal World Order: Internationalism After America," *Foreign Affairs* 90, no. 3 (May/June 2011).

131 *Scholars and policymakers feverishly debate:* For example, Council on Foreign Relations president Richard N. Haass argues that "the world needs an updated operating system—call it World Order 2.0—that takes into account new forces, challenges, and actors" because the current "trend is one of declining order." Other scholars like Princeton professor John Ikenberry argue that it is American leadership that is in crisis rather than the world order. Richard N. Haass, *A World in Disarray: American Foreign Policy and the Crisis of the Old Order* (New York: Penguin Press, 2017), 2–6. In contrast, G. John Ikenberry argues that "American leadership may be in crisis, but the world order is not." G. John Ikenberry, "American Leadership May Be in Crisis, but the World is Not," *Washington Post*, January 27, 2016, https://www.washingtonpost.com/news/in-theory/wp /2016/01/27/american-leadership-is-in-crisis-but-the-world-order-is -not/.

132 *It is now more comfortable:* See Chapter Ten and Michal Meidan, "The Structure of China's Oil Industry: Past Trends and Future Prospects," Oxford Institute for Energy Studies, University of Oxford, U.K., May 2016, 54, https://www.oxfordenergy.org/wpcms/wp-content/uploads/2016/05 /The-structure-of-Chinas-oil-industry-past-trends-and-future-pros pects-WPM-66.pdf.

133 *The goal was to establish:* These countries had good reasons to believe they could face persistent challenges and manipulation from OPEC in the years ahead. Kissinger allegedly advocated for an explicitly anti-OPEC body, but European countries and Japan demurred in the interest of maintaining relations with Arab countries. For more, see Dries Lesage and Thijs Van de Graaf, *Global Energy Governance in a Multipolar World* (London: Routledge, 2013), 59–60.

133 *In 2012, China began to clamor:* For example, in 2012, at an energy summit in Abu Dhabi, Chinese Premier Wen Jiabao advocated that a G20-like body be created to establish "fair, reasonable and binding" global rules to stabilize oil and gas markets; he proposed that stakeholders from consumer, supplier, and transit countries all be involved. David Worthington, "China Proposes Global Energy Governance, Development," *ZDNet*,

January 16, 2012, http://www.zdnet.com/article/china-proposes-global
-energy-governance-development/. Two years later, Chinese leaders were
involved in commissioning a study by the Grantham Institute of Imperial College London on whether a new energy order was needed and, if
so, how China could advance it. It emphasized the need for new, more
inclusive institutions. Neil Hirst et al., *Global Energy Governance Reform
and China's Participation* (Beijing and London: Energy Research Institute, NDRC and Grantham Institute, Imperial College London, Imperial College London, November 2014), https://workspace.imperial.ac.uk
/grantham/Public/publications/Global%20Energy%20Governance%20
and%20China's%20Participation%20-%20Consultation%20report%20
(English).pdf.

133 *It was not simply pushing back*: For a sense of the myriad of energy governance institutions, see David Goldwyn and Phillip Cornell, *Report of
the Atlantic Council Task Force on Reform of the Global Energy Architec-
ture* (Washington, D.C.: Atlantic Council Global Energy Center, April
2017), http://www.atlanticcouncil.org/images/publications/Reform_of
_the_Global_Energy_Architecture_web_0418.pdf.

133 *Most recently, China has focused*: Chinese officials were major drivers of
the G20 Principles on Energy Collaboration endorsed at the 2014 G20
summit in Brisbane, Australia. "G20 Principles on Energy Collaboration,"
G20 Australia 2014, November 16, 2014, www.g20australia.org/sites
/default/files/g20_resources/library/g20_principles_energy_collabora
tion.pdf; John J. Kirton, *China's G20 Leadership* (London: Routledge,
2016), 83.

133 *President Xi Jinping has reportedly said*: Ibid.

134 *While some believe that China*: The U.S. Congress in particular was reluctant to approve governance changes that would have increased China's influence in the International Monetary Fund to be more in accord with the
size of its economy. See Leonid Bershidsky, "IMF Reform is Too Little, Way
Too Late," *Bloomberg View*, December 18, 2015, https://www.bloomberg
.com/view/articles/2015-12-18/imf-reform-is-too-little-way-too-late.

134 *There, he explained that strengthening ties*: Dr. Fatih Birol, "Standing
Together: A New Era of IEA-China Co-operation," (speech, Chinese
Academy of Social Sciences, Beijing, September 9, 2015), 7, www.iea.org
/newsroomandevents/speeches/150909_CASS.pdf.

134 *China and other Asian countries*: See "Joint Ministerial Declaration on

the Occasion of the 2015 IEA Ministerial Meeting Expressing the Activation of Association," International Energy Agency, November 18, 2015, www.iea.org/media/news/2015/press/IEA_Association.pdf. For a nod of appreciation for these efforts, see the "G20 Energy Ministerial Meeting Beijing Communiqué," G20 2016 China, June 29, 2016, https://ec.europa.eu/energy/sites/ener/files/documents/Beijing%20Communique.pdf.

134 *Moreover, many of the steps:* China has taken advantage of low global oil prices to undertake an aggressive program to build strategic oil reserves. Some estimate these efforts boosted Chinese import demand in 2015 and 2016 by as much as 15%. See "Oil Bulls Beware Because China's Almost Done Amassing Crude," Bloomberg News, June 30, 2016, https://www.bloomberg.com/news/articles/2016-06-30/oil-bulls-beware-because-china-s-almost-done-amassing-crude.

135 *In March 2001, with the Bush Administration:* Graydon Carter, *What We've Lost* (London: Macmillan, 2004), 143.

135 *Rice didn't mince her words:* Condoleezza Rice, *No Higher Honor: A Memoir of My Years in Washington* (New York: Broadway Paperbacks, September 4, 2012), 41.

135 *If the reaction at the luncheon:* Jeffrey Kluger, "A Climate of Despair," *Time*, April 1, 2011, http://content.time.com/time/magazine/article/0,9171,104596,00.html.

135 *The French minister for the environment:* Edmund L. Andrews, "Bush Angers Europe by Eroding Pact on Warming," *New York Times*, April 1, 2001, http://www.nytimes.com/2001/04/01/world/bush-angers-europe-by-eroding-pact-on-warming.html.

135 *Romano Prodi, the former prime minister:* "Dead or Comatose?," *The Globalist*, July 12, 2001, www.theglobalist.com/dead-or-comatose/.

135 *In her 2011 memoir:* Condoleezza Rice, *No Higher Honor: A Memoir of My Years in Washington*, 41–46.

135 *Even though the Bush administration:* Bush administration objections were articulated in a presidential statement in June 2001. George W. Bush, "President Bush Discusses Global Climate Change," The White House Office of the Press Secretary, June 11, 2001, http://georgewbush-whitehouse.archives.gov/news/releases/2001/06/20010611-2.html. In 1997, in a 95–0 vote, the U.S. Senate made clear the conditions under which it would and would not sign a climate treaty. Byrd-Hagel Resolution,

S. RES. 98, 105th Cong. (July 25, 1997), https://www.nationalcenter.org /KyotoSenate.html.

136 *Speaking in Alaska in August 2015:* Colleen McCain Nelson, "Obama Calls for U.S. to Show Leadership in Fighting Climate Change," *Wall Street Journal*, August 31, 2015, www.wsj.com/articles/obama-calls -for-u-s-to-show-leadership-in-fighting-climate-change-1441074557 ?alg=y.

136 *Melanie Nakagawa, a State Department official:* Melanie Nakagawa, in-person conversation with author, Washington, DC, July 7, 2016.

136 *Environmental Protection Agency Administrator:* Jared Gilmour and David J. Unger, "EPA Chief: New Climate Rules Are Safe from Courts, Congress," *Christian Science Monitor*, July 7, 2015, www.csmonitor.com /Environment/Energy/2015/0707/EPA-chief-New-climate-rules-are-safe -from-courts-Congress.

136 *McCarthy pointed to India:* Ibid.

137 *In urging China—and eventually others:* "US Carbon Emissions Set to Fall to Lowest Level in Two Decades," *Guardian*, April 10, 2015, www .theguardian.com/environment/2015/apr/10/us-carbon-emissions-set -to-fall-to-lowest-level-in-two-decades; Christopher Martin, "U.S. Carbon Emissions Falling to Two-Decade Low in Coal Shift," Bloomberg, April 9, 2015, www.bloomberg.com/news/articles/2015-04-09/u-s-car bon-emissions-falling-to-two-decade-low-in-coal-shift.

137 *In private, administration officials:* Kelly Sims Gallagher, professor at Tufts University and former climate negotiator for the Obama Administration, in-person conversation with author, Cambridge, MA, August 4, 2015.

137 *Even in a low-cost-energy environment:* See, for instance, William Mauldin, "How Much Will the Paris Climate Deal Cost the U.S.?" *Wall Street Journal*, December 14, 2015, http://blogs.wsj.com/economics/2015/12/14 /how-much-will-the-paris-climate-deal-cost-the-u-s/.

137 *He lamented the costs:* Donald J. Trump, "Statement by President Trump on the Paris Climate Accord," The White House Office of the Press Secretary, June 1, 2017, https://www.whitehouse.gov/the-press-office /2017/06/01/statement-president-trump-paris-climate-accord.

137 *His administration could—and should:* See Meghan L. O'Sullivan, "How Trump Is Surrendering America's Soft Power," *Bloomberg View,* June 2, 2017, https://www.bloomberg.com/view/articles/2017-06-02/how-trump -is-surrendering-america-s-soft-power.

138 *Not only are sub-state actors:* For more on how the gas boom could further undermine coal, see "Future coal production depends on resources and technology, not just policy choices," U.S. Energy Information Administration, June 26, 2017, https://www.eia.gov/todayinenergy/detail .php?id=31792.

138 *But lurking behind these more tangible:* For a scene-setter memo written before the trip, see David Shambaugh, "The China Awaiting President Obama," The Brookings Institution, November 10, 2009, www.brookings .edu/research/opinions/2009/11/china-shambaugh.

138 *Just as the United States was seeking:* Shirley A. Kan, *China and Proliferation of Weapons of Mass Destruction and Missiles: Policy Issues* (CRS Report No. RL31555) (Washington, D.C.: Congressional Research Service, January 5, 2015), 10, http://fas.org/sgp/crs/nuke/RL31555.pdf; Marybeth Davis et al., *China-Iran: A Limited Partnership*, US-China Economic and Security Review Commission, www.uscc.gov/sites/default/files/Re search/China-Iran--A%20Limited%20Partnership.pdf.

139 *"The only reason the Chinese had any interest":* David Goldwyn, telephone conversation with author, July 15, 2015.

140 *The U.S. Commerce Department invited:* Ibid.

140 *Some received help assessing:* See Congressional Budget Justification: Foreign Operations (Washington, D.C.: United States of America Department of State, FY 2015), 234, www.state.gov/documents/organiza tion/224069.pdf.

141 *"These conversations provided the entry point":* David Goldwyn, telephone conversation with author, July 14, 2015.

142 *One cavern alone is easily large enough:* "SPR Storage Site," U.S. Department of Energy, Office of Fossil Energy, http://energy.gov/fe/services/pe troleum-reserves/strategic-petroleum-reserve/spr-storage-sites.

142 *The amount of oil held in them:* This ninety-day supply can include commercial stocks, not just government-held ones.

142 *Just as America's rising import dependence:* See "U.S. Ending Stocks of Crude Oil in SPR," U.S. Energy Information Administration, April 28, 2017, https://www.eia.gov/dnav/pet/hist/LeafHandler.ashx?n=PET&s =MCSSTUS1&f=M.

142 *Certainly, this has been the logic:* See Javier Blas, "U.S. Plans to Sell Down Strategic Oil Reserve to Raise Cash," Bloomberg, October 27, 2015, http:// www.bloomberg.com/news/articles/2015-10-27/u-s-plans-to-sell-down

-strategic-oil-reserve-to-raise-cash. The 2017 budget proposal of the Trump administration also included a proposal to reduce the SPR by half. Chris Mooney and Steven Mufson, "Trump seeks to sell off half of the Strategic Petroleum Reserve," *Washington Post*, May 23, 2017, https://www.washingtonpost.com/news/energy-environment/wp/2017/05/22/trump-seeks-to-sell-off-half-of-the-strategic-petroleum-reserve/?utm_term=.c7efbb27c01b.

143 *Initially, the president was authorized:* Robert Bamberger, "The Strategic Petroleum Reserve: History, Perspectives, and Issues" (CRS Report No. RL33341) (Washington, D.C.: Congressional Research Service, August 18, 2009), 3–4, https://www.fas.org/sgp/crs/misc/RL33341.pdf.

143 *In the wake of the 1990* Exxon Valdez *spill:* Ibid., 3.

143 *Subsequent revisions further loosened the criteria:* Ibid., 7–8.

143 *In fact, the SPR has only:* While the United States has only released stocks from the SPR in conjunction with others in three circumstances, it has on other occasions released stocks unilaterally. Sometimes, SPR releases occur in non-emergency settings. See "International Energy Agency Members Release Strategic Petroleum Stocks," U.S. Energy Information Administration, June 24, 2011, http://www.eia.gov/todayinenergy/detail.php?id=1950. For records of unilateral releases, see "SPR Quick Facts and FAQs," U.S. Department of Energy, Office of Fossil Energy, http://energy.gov/fe/services/petroleum-reserves/strategic-petroleum-reserve/spr-quick-facts-and-faqs.

143 *These arguments deserve deeper examination:* There is, in fact, a robust debate about whether the SPR is needed in this era of energy abundance. For varying viewpoints, see David L. Goldwyn and Robert McNally, "Seven Fat Years: The Importance of Preserving the U.S. Strategic Petroleum Reserve," Brookings, July 17, 2015, https://www.brookings.edu/blog/order-from-chaos/2015/07/17/seven-fat-years-the-importance-of-preserving-the-u-s-strategic-petroleum-reserve/; Nicolas Loris, "Why Congress Should Pull the Plug on the Strategic Petroleum Reserve," Heritage Foundation, August 20, 2015, http://www.heritage.org/research/reports/2015/08/why-congress-should-pull-the-plug-on-the-strategic-petroleum-reserve; "Does the U.S. Need a Large Strategic Petroleum Reserve?" *Wall Street Journal*, November 15, 2015, http://www.wsj.com/articles/does-the-u-s-need-a-large-strategic-petroleum-reserve-1447642801; for the position of the U.S. Department of Energy, see U.S. Department of Energy,

Long-Term Strategic Review of the U.S. Strategic Petroleum Reserve: Report to Congress (Washington, D.C.: U.S. Department of Energy, August 2016), https://www.energy.gov/sites/prod/files/2016/09/f33/Long-Term%20 Strategic%20Review%20of%20the%20U.%20S.%20Strategic%20Petro leum%20Reserve%20Report%20to%20Congress_0.pdf.

144 *Together, more responsive U.S. tight oil:* Another potential complication worth mentioning is that, were the United States to use the SR in this fashion, it would further reduce the incentives of Saudi Arabia to hold spare capacity.

144 *Should the United States find:* It takes approximately two weeks from the time of a decision by the U.S. president to tap into the SPR until the time the oil enters the market. Daniel Goldstein, "U.S. Adding 5 Million Barrels of Oil to Its Reserves Because of Cheap Prices," *Market Watch*, March 13, 2015, www.marketwatch.com/story/us-adding-5-million-bar rels-of-oil-to-its-reserves-because-of-cheap-prices-2015-03-13.

144 *President Luiz Inácio Lula da Silva:* Agence France-Presse, "God Is Brazilian: Lula," *Fin 24*, November 20, 2007, www.fin24.com/International /God-is-Brazilian-Lula-20071120.

144 *Few American political figures have been:* One exception is Pat Wood, a former chairman of the Federal Energy Regulatory Commission, who in 2012, declared that "God gave us a great big gift here with fracking." Shelley DuBois, "Former FERC Regulator: God Gave Us Fracking," *Fortune*, April 16, 2012, http://fortune.com/2012/04/16/former-ferc-regulator -god-gave-us-fracking/.

Seven: Energy Abundance, Climate, and the Environment

146 *The majors largely sold off:* In 2016, some majors began once again to make investments in renewable energy, although such investments were very small in relation to their overall capital budget. See Joe Ryan, "Big Oil Unexpectedly Backing Newest Non-Fossil Fuels," Bloomberg, May 10, 2016, https://www.bloomberg.com/news/articles/2016-05-10/big -oil-unexpectedly-backs-newest-non-fossil-fuels.

146 *Some explained, quite simply, that:* See, for instance, Lord John Browne, Executive Chairman of L1 Energy and former CEO of BP, "The New Energy Environment" (seminar, Harvard Kennedy School of Government, Cambridge, MA, March 10, 2016).

148 *One form of unconventional oil:* Richard K. Lattanzio, "Canadian Oil Sands: Life-Cycle Assessments of Greenhouse Gas Emissions," Congressional Research Service, March 10, 2014, 2, www.fas.org/sgp/crs/misc/R42537.pdf. Some studies calculated this percentage to be significantly lower. See, for instance, IHS Cambridge Energy Research Associates, *Oil Sands, Greenhouse Gases, and US Oil Supply: Getting the Numbers Right* (Cambridge, MA: IHS CERA, 2010), https://cdn.ihs.com/ihs/cera/Oil-Sands-Greenhouses-Gases-and-US-Oil-Supply.pdf.

148 *It is, however, useful to keep in mind:* International Energy Agency, *World Energy Outlook 2016* (Paris: OECD Publishing, 2016), November 16, 2016, 136, http://www.iea.org/newsroom/news/2016/november/world-energy-outlook-2016.html.

148 *Extra-heavy oil and oil shale:* See Adam R. Brandt, "Converting Oil Shale to Liquid Fuels: Energy Inputs and Greenhouse Gas Emissions of the Shell in Situ Conversion Process," *Environmental Science & Technology* 42, no. 19 (2008): 7489–95, http://pubs.acs.org/doi/pdf/10.1021/es800531f; Stefan Unnasch et al., *Assessment of Life Cycle GHG Emissions Associated with Petroleum Fuels* (Portola Valley, CA: Life Cycle Associates, LLC, February, 2009), 61, www.newfuelsalliance.org/NFA_PImpacts_v35.pdf.

148 *By contrast, the production of tight oil:* Global unconventional numbers are from International Energy Agency, *World Energy Outlook 2016*, 136. As for the United States, tight oil production constituted 53 percent of U.S. oil production as of January 2017. "Table: Oil and Gas Supply," *Annual Energy Outlook 2017*, U.S. Energy Information Administration, https://www.eia.gov/outlooks/aeo/data/browser/#/?id=14-AEO2017&region=0-0&cases=ref2017&start=2015&end=2018&f=A&linechart=ref2017-d120816a.23-14-AEO2017~ref2017-d120816a.8-14-AEO2017~ref2017-d120816a.10-14-AEO2017&ctype=linechart&sid=ref2017-d120816a.10-14-AEO2017~ref2017-d120816a.8-14-AEO2017~ref2017-d120816a.23-14-AEO2017&sourcekey=0.

148 *Second, tight oil in the United States:* Jesse Esparza et al., "Argentina Seeking Increased Natural Gas Production from Shale Resources to Reduce Imports," U.S. Energy Information Administration, February 10, 2017, https://www.eia.gov/todayinenergy/detail.php?id=29912.

149 *Some research outfits have calculated:* For instance, see IHS Energy, Comparing GHG Intensity of Oil Sands and the Average US Crude Oil

(Calgary, IHS, May 2014), 11, https://www.ihs.com/products/energy
-industry-oil-sands-dialogue.html?ocid=cera-osd:energy:print:0001.

149 *Stanford professor Adam Brandt:* Brandt is quoted in Tona Kunz, "Analysis Shows Greenhouse Gas Emissions Similar for Shale, Crude Oil," Argonne National Laboratory, October 15, 2015, https://www.anl
.gov/articles/analysis-shows-greenhouse-gas-emissions-similar-shale
-crude-oil. A more recent study concludes that life cycle greenhouse gas emissions in the Bakken are comparable to other crudes because flaring is "largely offset at the refinery due to the physical properties of this tight oil." Ian J. Laurenzi, Joule A. Bergerson, and Kavan Motazedi, "Life cycle greenhouse gas emissions and freshwater consumption associated with Bakken tight oil," *Proceedings of the National Academy of Sciences*, 113, no. 48 (2016): 11, http://www.pnas.org/content/113/48
/E7672.full.

149 *One scenario, the High Oil and Gas:* In the "High Oil and Gas Research and Technology Case," the United States produces 14 percent more oil overall (crude plus NGLs) and 25 percent more tight oil in 2020 than in the EIA's reference case. These numbers jump to 53 percent more oil overall and 90 percent more tight oil in the high-resource case than in the reference one when one looks out to 2040. "Table: Oil and Gas Supply," Annual Energy Outlook 2017, U.S. Energy Information Administration, https://
www.eia.gov/outlooks/aeo/data/browser/#/?id=14-AEO2017&region
=0-0&cases=ref2017~lowprice~highrt&start=2015&end=2020&f=A
&linechart=highrt-d120816a.23-14-AEO2017~ref2017-d120816a.23-14
-AEO2017~ref2017-d120816a.8-14-AEO2017~highrt-d120816a.8-14
-AEO2017~ref2017-d120816a.10-14-AEO2017~highrt-d120816a.10
-14-AEO2017&ctype=linechart&sid=highrt-d120816a.23-14-AEO
2017~ref2017-d120816a.23-14-AEO2017~ref2017-d120816a.8-14
-AEO2017~highrt-d120816a.8-14-AEO2017~ref2017-d120816a.10-14
-AEO2017~highrt-d120816a.10-14-AEO2017&sourcekey=0.

149 *In both the reference:* See U.S. Energy Information Administration, *Annual Energy Outlook 2017* (Washington, DC: U.S. Department of Energy, 2017), https://www.eia.gov/outlooks/aeo/data/browser/#/?id
=17-AEO2017&region=1-0&cases=ref2017~highrt&start=2015
&end=2050&f=A&linechart=ref2017-d120816a.40-17-AEO2017.1
-0~ref2013-d102312a.41-17-AEO2013.1-0~highrt-d120816a.40
-17-AEO2017.1-0&map=highrt-d120816a.3-17-AEO2017.1-0&c

type=linechart&sid=ref2017-d120816a.40-17-AEO2017.1-0~highrt
-d120816a.40-17-AEO2017.1-0~ref2013-d102312a.41-17-AEO2013.1
-0&sourcekey=0.

149 *But the high-resource case involves:* "Table: Energy-Related Carbon Dioxide Emissions by Sector and Source," Annual Energy Outlook 2017,
U.S. Energy Information Agency, https://www.eia.gov/outlooks/aeo
/data/browser/#/?id=17-AEO2017&region=1-0&cases=ref2017~high
price~lowprice~highrt&start=2015&end=2050&f=A&linechart
=ref2017-d120816a.43-17-AEO2017.1-0~highrt-d120816a.43-17
-AEO2017.1-0&map=highprice-d120816a.3-17-AEO2017.1-0&ctype
=linechart&sid=highrt-d120816a.43-17-AEO2017.1-0~~~~ref
2017-d120816a.43-17-AEO2017.1-0~~~~~&sourcekey=0.

149 *The data on the extent to which:* See "Global Liquid Fuels," Short-Term
Energy and Summer Fuels Outlook, U.S. Energy Information Administration, April 11, 2017, https://www.eia.gov/outlooks/steo/report
/global_oil.cfm; International Energy Agency, *Oil Market Report* (Paris:
OECD/IEA Publishing, 2017), https://www.iea.org/media/omrreports
/tables/2017-03-15.pdf.

149 *The Paris-based IEA:* Pierpaolo Cazzola et al., "Production Costs of
Alternative Transportation Fuels: Influence of Crude Oil Price and
Technology Maturity," International Energy Agency, 2013, 9, www.iea
.org/publications/freepublications/publication/FeaturedInsights_Alter
nativeFuel_FINAL.pdf.

150 *Of the twenty possible alternative:* Ibid., 7.

150 *While moving natural gas into transportation:* Several studies support this
point. See Hari Chandan Mantripragada, "CO_2 Reduction Potential of
Coal-to-Liquids (CTL) Plants," *Energy Procedia* 1, no. 1, 4331–38, http://
repository.cmu.edu/cgi/viewcontent.cgi?article=1067&context=epp;
Xunmin Ou, Xiaoyu Yan, and Xiliang Zhang, "Using Coal for Transportation in China: Life Cycle GHG of Coal-Based Fuel and Electric
Vehicle, and Policy Implications," *International Journal of Greenhouse
Gas Control* 4, no. 5 (September 2010): 878–87, www.researchgate.net
/publication/222952135_Using_coal_for_transportation_in_China
_Life_cycle_GHG_of_coal-based_fuel_and_electric_vehicle_and_policy
_implications; Paulina Jaramillo, W. Michael Griffin, and H. Scott Mat-
thews, "Comparative Analysis of the Production Costs and Life-Cycle
GHG Emissions of FT Liquid Fuels from Coal and Natural Gas," *Environ-*

mental Science & Technology 42, no. 20 (2008), www.cmu.edu/gdi/docs/comparative-analysis.pdf.

150 *Throughout the mid-2000s:* This method is officially called the Fischer-Tropsch process. It was developed by Germany during World War II as the country struggled to maintain adequate access to oil. For more information on coal to liquid technology, see James T. Bartis, Frank Camm, and David S. Ortiz, *Producing Liquid Fuels from Coal: Prospects and Policy Issues* (Santa Monica, CA: RAND Corporation, 2008), http://www.rand.org/pubs/monographs/MG754.html.

150 *One such effort was the Coal to Liquid:* Editorial, "Coal-to-Liquid Boondoggle: A Risky Solution to America's Energy Woes," *Washington Post*, June 18, 2007, www.washingtonpost.com/wp-dyn/content/article/2007/06/17/AR2007061700945.html; "S. 154 (is)-Coal-to-Liquid Fuel Energy Act of 2007," U.S. Government Publishing Office, https://www.gpo.gov/fdsys/pkg/BILLS-110s154is/content-detail.html.

150 *Enthusiasm for coal-to-liquids was not limited:* "Accelergy Unveils Pilot Plant and Signals Move Into Chinese Market," *Business Wire*, June 29, 2011, www.businesswire.com/news/home/20110629006179/en/Accelergy-Unveils-Pilot-Plant-Signals-Move-Chinese.

150 *Generally viewed as not being commercial:* This is $60 in 2007 dollars. Bartis, Camm, and Ortiz, Producing Liquid Fuels from Coal: Prospects and Policy Issues, xx.

150 *The advent of shale gas:* Christopher Martin, "U.S. Carbon Emissions Falling to Two-Decade Low in Coal Shift," Bloomberg, April 9, 2015, www.bloomberg.com/news/articles/2015-04-09/u-s-carbon-emissions-falling-to-two-decade-low-in-coal-shift.

150 *Between 2005 and 2015, U.S. CO_2 emissions:* "U.S. Energy-Related Carbon Dioxide Emissions, 2015," U.S. Energy Information Administration, March 16, 2017, https://www.eia.gov/environment/emissions/carbon/.

150 *In 2015, U.S. per capita:* "CO_2 emissions (metric tons per capita.") The World Bank, http://data.worldbank.org/indicator/EN.ATM.CO2E.PC?locations=US; "Inventory of U.S. Greenhouse Gas Emissions and Sinks," Environmental Protection Agency, April 14, 2017, 3-33, https://www.epa.gov/sites/production/files/2017-02/documents/2017_complete_report.pdf.

150 *As made clear by the International Panel:* Thomas Bruckner et al., "Energy Systems," in *Climate Change 2014: Mitigation of Climate Change.*

Contribution of Working Group III to the Fifth Assessment Report of the Intergovernmental Panel on Climate Change, ed. Ottmar Edenhofer et al. (Cambridge, U.K.: Cambridge University Press, 2014), 527, https://www.ipcc.ch/pdf/assessment-report/ar5/wg3/ipcc_wg3_ar5_chapter7.pdf.

151 *The IEA agreed, declaring in 2013:* International Energy Agency, *Redrawing the Energy-Climate Map: World Energy Outlook Special Report* (Paris: OECD Publishing, June 10, 2013), 28, www.worldenergyoutlook.org/media/weowebsite/2013/energyclimatemap/RedrawingEnergyClimateMap.pdf.

151 *Due to the shale gas bonanza:* "Henry Hub Natural Gas Spot Price," U.S. Energy Information Administration, November 2, 2016, https://www.eia.gov/dnav/ng/hist/rngwhhdd.htm.

151 *Utilities did what made:* See Nathan Hultman, Dylan Rebois, Michael Scholten, and Christopher Ramig, "The Greenhouse Impact of Unconventional Gas for Electricity Generation," *Environmental Research Letters* 6 (2011), http://iopscience.iop.org/article/10.1088/1748-9326/6/4/044008/pdf; Ernest J. Moniz et al., *The Future of Natural Gas: An Interdisciplinary MIT Study* (Cambridge: MIT, 2011), https://energy.mit.edu/wp-content/uploads/2011/06/MITEI-The-Future-of-Natural-Gas.pdf; Stephen P. A. Brown, Alan J. Krupnick, and Margaret A. Walls, "Natural Gas: A Bridge to a Low-Carbon Future?" (Issue Brief 09-11, Resources for the Future and National Energy Policy Institute, December 2009), www.rff.org/RFF/Documents/RFF-IB-09-11.pdf; Michael A. Levi, "Climate Consequences of Natural Gas as a Bridge Fuel," *Climate Change* 118, no. 3 (2013): 609–23, www.cfr.org/natural-gas/climate-consequences-natural-gas-bridge-fuel/p29772.

151 *After rising consistently since the 1960s:* "Table 1.3 Primary Energy Consumption by Source," U.S. Energy Information Administration, https://www.eia.gov/totalenergy/data/browser/?+b1=T01.03#/?f=A.

151 *Over the following decade:* Ibid.

151 *In 2005, half the electricity:* "Net Generation for All Sectors, Annual," Electricity Data Browser, U.S. Energy Information Administration, https://www.eia.gov/electricity/data/browser/#/topic/0?agg=2,0,1&fuel=vvg&geo=g&sec=g&linechart=ELEC.GEN.ALL-US-99.A~ELEC.GEN.COW-US-99.A~ELEC.GEN.NG-US-99.A&columnchart=ELEC.GEN.ALL-US-99.A&map=ELEC.GEN.ALL-US-99.A

&freq=A&start=2008&end=2016&ctype=linechart<ype=pin
&rtype=s&pin=&rse=0&maptype=0.

151 *A decade later, natural gas was virtually:* Ibid.

151 *The number of American coal mines:* "Annual Coal Report," U.S. Energy Information Administration, November 3, 2016, www.eia.gov/coal/annual/; Tyler Hodge, "Natural Gas Expected to Surpass Coal in Mix of Fuel Used for U.S. Power Generation in 2016," U.S. Energy Information Administration, March 16, 2016, www.eia.gov/todayinenergy/detail.cfm?id=25392.

151 *The industry lost 98 percent:* This assessment is based on the Dow Jones U.S. Coal Index. The index peaked in 2008 at $724. In January 2016, it hit a nadir of $12, a reduction of 98 percent. In April 2016, with the Trump administration taking a policy stance more supportive of coal, the index had bounced back to $42. "Dow Jones U.S. Coal Index (INDEXDJX:D JUSCL)," Google Finance, accessed April 9, 2017, https://www.google.com/finance?cid=4931635.

151 *Indicative of future expectations for coal:* Victor Reklaitis, "Bankrupt Peabody's stock plunge, in one chart," *MarketWatch,* April 14, 2016, www.marketwatch.com/story/bankrupt-peabodys-stock-plunge-to-around-1-in-one-chart-2016-04-13.

151 *According to David Victor:* Bjørn Lomborg, "A Fracking Good Story," *Project Syndicate,* September 15, 2012, www.slate.com/articles/health_and_science/project_syndicate/2012/09/thanks_to_fracking_u_s_carbon_emissions_are_at_the_lowest_levels_in_20_years_.html. Intent confirmed via David Victor, e-mail message to author, July 27, 2016.

152 *After a series of complex computations:* Richard G. Newell and Daniel Raimi, "Implications of Shale Gas Development for Climate Change," *Environmental Science & Technology* 48, no. 15 (April 22, 2014), http://pubs.acs.org/doi/abs/10.1021/es4046154. Also see Haewon McJeon et al., "Limited Impact on Decadal-Scale Climate Change from Increased Use of Natural Gas," *Nature* 514 (October 15, 2014): 482–85, www.nature.com/nature/journal/v514/n7523/full/nature13837.html.

153 *In a 2011 report discussed earlier:* International Energy Agency, *World Energy Outlook 2011 Special Report: Are We Entering a Golden Age of Gas?* (Paris: OECD Publishing, 2011), 37–38, www.worldenergyoutlook.org/media/weowebsite/2011/WEO2011_GoldenAgeofGasReport.pdf.

153 *But, similar to the other studies:* Ibid., 37–38.

153 *Talking with me on the margins:* Richard Newell, in-person conversation with author, Houston, Texas, March 9, 2017.

153 *Perhaps most importantly, it is essential:* Richard Newell also stressed the importance of minimizing the release of methane in natural gas production, as indicated by his study referenced earlier.

153 *With abundant natural gas:* Newell and Raimi, "Implications of Shale Gas Development for Climate Change."

153 *Ultimately, technologies such as carbon:* For more on carbon, capture, and storage, see Michael Gebert Faure and Roy A. Partain, *Carbon Capture and Storage: Efficient Legal Policies for Risk Governance and Compensation* (Cambridge: MIT Press, 2017).

153 *most scenarios depicting a global energy system:* The IEA makes this point in another way in its 2016 World Energy Outlook. "On the one hand, gas is too carbon intensive to take a long-term lead in the decarbonisation of the energy sector. Uncertainty over the extent of leakage of methane, a potent GHG, along the gas supply chain also cast a shadow over the fuel's environmental credentials. On the other hand, natural gas is the least carbon intensive of the fossil fuels and thus burning gas is a much more efficient way to use a limited carbon budget than combusting coal or oil. Gas is especially advantageous to the transition if it can help smooth the integration of renewables into power systems along the way." International Energy Agency, *World Energy Outlook 2016*, 163.

154 *While global investment in renewables:* Angus McCrone, *Global Trends in Renewable Energy Investment 2017: Key Findings* (Frankfurt: Frankfurt School–UNEP Centre/BNEF, 2017), 12, http://fs-unep-centre.org/sites/default/files/attachments/gtr_2017_-_key_findings.pdf.

154 *A closer look, however, suggests:* The report also notes that the timing of investments as well as a slowdown in renewable energy deployment in China and Japan also contributed to the 2016 investment drop. Ibid., 12–13.

154 *A report by the United National Environment:* Ibid., 11.

154 *For instance, in the United States:* Tom Randall, "What Just Happened in Solar is a Bigger Deal Than Oil Exports," Bloomberg, December 17, 2015, https://www.bloomberg.com/news/articles/2015-12-17/what-just-happened-to-solar-and-wind-is-a-really-big-deal.

155 *With an air of excitement in the room:* Donald J. Trump, "Presidential Executive Order on Promoting Energy Independence and Economic

Growth," The White House Office of the Press Secretary, March 28, 2017, https://www.whitehouse.gov/the-press-office/2017/03/28/presiden tial-executive-order-promoting-energy-independence-and-economi-1.

155 *The order focused on repealing regulations:* "Remarks by President Trump at Signing of Executive Order to Create Energy Independence," The White House Office of the Press Secretary, March 28, 2017, www.white house.gov/the-press-office/2017/03/28/remarks-president-trump-signing -executive-order-create-energy.

155 *this bilateral accord was a springboard:* It is important to note two things, however significant this agreement was in catalyzing further action. First, China agreed to reduce its carbon intensity and to cap total CO_2 emissions, but it did not agree to decrease emissions. Second, even if all countries keep to the pledges they made in the Paris Agreement, the effort will still fall short of what is needed to avert "catastrophic" climate change.

156 *Howard Rogers, a scholar at the Oxford Institute:* Howard Rogers, "The Forthcoming LNG Supply Wave: A Case of 'Crying Wolf'?" (Energy Insight: 4, Oxford Institute for Energy Studies, University of Oxford, Oxford, U.K., February 2017), 8-9, https://www.oxfordenergy.org /wpcms/wp-content/uploads/2017/02/The-Forthcoming-LNG-Supply -Wave-OIES-Energy-Insight.pdf.

157 *Yet over the course:* For instance, a contentious piece of legislation in Colorado would introduce significant setback provisions that some see as effective bans on fracking. See John Fryer, "Oil, gas school setbacks bill clears Colorado House in party-line vote, "*Longmont Times-Call,* March 29, 2017, www.timescall.com/longmont-local-news/ci_30887589/colo rado-house-approves-lafayette-rep-mike-footes-oil.

158 *Yet, he did express concern to me:* The major environmental impacts of which Deutch spoke are air quality, water quality, community impact, land use, and induced seismicity. John Deutch, in-person conversation with author, Cambridge, MA, April 19, 2017. Confirmed via John Deutch, e-mail message to author, April 20, 2017.

158 *Just ask the residents of Sparks:* John Daly, "U.S. Government Confirms Link Between Earthquakes and Hydraulic Fracturing," Oilprice.com, November 8, 2011, http://oilprice.com/Energy/Natural-Gas/U.S.-Government -Confirms-Link-Between-Earthquakes-and-Hydraulic-Fracturing.html. Also see National Research Council et al., *Induced Seismicity Potential in Energy Technologies* (Washington, D.C.: National Academies Press, 2013).

158 *Or solicit the views of Hugh Fitzsimons:* Kate Galbraith, "As Fracking Increases, So Do Fears About Water Supply," *New York Times,* March 7, 2013, www.nytimes.com/2013/03/08/us/as-fracking-in-texas-increases-so -do-water-supply-fears.html?pagewanted=all.

158 *Or you may wish to talk:* John Fryar, "Weld County Resident Says Fracking Is Costing Him Sleep," *Times-Call Local News,* August 16, 2014, www .timescall.com/longmont-local-news/ci_26350144/weld-county-resident -says-fracking-is-costing-him.

158 *Or even chat with the residents:* Abrahm Lustgarten, "Feds Warn Residents Near Wyoming Gas Drilling Sites Not to Drink Their Water," *ProPublica,* September 1, 2010, www.propublica.org/article/feds-warn-residents-near -wyoming-gas-drilling-sites-not-to-drink-their-water.

159 *There is certainly room to remove:* For instance, deregulation makes sense where federal regulation duplicate state regulation of fracking.

160 *Halliburton claims it can create:* Patrick J. Kiger, "Green Fracking? 5 Technologies for Cleaner Shale Energy," *National Geographic,* March 21, 2014, http://news.nationalgeographic.com/news/energy/2014 /03/140319-5-technologies-for-greener-fracking/.

160 *Others have started to use:* Roger Real Drouin, "On Fracking Front, a Push to Reduce Leaks of Methane," *Yale Environment 360,* April 7, 2014, http://e360.yale.edu/mobile/feature.msp?id=2754.

160 *The report's bottom line:* Secretary of Energy Advisory Board, *Shale Gas Production Subcommittee Second Ninety Day Report* (Washington, D.C.: U.S. Department of Energy, November 18, 2011), https://energy.gov /sites/prod/files/90day_Report_Second_11.18.11.pdf.

160 *Other organizations such as the IEA:* See International Energy Agency, *Golden Rules for a Golden Age of Gas: World Energy Outlook Special Report on Unconventional Gas* (Paris: OECD Publishing, November 12, 2012), www.worldenergyoutlook.org/media/weowebsite/2012/goldenrules /weo2012_goldenrulesreport.pdf; "Performance Standards," Center for Responsible Shale Development, April 7, 2016, http://www.responsible shaledevelopment.org/wp-content/uploads/2016/09/Performance-Stan dards-v.-1.4-2.pdf.

160 *Concerns that the implementation:* Some technological innovations will be more expensive to deploy. For instance, David Burnett, a professor at Texas A&M, estimates that producing shale gas with waterless fracking could be 25 percent more expensive. (Waterless fracking, however, is not

considered among the "best practices" as the technology is still nascent.) This number, however, is likely to fall as the technologies are perfected—and could seem less burdensome if policy changes increased the now extremely low cost of freshwater in many states. See Patrick J. Kiger, "Green Fracking? 5 Technologies for Cleaner Shale Energy."

161 *In a 2016 presentation:* Michael Porter, "Realizing America's Unconventional Energy Opportunity" (presentation, GLOBE 2016, Vancouver, March 3, 2016), Slide 12.

161 *A drive to keep oil and gas:* For an interesting preliminary study that concludes the costs of air pollution and CO_2 emissions are significantly greater for rail than pipeline, see Karen Clay et al., "Economics and Externalities of Moving Crude Oil by Pipelines and Railroads: Evidence From the Bakken Formation" (presentation, American Economic Association, Chicago, IL, January 8, 2017).

161 *Moreover, whereas some deregulation:* See "Innovation through Regulation," *Economist*, June 2, 2009, www.economist.com/node/13766329.

161 *In a world increasingly cognizant:* See, for example, Trefor Moss, "Ford to Make Electric Cars in China Amid Green Drive," *Wall Street Journal*, April 7, 2017, https://www.wsj.com/articles/ford-to-make-electric-cars -in-china-amid-green-drive-1491475032.

162 *Several U.S. companies:* Many companies favor a tax on carbon in order to provide certainty around investment decisions. Tim Puko, "Big Oil Steps Up Support for Carbon Tax," *Wall Street Journal*, June 20, 2017, https://www.wsj.com/articles/big-oil-steps-up-support-for-carbon-tax -1497931202.

162 *Some environmentalist groups—such as:* The Environmental Defense Fund openly states that part of its ability to make an impact depends on its partnerships with corporate actors. See "Partnerships: The Key to Scalable Solutions," Environmental Defense Fund (EDF), https://www .edf.org/approach/partnerships.

Eight: Europe:—Catching a Break

165 *Calling European sanctions:* See video of "Brussels Forum: Europe in Transition," YouTube, 1:10:25, posted by "GermanMarshallFund," March 21, 2014, https://www.youtube.com/watch?v=X9zTuJ4piGM.

165 *Foremost among them was the fact:* Some claim that Kosovo was another

instance of changing borders through force. For a good exposition of the differences, see David L. Phillips, "Crimea Is Not Kosovo," *World Post*, March 10, 2014, www.huffingtonpost.com/david-l-phillips/crimea-is-not -kosovo_b_4936365.html.

166 *This view prevailed among many:* Russia has been able to wield influence by providing certain countries with advantageous pricing, so the threat of normal market pricing is a real and potent one. This has largely been the case with Ukraine in the crises of 2006 and 2009, when Russia decided that pro-Western political events in Kiev removed the rationale for Russia to continue to provide low-cost gas to Ukraine. Disputes over raising the price led Russia to suspend gas supply to Ukraine. In 2009, Gazprom—frustrated by Ukraine's siphoning off of gas intended for other countries—stopped supply to the rest of Europe for thirteen days given that 80 percent of Russian gas heading to Europe transited Ukraine at the time.

166 *Many argued that these crises:* See "Remarks by Paolo Scaroni," Chief Executive Officer, Eni, Delivered at CERA Week, Houston, TX, March 7, 2010. https://www.eni.com/en_IT/attachments/media/news/speech_Scaroni %20.pdf.

166 *In a regretful tone, Putin explained:* See "Message from the President of Russia to the Leaders of Several European Countries," President of Russia, April 10, 2014, http://en.kremlin.ru/events/president/news/20751.

166 *Russian gas accounted for a third:* If Turkey is included, this percentage is even higher. BP, *BP Statistical Review of World Energy 2016*, 23–28, http:// www.bp.com/content/dam/bp/pdf/energy-economics/statistical-re view-2016/bp-statistical-review-of-world-energy-2016-full-report.pdf.

167 *Instead, for years, the energy focus:* For a full exploration of the evolution of the "securitization" of Europe's energy policy, see Svein S. Andersen, Andreas Goldthau, and Nick Sitter, eds., *Energy Union: Europe's New Liberal Mercantilism?* (London: Palgrave, 2017).

167 *New reverse-flow pipeline capabilities:* At the time, Ukraine was able to import 5 percent of its consumption through these countries. BP, *BP Statistical Review of World Energy 2016*, 23, http://www.bp.com/content /dam/bp/pdf/energy-economics/statistical-review-2016/bp-statistical -review-of-world-energy-2016-full-report.pdf; Frank Umbach, "The Energy Dimensions of Russia's Annexation of Crimea," *NATO Review Magazine*, May 27, 2014, www.nato.int/docu/review/2014/nato-energy

-security-running-on-empty/Ukraine-energy-independence-gas-depen
dence-on-Russia/EN/index.htm.

167 *For instance, the Nabucco Pipeline:* See also Friedel Taube, "Nabucco Pipeline Future Uncertain as Hungary Backs Russian Rival," Deutsche Welle,
April 26, 2012, www.dw.de/nabucco-pipeline-future-uncertain-as-hun
gary-backs-russian-rival/a-15910599.

167 *In the trio of priorities:* At the national level, not all European countries
have ranked energy security last. As indicated later in the chapter, European countries have very different levels of import dependence and con-
sequently different national priorities.

167 *In the words of Richard Morningstar:* Richard Morningstar, in-person
conversation with author, Abu Dhabi, UAE, January 12, 2017.

168 *Reagan dismissed the ongoing détente:* Richard C. Thornton, *The Reagan
Revolution II: Rebuilding the Western Alliance* (Bloomington, IN: Trafford
Publishing, 2006), 142–44.

168 *Even more telling is the proposed:* As of mid-2017, the list of proposed concessions by Gazprom is out for comment among European states. Some,
such as Lithuania and Poland, are urging the EU to take an even tougher
stance against Gazprom. See Natalia Drozdiak, "EU Regulator to Test
Gazprom's Antitrust Remedies in Eastern Europe," *Wall Street Journal*,
March 13, 2017, https://www.wsj.com/articles/eu-regulator-to-test-gaz
proms-antitrust-remedies-in-eastern-europe-1489404872; Simone Tagliapietra, "The EU antitrust case: no big deal for Gazprom," *Bruegel*,
March 15, 2017, bruegel.org/2017/03/the-eu-antitrust-case-no-big-deal
-for-gazprom/.

168 *Under the proposed settlement:* In the absence of a settlement, the fines
imposed on Gazprom could amount to as much as 10 percent of its global
annual turnover. Tagliapietra, "The EU antitrust case: no big deal for
Gazprom."

169 *For years, Ukraine depended on:* This proportion varied from two-thirds
to nearly all of Ukraine's gas imports. BP, *BP Statistical Review of World
Energy 2016*, http://www.bp.com/content/dam/bp/excel/energy-eco
nomics/statistical-review-2016/bp-statistical-review-of-world-ener
gy-2016-workbook.xlsx; Leonid Bershidsky, "How Ukraine Weaned
Itself Off Russian Gas," Bloomberg, January 12, 2016, https://www
.bloomberg.com/view/articles/2016-01-12/how-ukraine-weaned-it
self-off-russian-gas.

169 *But, thanks to gas market reforms:* Tim Daiss, "Ukraine Celebration: One Year Without Russian Gas," *Forbes,* November 27, 2016, https://www.forbes.com/sites/timdaiss/2016/11/27/ukraine-celebration-one-year-without-russian-gas/#7d7d34f462f4; Roman Olearchyk, "Ukrainians will finally feel benefit of reforms this year, says PM," *Financial Times,* April 5, 2017, https://www.ft.com/content/80e6b64a-1927-11e7-9c35-0dd2cb31823a.

169 *"Gas is so abundant now":* Andriy Kobolyer, in-person conversation with author, Kiev, Ukraine, October 17, 2016.

170 *The U.S. EIA estimated:* The U.S. Energy Information Administration looked at eleven European countries and estimated them to hold 470 tcf of shale gas or 6.5 percent of the global shale gas reserves estimated in 2013. Shale gas in Ukraine would add another 128 tcf, or 1.8 percent of estimated global shale reserves. Russia holds 287 tcf, according to this source. These figures refer to unproved resources, meaning they cannot necessarily be produced commercially with today's technology. U.S. Energy Information Administration, *Technically Recoverable Shale Oil and Shale Gas Resources: An Assessment of 137 Shale Formations in 41 Countries Outside the United States* (Washington, DC: U.S. Department of Energy, June 2013), 6–7, www.eia.gov/analysis/studies/worldshalegas/pdf/fullreport.pdf.

170 *A 2013 study by Germany's Federal Institute:* Andreas Bahr et al., *BGR Energy Study 2013: Reserves, Resources and Availability of Energy Resources* (Hannover: Federal Institute for Geosciences and Natural Resources [BGR], December 2013), 39, www.bgr.bund.de/EN/Themen/Energie/Downloads/energiestudie_2013_en.pdf;jsessionid=04547424D-826C776449087E66D430EB4.1_cid284?__blob=publicationFile&v=2.

170 *If all European countries:* International Energy Agency, *Golden Rules for a Golden Age of Gas: World Energy Outlook Special Report on Unconventional Gas* (Paris: OECD Publishing, November 12, 2012), 130, www.worldenergyoutlook.org/media/weowebsite/2012/goldenrules/weo2012_goldenrulesreport.pdf.

170 *This latter number is impressive:* In 2016, Europe (including Turkey) imported 166 bcm or 5.9 tcf of natural gas from Russia. BP, *BP Statistical Review of World Energy 2017,* 34, http://www.bp.com/content/dam/bp/en/corporate/pdf/energy-economics/statistical-review-2017/bp-statistical-review-of-world-energy-2017-full-report.pdf.

170 *International investors flocked:* Andrew Kureth, "Polish Shale Gas Hits a Dry Well," *Politico*, June 16, 2015, www.politico.eu/article/polish-shale -gas-hits-a-dry-well/.

171 *government awarded more than ninety:* "Exxon to Start Fracking 2nd Polish Shale Gas," *Reuters*, September 28, 2011, http://www.reuters.com /article/poland-shale-exxon-idUSL5E7KS0QK20110928; Neil Buckley, "Eastern European shale exploration on ice as boom turns to bust," *Financial Times*, October 28, 2015, https://www.ft.com/content/72a0fbd4 -7cae-11e5-a1fe-567b37f80b64.

171 Yet, *in March 2012, the Polish Geological Institute:* In 2013, the U.S. Energy Information Administration assessed—based on historical factors, rather than new exploration—that Poland had as much as 146 tcf of shale gas reserves. "World Shale Resource Assessments," U.S. Energy Information Administration, September 24, 2015, https://www.eia.gov/analysis/studies /worldshalegas/. The Polish Geological Institute estimated this amount to be between 12.2 tcf and 27.1 tcf. See Polish Geological Institute National Research Institute, *Assessment of Shale Gas and Shale Oil Resources of the Lower Paleozoic Baltic-Podlasie-Lublin Basin in Poland* (Warsaw: Polish Geological Institute, March 2012), 5, https://www.pgi.gov.pl/en/docman-dokumen ty-pig-pib/docman/aktualnosci-2012/zasoby-gazu/769-raport-en/file.html.

171 *One executive from a major U.S.:* Buckley, "Eastern European Shale Exploration on Ice as Boom Turns to Bust."

171 *Far from turning Poland:* Ibid.

171 *In Austria, simply complying:* Brad Plumer, "How Long Before Fracking Spreads to Europe? A Decade, at Least," *Washington Post*, February 7, 2013, https://www.washingtonpost.com/news/wonk/wp/2013/02/07/will -fracking-ever-spread-to-europe-maybe-in-a-decade/.

171 *As described by Oxford scholar:* Jonathan Stern, "The Future of Gas in Decarbonising European Energy Markets: The Need for a New Approach" (The Oxford Institute for Energy Studies, Oxford University, U.K., January 2017), 2–4, https://www.oxfordenergy.org/wpcms/wp-content/up loads/2017/01/The-Future-of-Gas-in-Decarbonising-European-Energy -Markets-the-need-for-a-new-approach-NG-116.pdf.

171 *France, Belgium, the Netherlands:* Brigitte Osterath, "What Ever Happened with Europe's Fracking Boom?," Deutsche Welle, July 20, 2016, www.dw.com/en/what-ever-happened-with-europes-fracking-boom /a-18589660.

171 *Germany has moved to the left:* Critics of the new law have been largely from the left, arguing that the ban is more of a "fracking permission" law than the ban they desire. "Die Beschlüsse des Bundestages vom 12. bis 14. Dezember," *Deutscher Bundestag*, December 14, 2012, https://www.bundestag.de/dokumente/textarchiv/2012/42036947_kw50_angenommen_abgelehnt/210232; "Bundestag beschließt weitgehendes Fracking-Verbot," *Zeit Online*, June 24, 2016, www.zeit.de/wirtschaft/2016-06/erdgasfoerderung-fracking-bundestag-verbot.

172 *Restrictions exist on fracking:* "RPT-Opposition, Disappointing Data Wither Europe's Shale Gas Prospects," Reuters, February 3, 2015, www.reuters.com/article/europe-shalegas-idUSL6N0VD1NE20150203.

173 *Long before there was talk:* Fiona Harvey, "Russia's 'Secretly Working with Environmentalists to Oppose Fracking,'" *Guardian*, June 19, 2014, www.theguardian.com/environment/2014/jun/19/russia-secretly-working-with-environmentalists-to-oppose-fracking; Arthur Herman, "Is Russia Funding Europe's Anti-fracking Green Protests?," *Hudson Institute*, July 21, 2014, www.hudson.org/research/10461-is-russia-funding-europe-s-anti-fracking-green-protests-.

173 *Given Europe's current trajectory, no rational actor:* International Energy Agency, *Golden Rules for a Golden Age of Gas: World Energy Outlook. Special Report on Unconventional Gas* (Paris: OECD Publishing, November 12, 2012), www.worldenergyoutlook.org/media/weowebsite/2012/goldenrules/weo2012_goldenrulesreport.pdf.

173 *In fact, even in the most optimistic:* Imports go up in part because gas is cheaper, generating more demand. See ibid., 129.

174 *She declared, "The liquefied:* "Lithuania Welcomes 'Game Changing' LNG Vessel," *LNG Industry*, October 27, 2014, www.lngindustry.com/liquid-natural-gas/27102014/Lithuania-welcomes-FSRU-Independence-1675/.

174 *U.S. Secretary of State John Kerry:* Georgi Kantchev, "With U.S. Gas, Europe Seeks Escape from Russia's Energy Grip, *Wall Street Journal*, February 25, 2016, www.wsj.com/articles/europes-escape-from-russian-energy-grip-u-s-gas-1456456892; "Let There Be Gas!," *Baltic Times*, November 5, 2014, www.baltictimes.com/news/articles/35728/.

174 *Secretary Kerry's letter did not specifically:* "U.S. Secretary John Kerry Congratulates Lithuania on LNG Terminal Opening," Embassy of the United States, Vilnius, Lithuania, October 27, 2013, http://vilnius.usembassy

.gov/press_releases/2014/10/27/2014--u.s.-secretary-john-kerrys-con
gratulates-lithuania-on-lng-terminal-opening.

174 *Just seven months earlier:* John Boehner, "Counter Putin by Liberating
U.S. Natural Gas," *Wall Street Journal,* March 6, 2014, http://online.wsj
.com/news/articles/SB10001424052702303824204579421024172546260.

174 *Four foreign ambassadors:* "In Response to Russian Aggression, Key
Central European Nations Plead for U.S. Natural Gas Exports," Speaker
Boehner's Press Office, March 8, 2014, www.speaker.gov/press-release/re
sponse-russian-aggression-key-central-european-nations-plead-us-nat
ural-gas-exports.

175 *Most analysts scoffed:* See, for example, Steve Mufson, "Boehner's plan
to save Ukraine: It's full of gas," *Washington Post,* March 7, 2014, https://
www.washingtonpost.com/news/wonk/wp/2014/03/07/boehners-plan
-to-save-ukraine-its-full-of-gas/?utm_term=.82520b1125b1.

175 *Yet, in the years since:* "Henry Hub Natural Gas Spot Price," U.S. Energy
Information Administration, https://www.eia.gov/dnav/ng/hist/rngwh
hdd.htm.

175 *By February 2017, Italy, Malta, Portugal:* "U.S. Natural Gas Exports
and Re-Exports by Country," U.S. Energy Information Administration,
https://www.eia.gov/dnav/ng/ng_move_expc_s1_m.htm.

175 *A report from Columbia University:* Jason Bordoff and Trevor Hauser,
"Amerian Gas to the Rescue? The Impact of US LNG Exports on European Security and Russian Foreign Policy," Columbia SIPA Center on
Global Energy Policy, September 2014, 29, http://energypolicy.colum
bia.edu/sites/default/files/energy/CGEP_American%20Gas%20to%20
the%20Rescue%3F.pdf.

176 *Moreover, Gazprom is not a static player:* Thierry Bros, "Putting a Price
on Gas or Putin's Gas Price?" Research Centre for Energy Management
(RCEN), June 2, 2014, www.rcem.eu/posts/2014/june/02/putting-a-price
-on-gas-or-putin%E2%80%99s-gas-price.aspx.

176 *In 2016, one consultancy projected:* Anna Shiryaevskaya, "More than
Half of U.S. LNG Is Destined for Europe, WoodMac Says," Bloomberg,
January 15, 2016, www.bloomberg.com/news/articles/2016-01-15/more
-than-half-of-u-s-lng-is-destined-for-europe-woodmac-says.

176 *Releasing another report at the same time:* Kantchev, "With U.S. Gas, Europe Seeks Escape from Russia's Energy Grip."

177 *In 2013, Europe was only utilizing:* Guy Chazan, "Europe Seeks

Alternative Gas Supplies," *Financial Times*, April 26, 2014, www.ft.com/intl /cms/s/0/b943b2c4-b8ed-11e3-98c5-00144feabdc0.html#axzz3HBrbA jsU; www.ft.com/intl/cms/s/0/b943b2c4-b8ed-11e3-98c5-00144feabdc0 .html,Äîaxzz3HBrbAjsU; King & Spalding, *LNG in Europe: An Overview of European Import Terminals in 2015*, 4, https://kslawemail.com/77/574/ uploads/lng-in-europe---v5.pdf.

177 *At the time, a third of Europe's:* In 2014, Europe's regasification capacity was 201 bcm or 7 tcf, which was equivalent to 40 percent of Europe's demand for gas. Ibid., 4, 10.

177 *Asian consumers, particularly after Japan:* The trajectory of European LNG imports was in fact down from 8.7 tcf a year in 2010 to 4.3 tcf a year in 2014. "Chapter 3. Natural Gas," International Energy Outlook 2016, May 11, 2016, https://www.eia.gov/outlooks/ieo/nat_gas.cfm.

177 *Had it, one can be sure:* "Regional Price Spreads: Predicting the Future of LNG," *S&P Global Platts*, August 26, 2014, http://blogs.platts .com/2014/08/26/lng-price-spreads/.http://blogs.platts.com/2014/08/26 /lng-price-spreads/.

178 *The shale boom has both directly:* See Chapter Three for a fuller discussion of these developments.

179 *In the end, Lithuania received:* "Lithuania to Pay More for Norwegian LNC than Russian Gas," Reuters, November 13, 2014, www.reuters.com/ article/lithuania-lng-idUSL6N0T268X20141113.

179 *Despite insisting publicly that maintaining:* Tatiana Mitrova, "The Geopolitics of Russian Natural Gas," Belfer Center for Science and International Affairs, Kennedy School of Government, Harvard University, February 21, 2014, 61–62, http://www.belfercenter.org/sites/default/files /files/publication/CES-pub-GeoGasRussia-022114.pdf.

180 *Although the population of Europe:* Oystein Noreng, "The Global Dimension of EU Energy Policy," in *Energy Union: Europe's New Liberal Mercantilism?* ed. Svein S. Andersen, Andreas Goldthau, and Nick Sitter (London: Palgrave, 2017), 65.

180 *As of 2014, natural gas prices:* Ibid., 69.

180 *In allowing greater access:* Thomas Raines and Shane Tomlinson, *Europe's Energy Union: Foreign Policy Implications for Energy Security, Climate, and Competitiveness* (London: Chatham House) March 2016, 21, https://www.chathamhouse.org/sites/files/chathamhouse/

publications/research/2016-03-31-europe-energy-union-raines-tom linson.pdf.

Nine: Russia—More Petulant, Less Powerful

182 *A podgy academic from Moscow:* It is worth noting, however, that Gaidar's grandfather was a Bolshevik revolutionary.

182 *As the chief architect of Russia's:* Andrew E. Kramer, "Russia's Market Reform Architect Dies at 53," *New York Times*, December 16, 2009, http://www.nytimes.com/2009/12/17/world/europe/17gaidar.html.

182 *Before his death, Gaidar wrote:* See Yegor Gaidar, *Collapse of an Empire: Lessons for Modern Russia* (Washington, DC: Brookings Institution Press, 2010).

183 *Some have suggested that economic calamity:* See Nikolay Petrov, "Putin's Downfall: The Coming Crisis of the Russian Regime," European Council on Foreign Relations, April 19, 2016, www.ecfr.eu/publications/summary/putins_downfall_the_coming_crisis_of_the_russian_regime7006.

183 *One expert, Dmitri Trenin:* See Dmitri Trenin, "Russia Needs a Plan C," Carnegie Moscow Center, December 15, 2015, http://carnegie.ru/commentary/2015/12/15/russia-needs-plan-c/in4j.

183 *Other analysts see the plunge:* For instance, see Matt O'Brian, "Sorry, Putin. Russia's Economy Is Doomed," *Washington Post*, December 15, 2014, https://www.washingtonpost.com/news/wonk/wp/2014/12/15/russias-economy-is-doomed-its-that-simple/.

183 *Optimistic commentators see the period:* See Daniel Treisman, "Oil and Democracy in Russia" (Working Paper 15667, The National Bureau of Economic Research, Cambridge, MA, January 2010), http://www.nber.org/papers/w15667; Anders Åslund, "What Falling Oil Prices Mean for Russia and Ukraine," Atlantic Council, January 22, 2016, www.atlanticcouncil.org/blogs/ukrainealert/what-the-falling-oil-price-means-for-russia-and-ukraine.

184 *Yekaterina, from the Siberian city:* For a full transcript of the broadcast, see "Direct Line with Vladimir Putin," President of Russia, April 14, 2016, http://en.special.kremlin.ru/events/president/transcripts/51716.

184 *Over the course of the next four hours:* Ibid.

184 *The questioner responded:* Ibid.

184 *By the time of the broadcast:* "Russia, Annual data and forecast," The Economist Intelligence Unit, July 19, 2016, country.eiu.com/article .aspx?articleid=294438413&Country=Russia&topic=Economy&sub topic=Charts+and+tables&subsubtopic=Annual+data+and+forecast.

184 *The economy had shrunk by 3.6:* "GDP growth (annual %)," The World Bank, data.worldbank.org/indicator/NY.GDP.MKTP.KD.ZG?locations=RU.

184 *Gross national income per capita:* "Russian Federation," The World Bank, 2016, http://data.worldbank.org/country/russian-federation.

184 *The number of Russians living in poverty:* Ibid.

184 *But sanctions are estimated:* See Robin Emmott, "Sanctions Impact on Russia to Be Longer Term, U.S. Says," Reuters, January 12, 2016, www.reuters.com/article/us-ukraine-crisis-sanctions-idUSKCN0U Q1ML20160112. For more on the impact of sanctions on Russian eco-nomic growth, see "IMF Survey: Cheaper Oil And Sanctions Weigh On Russia's Growth Outlook," The International Monetary Fund, August 3, 2015, www.imf.org/external/pubs/ft/survey/so/2015/car080315b.htm.

185 *Russian finance minister Anton Siluanov:* Andrey Gurkov, "Sanctions Aren't the Key Source of Russia's Woes," Deutsche Welle, January 30, 2015, www.dw.com/en/sanctions-arent-the-key-source-of-russias-woes /a-18225289.

185 *he estimated that Russia loses:* Olga Tanas, "Russia Sees $140 Billion An-nual Loss from Oil, Sanctions," Bloomberg, November 24, 2014, www .bloomberg.com/news/articles/2014-11-24/russia-sees-140-billion-an nual-loss-from-oil-sanctions.

185 *In 2015, the energy sector accounted:* A. A. Makarov et al., ed. *Global and Russian Energy Outlook 2016* (Moscow: The Energy Research Institute of the Russian Academy of Science [ERI RAS] and Analytical Center of the Government of the Russian Federation [ACRF], 2016), 172, https://www .eriras.ru/files/forecast_2016.pdf.

185 *Energy sales and taxes provided:* Russia's export revenue and approxi-mately half of the revenues feeding the total federal budget came from energy in 2013. The total federal budget in 2014 was $432 billion. U.S. Energy Information Administration, *Annual Energy Outlook 2015: With Projections to 2040* (Washington, DC: U.S. Department of Energy, 2015), www.eia.gov/forecasts/aeo/pdf/0383(2015).pdf.

185 *Few economies outside the Gulf Cooperation Council:* On average,

hydrocarbon-related revenues make up 46 percent of the GDP of GCC countries and 75 percent of their total export revenues. Nicolas Parasie, "GCC Countries Still Too Reliant on Oil and Gas Revenues, S&P Says," *Wall Street Journal*, June 30, 2014, http://blogs.wsj.com/middle east/2014/06/30/gcc-countries-still-too-reliant-on-oil-and-gas-reve nues-sp-says/.

185 *According to Russian economist Tatiana Mitrova:* Tatiana Mitrova, in-person conversation with author, Moscow, Russia, May 26, 2014.

185 *In 1999, when Putin first became:* Putin was appointed to be "acting president" by President Boris Yeltsin in December 1999; he was first elected to the office in March 2000. The GDP per capita data used in the text are in current US dollars. "GDP per capita, PPP (current international $)," The World Bank, 2016, http://data.worldbank.org /indicator/NY.GDP.PCAP.PP.CD; "GDP per capita (current US$)," The World Bank, 2016, http://data.worldbank.org/indicator/NY.GDP .PCAP.CD?locations=RU.

185 *Without the propeller of steadily rising prices:* Russian GDP growth was 0.7 percent in 2014 and 1.3 percent in 2013. "GDP Growth (Annual %)," The World Bank, 2016, http://data.worldbank.org/indicator/NY.GDP .MKTP.KD.ZG?locations=RU.

185 *Had the unconventional energy revolution:* An American Petroleum Institute study conducted in October 2014 in collaboration with the consultancy ICF International found "that international Brent crude oil prices would have averaged $122 to $150 per barrel in 2013 without U.S. HMSHF crude oil and condensate production increases." HMSHF stands for horizontal multi-stage hydraulic fracturing, or fracking as it is better known. "U.S. Oil Impacts: The Impacts of Horizontal Multi-stage Hydraulic Fracturing Technologies on Historical Oil Production, International Oil Costs, and Consumer Petroleum Product Costs," ICF International, Presentation to The American Petroleum Institute, October 30, 2014, Fairfax, VA, www.api.org/~/media/Files/Policy/Hydraulic_Frac turing/ICF-Hydraulic-Fracturing-Oil-Impacts.pdf; "Europe Brent Spot Price FOB," U.S. Energy Information Administration, https://www.eia .gov/dnav/pet/hist/LeafHandler.ashx?n=pet&s=rbrte&f=a.

186 *Russia had several advantages:* See "Russian Federation: Overview," The World Bank, October 6, 2016, www.worldbank.org/en/country/russia/ overview.

186 *First, Russia had the ability to increase:* In a 2016 interview, Putin acknowledged, "We suffer dangerous revenue losses in our export of oil and gas." Nicolaus Blome and Kai Diekmann, "Putin—The Interview: 'For Me, It Is Not Borders That Matter,'" *Bild*, November 1, 2016, www.bild.de/politik/ausland/wladimir-putin/russian-president-vladimir-putin-the-interview-44092656.bild.html.

186 *Developing Russia's own vast unconventional:* By all accounts, Russia's own unconventional oil deposits are the largest in the world, estimated by the EIA to be 75 billion barrels. See James Henderson, "Tight Oil Developments in Russia," (WPM 52, Oxford Institute for Energy Studies, University of Oxford, U.K., October 2013), www.oxfordenergy.org/wpcms/wp-content/uploads/2013/10/WPM-52.pdf.

186 *it would require the lifting of sanctions:* Experts point out that Russia's unconventional fields—such as the Bazhenov formation in Siberia—lack the clean "wedding cake" structure found in the American Bakken fields and elsewhere. Walter Russell Mead quoted in Patrice Hill, "Siberian Shale Find Fuels Russia's Fracking Future," *Washington Times*, February 18, 2014, www.washingtontimes.com/news/2014/feb/18/siberian-shale-find-fuels-russias-fracking-future/?page=all. In 2013, Russian energy minister Alexander Novak estimated that, even with incentives from the government for exploration and production, tight oil would eventually add only 300,000 to 400,000 barrels per day—or less than 4 percent of Russia's overall 2013 oil production. "Russia Shuns OPEC, Plans Increased Oil Production," *Moscow Times*, December 3, 2013, www.themoscowtimes.com/news/article/russia-shuns-opec-plans-increased-oil-production/490833.html.

186 *One of the successor accounts:* In 2008, this $157 billion fund was split into the Reserve Fund and the National Welfare Fund. "Economic policy: Stabilisation Fund will be split into two funds in 2008," The Economist Intelligence Unit, September 1, 2007.

187 *Its size dropped:* The fund fell from a 2014 high of $92 billion to $16 billion in April 2017. "Sovereign wealth fund declines sharply in December," The Economist Intelligence Unit, January 11, 2017.

187 *Finally, and most importantly, the government:* Dmitry Dokuchaev, "Why a Free-Floating Ruble Is Not in Freefall," *Russia Direct*, November 18, 2015, www.russia-direct.org/analysis/why-free-floating-ruble-not-freefall.

187 *Almost instantly, the ruble weakened:* The weakened ruble also makes imported goods more expensive, having the effect of making Russians relatively "poorer." Vladimir Kuznetsov, "Ruble Drops to Record in Worst Month in Five Years as Bonds Fail," Bloomberg, November 28, 2014, https://www.bloomberg.com/amp/news/articles/2014-11-28/ruble -heads-for-weakest-month-in-five-years-after-oil-slump.

187 *In January 2016, Finance Minister:* Olga Tanas, Anna Andrianova, and Ryan Chilcote, "Russia on War Path Leaves Siluanov to Scrape By as Funds Ebb," Bloomberg, January 13, 2016, https://www.bloomberg.com /news/articles/2016-01-13/russia-seeks-20-billion-budget-boost-as-silu anov-warns-on-oil.

187 *He warned that a failure to do so:* Ibid.

188 *Putin, speaking in December 2015:* James Marson and Andrey Ostroukh, "Vladimir Putin Says Russia's Economic Crisis Has Peaked," *Wall Street Journal,* December 17, 2015, https://www.wsj.com/articles /vladimir-putin-warns-government-may-adjust-budget-over-oil-price -fall-1450346139.

188 *However, as Andrey Movchan:* Andrey Movchan, in-person conversation with author, Moscow, Russia, October 21, 2016.

188 *Such moves could buy Putin:* One independent analyst sought to estimate the price of oil Russia would need to balance its economy in 2017. He found $68 to be the magic number, in contrast to $82 cited by the Russian finance minister in January 2016. See Jacob Shapiro, "Here's Where Oil Needs to be for Russia to Break Even," *Business Insider,* December 28, 2016, http://www.businessinsider.com/russia-needs-higher-oil-prices-to -break-even-2016-12.

188 *The publication asked dozens of American:* "Will the Putin Regime Crumble? Foreign Affairs' Brain Trust Weighs In," *Foreign Affairs,* April 17, 2016, https://www.foreignaffairs.com/articles/russian-federation/2016-04-17 /will-putin-regime-crumble.

189 *At the Pushkin Café:* Dmitri Trenin, in-person conversation with author, Moscow, Russia, October 21, 2016.

189 *According to* Forbes *magazine, the number:* Kerry A. Dolan and Luisa Kroll, "Inside the 2014 Forbes Billionaires List: Facts and Figures," *Forbes,* March 3, 2014, http://www.forbes.com/sites/luisakroll/2014/03/03/inside -the-2014-forbes-billionaires-list-facts-and-figures/; Katie Sola, "The 25

Countries With The Most Billionaires," *Forbes*, March 09, 2016, http://www.forbes.com/sites/katiesola/2016/03/08/the-25-countries-with-the-most-billionaires/6/#1bd5e6dd461b.

189 *Over dinner at Moscow's:* Western journalist, in-person conversation with author, Moscow, Russia, May 28, 2014.

190 *According to David Szakonyi:* David Szakonyi, "Putin Is Still Standing: The Elites That Keep the President in Power," *Foreign Affairs*, July 26, 2016, https://www.foreignaffairs.com/articles/russian-federation/2016-07-26/putin-still-standing.

190 *The execution of import substitution:* Two contracts were reportedly awarded to the company of Arkady Rotenberg: one valued at $285 million to build a railway and the other, valued at $4 billion, to build a bridge to Crimea. See "Putin's Judo Partner Awarded $285M Contract For Crimea Railway," *Moscow Times*, January 16, 2017, https://themoscowtimes.com/news/putins-friend-to-construct-russo-crimean-railway-56826.

190 *Stephen Kotkin of Princeton University:* "Will the Putin Regime Crumble? Foreign Affairs' Brain Trust Weighs In," *Foreign Affairs*. Brezhnev had several strokes and heart attacks from 1974 to 1976. Although he was clinically dead after his 1976 stroke, his inner circle allegedly revived him and, for six years until his death in 1982, kept him in power although he had no cognitive ability to govern. For more details, see Stephen Kotkin, *Armageddon Averted: The Soviet Collapse, 1970–2000* (Oxford: Oxford University Press, 2001), 49–50.

191 *Spanning eleven time zones, Russia is composed:* Russia has forty-six provinces, 21 republics, 9 territories, 4 autonomous districts, an autonomous province, and 2 federal cities. The Republic of Crimea and the federal city of Sevastopol are still internationally recognized as part of Ukraine.

191 *A very important source of budget revenues:* Thomas F. Remington, "Here's how Alexander Hamilton would understand Russia's regional debt crisis," *Washington Post*, March 24, 2016, https://www.washingtonpost.com/news/monkey-cage/wp/2016/03/24/heres-how-alexander-hamilton-would-understand-russias-regional-debt-crisis/.

191 *Now, with less largesse to distribute:* Moscow is aware of the challenge and debating measures to help deeply indebted federal entities. See Olga Tanas, "Time Expiring for Russian Regions to Fix Budgets, Fitch Says," Bloomberg, April 28, 2016, http://www.bloomberg.com/news/articles/2016-04-28/fitch-ticks-off-the-time-for-russian-regions-to-find-budget-fix.

191 *These limitations could result:* Ariel Cohen, "A Threat to the West: The Rise of Islamist Insurgency in the Northern Caucasus and Russia's Inadequate Response," The Heritage Foundation, March 26, 2012, www .heritage.org/research/reports/2012/03/a-threat-to-the-west-the-rise-of -islamist-insurgency-in-the-northern-caucasus.

191 *A sampling of Russian experts:* Holly Ellyatt, "Why Russia Must Reform its Energy Sector," CNBC, June 21, 2013, http://www.cnbc.com /id/100834136.

192 *In the final months of that year:* Novatek was given permission to export LNG from its Yamal project to Asia. "Russia: Putin Approves LNG Exports for Novatek and Rosneft," *LNG World News*, December 3, 2013, https://www.lngworldnews.com/russia-putin-approves-lng-exports-for -novatek-and-rosneft/.

192 *On a sweltering Moscow day:* Advisor to the Russian government, in-person conversation with author, Moscow, Russia, May 28, 2014.

193 *Reportedly, previous ministers submitted plans:* Kathrin Hille, "Kudrin's Resurrection Fails to Damp Doubts on Russian Reforms," *Financial Times*, May 8, 2016, https://next.ft.com/content/7f30cba8-13ae-11e6 -91da-096d89bd2173.

193 *"The proposals Kudrin has advocated":* Ibid.

193 *In the* Moscow Times, *Rogov wrote:* Kirill Rogov, "Putin's Economist: Why Kudrin Tapped to Write New Reforms Plan," *Moscow Times*, April 20, 2016, https://themoscowtimes.com/articles/putins-economist -why-kudrin-tapped-to-write-new-reforms-plan-op-ed-52621.

194 *Similarly, at the end of 2004:* For perceptions of a slowing economy at the end of 2004, see Gérard Roland, "The Russian Economy in the Year 2005," University of Berkeley, 2006, 3–6, http://eml.berkeley.edu/~groland/pubs /The_Russian_Economy_in_the_Year_2005.pdf.

194 *Putin, himself a moderate nationalist:* For a detailed examination of the many facets of Putin's persona, see Fiona Hill and Clifford G. Gaddy, *Mr. Putin: Operative in the Kremlin* (Washington, DC: Brookings), 2013.

194 *The Kremlin has even reportedly:* See Maria Snegovaya, "How Putin's Worldview May Be Shaping His Response in Crimea," *Washington Post*, March 2, 2014, www.washingtonpost.com/blogs/monkey-cage /wp/2014/03/02/how-putins-worldview-may-be-shaping-his-response -in-crimea/.

195 *Some believe the annexation:* Those espousing this view often point to

Putin's public call for Russians to "acknowledge that the collapse of the Soviet Union was a major geopolitical disaster of the century" in a state of the nation address nearly a decade earlier. See Vladimir Putin, "Annual Address to the Federal Assembly of the Russian Federation," The Kremlin, Moscow, Russia, April 25, 2005, http://en.kremlin.ru/events/president/transcripts /22931.

195 *As evidence, some even pointed:* "Vladimir Putin's Interview with the BBC (March 13, 2000)," *Gazeta,* March 7, 2001, https://www.gazeta .ru/2001/02/28/putin_i_bbc.shtml.

195 *These elites also lamented the perceived:* See Vladimir Putin, "A New Integration Project for Eurasia—The Future Is Born Today," Center for Strategic Assessment and Forecasts, October 4, 2011, http://csef.ru/en /politica-i-geopolitica/223/novyj-integraczionnyj-proekt-dlya-evrazii -budushhee-kotoroe-rozhdaetsya-segodnya-1939; "The Article by Vladimir Putin '50 Years of the European Integration and Russia' is Published Today in the European Media," President of Russia, March 25, 2007, http://en.kremlin.ru/events/president/news/37692.

196 *As one leader of a Gulf country:* Anonymous Gulf state leader, in-person conversation with author, the Middle East, January 2016.

196 *So far, this approach has served:* See Joshua Tucker, "Why We Should Be Confident That Putin Is Genuinely Popular in Russia," *Washington Post,* November 24, 2015, https://www.washingtonpost.com/news/mon key-cage/wp/2015/11/24/why-we-should-be-confident-that-putin-is -genuinely-popular-in-russia/.

196 *According to Yegor Gaidar's book:* Yegor Gaidar, *Collapse of an Empire: Lessons for Modern Russia* (Washington, DC: Brookings, 2007), 106–7.

197 *one study by the Oxford Institute:* Andreas Economou et al., "Oil Price Paths in 2017: Is a Sustained Recovery of the Oil Price Looming?" (Energy Insight: 1, Oxford Institute for Energy Studies, University of Oxford, Oxford, U.K., January 2017), 10, https://www.oxfordenergy.org/wpcms /wp-content/uploads/2017/01/Oil-Price-Paths-in-2017-Is-a-Sustained -Recovery-of-the-Oil-Price-Looming-OIES-Energy-Insight.pdf.

197 *With the removal of many sanctions:* Mitchell A. Orenstein and George Romer, "Putin's Gas Attack: Is Russia Just in Syria for the Pipelines," *Foreign Affairs,* October 14, 2015.

197 *Finally Russian interest in cooperation:* For example, the 2016 deal

between OPEC and non-OPEC countries translated into significant gains for Russia. Russia exported an average of 5 million barrels per day from 2014–2016. This meant that for every dollar increase in the price of oil, Russia gained approximately $5mn on a daily basis or $1.8bn annually. There was an $8 increase in the Brent Price of Oil from November to April. That would roughly translate into an increase of $15bn in Russia's export revenues. "JODI-Oil, Primary (all data)," Joint Organizations Data Initiative, www.jodidb.org/ReportFolders/reportFolders.aspx?sCS_refer er=&sCS_ChosenLang=en; "Europe Brent Spot Price FOB," U.S. Energy Information Administration, https://www.eia.gov/dnav/pet/hist/Leaf Handler.ashx?n=pet&s=rbrte&f=m.

198 *In 2012, long before Russia's confrontation:* Denis Pinchuk and Alexei Anishchuk, "Putin Tells Russian Gas Exporters to Look East," Reuters, October 23, 2012, www.reuters.com/article/us-russia-putin-energy-idUS BRE89M0RR20121023.

199 *A great power should not:* Russian academic, in-person conversation with author, Moscow, Russia, May 27, 2014.

199 *Even the mouse's corpus:* "The Resident Population in an Average Year," EMISS Government Statistics, Ministry of Communications of Russia, https://fedstat.ru/indicator/31556?#.

199 *Yevgenii Nazdratenko, the recalcitrant:* Yevgenii Nazdratenko, "Russia: Far Eastern Governor Wants to Import More Russians," *Info-Prod Research*, June 5, 2000, www.highbeam.com/doc/1G1-62504369.html.

199 *Shaikin likely exaggerates when he claims:* "Russia Fears Flood of Illegal Chinese Entering Russian Far East," *Peace and Freedom*, July 6, 2013, https://johnib.wordpress.com/tag/alexander-shaikin/. According to the Woodrow Wilson Center in Washington, D.C., "The number of Chinese visiting, working, or living in Russia has been among the most wildly abused data points in a country known for statistical anomalies." See Maria Repnikova and Harley Balzer, "Chinese Migration to Russia: Missed Opportunities," Woodrow Wilson International Center for Scholars, 2009, 13, https://www.wilsoncenter.org/sites/default/files/No3_Chi neseMigtoRussia.pdf.

199 *But the Russian Federal Migration:* Peter Zeihan, "Analysis: Russia's Far East Turning Chinese," *ABC News*, July 14, http://abcnews.go.com/Inter national/story?id=82969.

199 *It is perhaps no surprise that:* Pepe Escobar, "The Roving Eye: The Future Visible in St Petersburg," *Asia Times*, May 29, 2014, www.atimes.com/atimes/Central_Asia/CEN-01-290514.html.

200 *More recently, a 2015 deal:* "How Russia Could Become a Food Superpower," *Wall Street Journal*, July 15, 2015, https://blogs.wsj.com/experts/2015/07/15/how-russia-could-become-a-food-superpower/.

200 *Russians declared, "China's creeping:* John Xenakis, "World View: Russia Makes a Controversial Deal to Lease Siberian Land to China," *Breitbart News*, June 21, 2015, http://www.breitbart.com/national-security/2015/06/21/world-view-russia-makes-a-controversial-deal-to-lease-siberia-land-to-china/.

200 *First, the crisis over Ukraine and Crimea:* In speaking to the Valdai Club—an annual gathering in Russia of international experts—in 2014, Putin explained, "Asia is playing an ever greater role in the world, in the economy and in politics, and there is simply no way we can afford to overlook these developments. Let me say again that everyone is doing this, and we will do so too, all the more so as a large part of our country is geographically in Asia. Why should we not make use of our competitive advantages in this area?" See full transcript at "Meeting of the Valdai International Discussion Club," President of Russia, October 24, 2014, http://en.kremlin.ru/events/president/transcripts/copy/46860.

200 *Today, Russia is desperately searching:* For instance, a decade ago, prospects for greater gas usage in Europe were on the rise as analysts and policymakers perceived gas as a bridge to a more low carbon fueled economy. Today, the combination of stagnant economic growth on the continent and a surfeit of European renewable energy has dulled this outlook. See Chapter Eight.

200 *In 2009, Gazprom anticipated:* Guy Chazan and Catherine Belton, "Gazprom Freezes Arctic Gas Project," *Financial Times*, August 29, 2012, www.ft.com/intl/cms/s/0/ab331568-f1d8-11e1-bba3-00144feabdc0.html#axzz3G3UL4vj6.

200 *These grandiose plans were:* Ibid.

201 *The room in Shanghai seemed far:* "Video: Russia, China sign 'gas deal of the century,' " YouTube video, 1:59, posted by "RT," May 21, 2014, https://www.youtube.com/watch?v=7MJJ-YzuKNs.

201 *The deal centered on the development:* "Power of Siberia," Gazprom, www.gazprom.com/about/production/projects/pipelines/ykv/.

202 *Under the thirty-year arrangement:* Nikos Tsafos, "The Russia-China Gas Deal: A \$400 Billion Marriage," *The National Interest*, May 29, 2014, http://nationalinterest.org/feature/the-russia-china-gas-deal-400 -billion-mirage-10556; the calculations in this sentence are based on the assumption that Chinese natural gas consumption is 360 bcm or 12.7 tcf in 2020 and on statistics provided in "Dry Natural Gas Consumption 2014," U.S. Energy Information Administration, www.eia.gov /cfapps/ipdbproject/iedindex3.cfm?tid=3&pid=26&aid=2&cid=CH, &syid=2010&eyid=2014&unit=BCF.

202 *During this time, China's demand:* Candace Dunn, "Natural Gas Serves a Small, but Growing, Portion of China's Total Energy Demand," U.S. Energy Information Administration, August 18, 2014, www.eia.gov/today inenergy/detail.cfm?id=17591.

202 *In 2012, CNPC estimated:* The 2012 CNPC estimate is sourced from Keun-Wook Paik, "Sino-Russian Gas Breakthrough and the Implications towards Regional and Global Gas Trading" (Presentation, Harvard Kennedy School Workshop, held in Washington D.C., Washington, D.C., January 29, 2014). In 2010, total Asian consumption of natural gas equaled 19.2 tcf. "Global natural gas consumption doubled from 1980 to 2010," U.S. Energy Information Administration, April 12, 2012, https://www .eia.gov/todayinenergy/detail.php?id=5810.

202 *Not only did China prefer:* Dunn, "Natural Gas Serves a Small, but Growing, Portion of China's Total Energy Demand."

203 *China—intrigued by the 2011:* U.S. Energy Information Administration, "World Shale Gas Resources: An Initial Assessment of 14 Regions Outside the United States" (Washington, D.C., U.S. Department of Energy, April 2011), 4, https://www.eia.gov/analysis/studies/worldshalegas/ar chive/2011/pdf/fullreport_2011.pdf.

203 *Talk began of how the United States:* Edward McAllister and Martin Roberts, "Analysis: U.S. LNG Exports a Real Possibility; Obstacles Remain," Reuters, December 3, 2010, www.reuters.com/article/us-lng-usa-export -idUSTRE6B24TN20101203.

203 *"Whether Gazprom slept through":* Stepan Kravchenko and Anna Shiryaevskaya, "Putin Dismisses Concern Gazprom 'Slept Through' Shale Gas Boom," Bloomberg, April 25, 2013, www.bloomberg.com/news/2013-04 -25/putin-dismisses-concern-gazprom-slept-through-shale-gas-boom .html.

203 *This figure is lower than Russia:* Although the final price of the gas remained shrouded in secrecy, most commentators believe China agreed to pay a price close to what Europe was paying for Russian gas at the time, leading to the valuation of $400bn. Author interviews in Moscow, Russia, May 2014.

203 *He was quick to tweet:* Louise Watt and Vladimir Isachenkov, "Gazprom Deal: China, Russia Natural Gas Pact Worth $400 Billion," *Huffington Post Business*, May 21, 2014, www.huffingtonpost.com/2014/05/21/china-russia-gas-deal-gazprom_n_5364004.html.

204 *Pushkov's counterpart in Russia's upper house:* Associated Press, "China Signs Giant 30-Year Deal with Russia for Natural Gas," *Washington Times*, May 21, 2014, www.washingtontimes.com/news/2014/may/21/china-signs-giant-30-year-deal-russia-natural-gas/?page=all.

204 *One Washington commentator asked:* Stephen Moore, "The Russia-China Pipeline," *National Review*, May 22, 2014, www.nationalreview.com/article/378555/russia-china-pipeline-stephen-moore.

204 *Others saw no occasion for levity:* Quote found in Dina Gusovsky, "Should America Worry About a China-Russia Axis?" *CNBC*, October 22, 2014, http://www.cnbc.com/2014/10/22/should-america-worry-about-a-china-russia-axis.html.

204 *One newspaper subsequently asked:* Emma Graham-Harrison et al., "China and Russia: The World's New Superpower Axis?," *Guardian*, July 7, 2015, www.theguardian.com/world/2015/jul/07/china-russia-superpower-axis.

204 *Predictions that "the US could find":* See, for example, Gordon G. Chang, "China and Russia: An Axis of Weak States," *World Affairs*, March/April 2014, www.worldaffairsjournal.org/article/china-and-russia-axis-weak-states.

204 *Reinforcing this sense of a burgeoning:* Carrie Gracie, "Brothers Again? How Deep Is the Xi-Putin Bromance?," *BBC News*, April 24, 2015, www.bbc.com/news/world-asia-china-32409409.

205 *But the two countries were slow:* See Philip Andrews-Speed and Roland Dannreuther, *China, Oil, and Global Politics* (London: Routledge, 2013), 119–22.

205 *They finally settled most of their border:* In 2009, China and Russia reached an agreement in which Russia would sell 300,000 barrels a day to China for twenty years in exchange for a $25 billion loan to Russian companies to

develop oil fields and build pipelines to deliver this oil to China. A spur of the Eastern Siberia–Pacific Ocean pipeline began delivering this oil in 2011.

205 *In addition to the natural gas deals:* Alexander Gabuev, "A 'Soft Alliance'? Russia-China Relations After the Ukraine Crisis," European Council on Foreign Relations, February, 2015, 4, www.ecfr.eu/page/-/ECFR126_-_A_Soft_Alliance_Russia-China_Relations_After_the_Ukraine_Crisis.pdf.

205 *During President Xi's 2015 visit:* "China, Russia Call for Proper Settlements of Major International Issues," *Xinhuanet,* May 9, 2015, http://news.xinhuanet.com/english/2015-05/09/c_134222910.htm.

205 *Perhaps most significantly, China has stepped:* In 2015, China Construction Bank Corp. pledged to provide as much as $25 billion as loan guarantees to struggling Russian companies. In 2016, the Bank of China agreed to a $2.2 billion, five year loan to Gazprom, the largest the company ever received from one bank. See Anna Baraulina, "Russia Wealth Fund to Team With China for $25 Billion of Lending," Bloomberg, May 8, 2015, https://www.bloomberg.com/news/articles/2015-05-08/russia-wealth -fund-to-team-with-china-for-25-billion-of-lending; "Gazprom secures large loan from China," Economist Intelligence Unit, March 3, 2016, http:/www.eiu.com/industry/article/1754001159/gazprom-secures -large-loan-from-china/2016-03-04.

206 *The combination of continued sanctions:* Morena Skalamera and Andreas Goldthau suggest that Russia's shift to Asia has been reduced to a more strategically limited "pivot to China." See Morena Skalamera and Andreas Goldthau, "Russia: Playing Hardball or Bidding Farewell to Europe?" (discussion paper 2016-03, Geopolitics of Energy Project, Belfer Center for Science and International Affairs, Harvard University, Cambridge, MA, June 2016), 15–20, http://www.belfercenter.org/sites/default/files/files /publication/Russia%20Hardball%20-%20Web%20Final.pdf.

206 *Ambitious Russian plans to meet:* Despite the slowing of LNG projects, Japanese prime minister Shinzo Abe more recently launched specific initiatives to forge closer economic ties with Russia. Abe has presented a package of business deals to Russia, with the understanding the Japanese companies will be more involved in infrastructure and energy development projects in Russia. One initiative was a $400 million loan from the state-owned Japan Bank for International Co-operation to Novatek to help develop the Yamal LNG project, despite the fact that Novatek is under sanctions. Abe and Putin have also engaged in sporadic discussions aimed at

the settlement of a longstanding territorial dispute surrounding the Kuril Islands. See "Japan Strengthens Outreach to Russia," Economist Intelligence Unit, September 13, 2016, http://country.eiu.com/article.aspx?articleid=1994605583&Country=Japan&topic=Politics. For the state of cooperation with South Korea, also see Younkyoo Kim, "The Impact of Low Oil Prices on South Korea," National Bureau of Asian Research, May 14, 2015, www.nbr.org/research/activity.aspx?id=562.

206 *The gas once expected to flow:* Stuart Radnedge, "Gazprom is postponing construction of the Vladivostok LNG plant on the country's Pacific Coast," *Gasworld*, June 29, 2015, https://www.gasworld.com/vladivostok-lng-plant-construction-shelved/2007743.article.

206 *Gazprom's CEO, Alexei Miller:* Ibid.

206 *While Russia's reliance on China:* See Meghan L. O'Sullivan, "Asia: A Geopolitical Beneficiary of the New Energy Environment," in *Asia's Energy Security Amid Global Market Change*, Muhamad Izham Abd. Shukor et al. (Washington, D.C.: The National Bureau of Asian Research, December 2016), 22–24, http://nbr.org/publications/special-report/pdf/Free/02172017/SR63_AsiasEnergySecurity_December2016.pdf.

207 *Specifically, the new price environment:* See "Russia-China Pipeline Plans in Crisis," Economist Intelligence Unit, February 4, 2016.

207 *Scholars from the U.K.-based International:* Samuel Charap, John Drennan and Pierre Noël, "Russia and China: A New Model of Great-Power Relations," *Survival* 59, no. 1, January 31, 2017.

207 *But the fact remains that, if China:* Author's own calculations from various Chinese sources. Also see Christine Forster and Stephanie Wilson, "China's lower gas demand leaves NOCs facing oversupply: WoodMac," Platts, ed. Megan Gordon, June 10, 2015, https://www.platts.com/latest-news/naturalgas/sydney/chinas-lower-gas-demand-growth-leaves-nocs-facing-27495854; Shi Xunpeng, Hari Malamakkavu Padinjare Variam, and Jacqueline Tao, "Global impact of uncertainties in China's gas market," *Energy Policy* 104 (2017), http://www.australiachinarelations.org/sites/default/files/2017%20Shi%20Variam%20Tao%20China%20gas%20market%20uncertainties.pdf.

207 *This dynamic will accentuate:* Linking overseas finance with Chinese construction companies is one of the ways in which the Chinese government hopes to address the overcapacity that has emerged as a result of the

slowdown of China's economy after decades of boom. See Bo Kong and Kevin P. Gallagher, "The Globalization of Chinese Energy Companies: The Role of State Finance," (Global Economic Governance Initiative, Boston University, 2016), https://www.bu.edu/pardeeschool/files/2016/06 /Globalization.Final_.pdf.

207 *If China gets involved in constructing*: Interestingly, China reportedly wanted to keep the Altai pipeline project alive, conditional upon holding an open tender and allowing Chinese companies to build the pipeline. Aleksandr Medvedev, the deputy director of Gazprom, rejected this proposal, stating that "we do not need Chinese men and equipment here, we never did and we never will." "Russia-China Pipeline Plans in Crisis," Economist Intelligence Unit, February 4, 2016.

208 *Negotiations between the Turkmen*: Andreas Heinrich and Heiko Pleines, eds., *Export Pipelines from the CIS Region: Geopolitics, Securitization, and Political Decision-Making* (Stuttgart: ibidem Press, 2014), 163–68.

208 *After a tense 2009 summit between*: Bruce Pannier, "Pipeline Explosion Raises Tensions Between Turkmenistan, Russia," RadioFreeEurope Radio Liberty, April 14, 2009, https://www.rferl.org/a/Pipeline_Explosion_Stokes _Tensions_Between_Turkmenistan_Russia/1608633.html.

209 *An agreement was then struck*: Isabel Gorst, "Russia welcomes end to Turkmen gas dispute," *Financial Times*, December 23, 2009, http://www .ft.com/cms/s/0/20dfe82e-ef69-11de-86c4-00144feab49a.html?ft_site =falcon&desktop=true#axzz4nrjqySIQ.

209 *The Russian-Turkmen gas trade resumed*: Russian-Turkmen gas trade stopped once again in 2016. "UPDATE 2-Russia resumes Turkmen gas imports, slashes volumes," Reuters, December 22, 2009, www.reuters .com/article/russia-turkmenistan-idUSLDE5BL0BX20091222; Catherine Putz, "Russia's Gazprom Stops Buying Gas from Turkmenistan," *The Diplomat*, January 6, 2016, thediplomat.com/2016/01/russias-gazprom -stops-buying-gas-from-turkmenistan/.

209 *Energy will help it consolidate*: Armenia, Belarus, Kazakhstan, Kyrgyzstan, and Russia are all members of the EEU; Tajikistan was still contemplating membership as of 2016.

209 *Russia, in the words*: Tatiana Mitrova, in-person conversation with author, Aspen, CO, July 6, 2016.

209 *However, there is little competition*: "The New Great Game in Central Asia," European Council on Foreign Relations, www.ecfr.eu/page

/-/China Analysis_The new Great Game in Central Asia_September2011
.pdf.

210 *Critical—and delicate—energy deals followed:* A pipeline delivering Ka-
     zakh oil to China delivered its first oil in 2006. That same year, negotia-
     tions between the Chinese and Turkmen governments bore fruit with the
     signing of an agreement related to the development and sale of natural
     gas. In 2007, China completed transit agreements with neighboring Uz-
     bekistan and Kazakhstan. Just two years later, the Central Asian-China
     Pipeline was inaugurated and Turkmen gas began to flow 7,000 km across
     four countries to Chinese markets, or almost one and a half the length of
     the United States from coast to coast.

210 *Whether he actually wrote the thesis:* "Researchers Peg Putin as a Pla-
     giarist Over Thesis," *Washington Times*, March 24, 2006, http://www
     .washingtontimes.com/news/2006/mar/24/20060324-104106-9971r.
     See "Putin's Thesis (Raw Text)," *Atlantic*, August 20, 2008, https://
     www.theatlantic.com/daily-dish/archive/2008/08/putins-thesis-raw
     -text/212739/.

210 *Released in May 2009, Russia's:* This quote is sourced from Keir Giles,
     "Russia's National Security Strategy to 2020," (Research Division, NATO
     Defense College, Rome, Italy, June 2009), 6, http://conflictstudies.co.uk
     /files/RusNatSecStrategyto2020.pdf.

Ten: China—Greater Degrees of Freedom

212 *In February 1960, in a drive:* Lai-Ha Chan et al., *China at 60: Global-Local
     Interactions* (Singapore: World Scientific, 2011), 235.

212 *As depicted in a 2009 Chinese film:* Xinhua, "China's 'Iron Man' an Undy-
     ing Legend," *People's Daily Online*, September 17, 2009, http://en.people
     .cn/90001/90776/90882/6760061.html; Henry M. Paulson, *Dealing with
     China: An Insider Unmasks the New Economic Superpower* (New York:
     Twelve, 2015).

212 *Legend holds that, without any other infrastructure:* Xinhua, "China's 'Iron
     Man,' an Undying Legend."

212 *Equal quantities of water:* Ibid.

213 *Foreshadowing his own early death:* Ibid.

213 *Yet despite Daqing's success:* In 2014, Daqing was producing about 850,000
     barrels a day, after declining from 1 mnb/d. The China National Petroleum

Corporation, Daqing's operator, reportedly intends to decrease production to 640,000 barrels a day by 2020 due to high costs of production, lower oil prices, and limited reserves. See "China: Overview," U.S. Energy Information Administration, May 14, 2015, www.eia.gov/beta/international/analysis.cfm?iso=CHN.

213 *The country's energy self-sufficiency:* For a history of China's oil dependencies, see Hong-Pyo Lee, "China's Petroleum Trade," *The Journal of East Asian Affairs* 4, no. 1 (1990).

213 *Whereas it had taken more:* BP p.l.c., *BP Statistical Review of World Energy 2016* (London: BP, June 2016), www.bp.com/content/dam/bp/excel/energy-economics/statistical-review-2016/bp-statistical-review-of-world-energy-2016-workbook.xlsx.

213 *As of 2014, it accounted:* Nick Butler, "China: The World's Energy Superpower," *Financial Times*, September 21, 2014, http://blogs.ft.com/nick-butler/2014/09/21/china-the-worlds-energy-superpower/.

214 *The Chinese government would have preferred:* For a longer exposition on Deng's approach, see Aaron L. Friedberg, *A Contest for Supremacy: China, America, and the Struggle for Mastery in Asia* (New York: W.W. Norton & Company, 2012), chapter 6.

214 *In 2009, two Chinese officials:* The central planning department of the Chinese government predicted that China would need to import 60 percent of its oil by 2020. Xiaoli Liu and Xinmin Jiang, "China's Energy Security Situation and Countermeasures," *International Journal of Energy Sector Management* 3, no. 1 (2009): 83–92.

214 *the following year in 2010:* In 2010, the IEA projected that 78 percent of China's oil needs would be met by imports in 2030. International Energy Agency, *World Energy Outlook 2010* (Paris: OECD Publishing, 2010), 105, 128, www.worldenergyoutlook.org/media/weo2010.pdf.

216 *The transformation of China:* Michael Forsythe and Jonathan Ansfield, "Fading Economy and Graft Crackdown Rattle China's Leaders," *New York Times*, August 22, 2015, www.nytimes.com/2015/08/23/world/asia/chinas-economy-and-graft-crackdown-rattle-leaders.html.

216 *In 1978, when China began to open:* Percentages are derived from "GDP (current US$)," The World Bank, 2016, http://data.worldbank.org/indicator/NY.GDP.MKTP.CD.

216 *In less than forty years, China has grown:* Forsythe and Ansfield, "Fading Economy and Graft Crackdown."

216 *It now constitutes approximately*: Percentage derived from "GDP (current US$)," The World Bank.

216 *In the mere three years*: Vaclav Smil, *Making the Modern World: Materials and Dematerialization* (West Sussex: John Wiley & Sons, 2014), 91.

216 *Yet China is now the world's largest economy*: If the size of the economies are not adjusted for Purchasing Power Parity, China's economy is still smaller than that of United States.

216 *It is also the planet's biggest manufacturer*: Wayne M. Morrison, *China's Economic Rise: History, Trends, Challenges, and Implications for the United States* (CRS Report No. RL33534) (Washington, DC: Congressional Research Service, October 21, 2015), 1, https://www.fas.org/sgp/crs/row/RL33534.pdf.

216 *During the first decade of*: Rakesh Kochhar, "A Global Middle Class Is More Promising than Reality," Pew Research Center, July 8, 2015, www.pewglobal.org/2015/07/08/a-global-middle-class-is-more-promise-than-reality/.

216 *Had she visited five years later*: In 2005, there were 2.6 million cars in Beijing; by January 2011, this number had risen to 4.8 million vehicles. "Beijing Unveils Measures to Ease Traffic Flow," *Beijing International*, www.ebeijing.gov.cn/BeijingInformation/BeijingNewsUpdate/t1146155.htm.

216 *She would have noticed how many*: Jim Gorzelany, "The Worst Traffic Jams in History," *Forbes*, May 21, 2013, www.forbes.com/sites/jimgorzelany/2013/05/21/the-worst-traffic-jams-in-history/.

217 *In the words of China scholar*: Susan L. Shirk, *China: Fragile Superpower* (New York: Oxford University Press, 2007), 54.

217 *Just twenty years earlier, China*: For harrowing accounts of this famine, see Frank Dikötter, *Mao's Great Famine: The History of China's Most Devastating Catastrophe* (London: Bloomsbury, 2011); Tania Branigan, "China's Great Famine, The True Story," *Guardian*, January 1, 2013, www.theguardian.com/world/2013/jan/01/china-great-famine-book-tombstone.

217 *With more disposable income*: Jikun Huang and Scott Rozelle, "Agricultural Development and Nutrition: The Policies Behind China's Success," World Food Programme, November 2009, http://documents.wfp.org/stellent/groups/public/documents/newsroom/wfp213339.pdf, 27.

217 *In 2002, both the Chinese government*: Daniel H. Rosen and Trevor

Houser, "What Drives China's Demand for Energy (and What It Means for the Rest of Us)," *The China Balance Sheet in 2007 and Beyond* (Washington, D.C.: CSIS and The Peterson Institute), 29, web.archive.org/web/20110220195805/http://csis.org/files/media/csis/pubs /090212_02what_drives_china_demand.pdf.

217 *They also predicted annual energy demand:* Ibid.

217 *China's economy grew rambunctiously:* China's GDP grew between 8.4 percent and 14.2 percent over the years 2000–2010 (inclusive). "GDP Growth (Annual %)," World Bank, 2016, http://data.worldbank.org/indicator/NY.GDP.MKTP.KD.ZG?locations=CN.

217 *Overall energy consumption grew four times:* Rosen and Houser, "What Drives China's Demand for Energy."

218 *By the mid-2000s, almost half:* According to C. Fred Bergsten, "China now accounts for 48 percent of global cement production, 49 percent of global flat glass production, 35 percent of global steel production, and 28 percent of global aluminum production." C. Fred Bergsten et al., *China's Rise: Challenges and Opportunities* (Chicago: Peterson Institute, 2009), 142–43.

219 *As China grew at an average growth:* Chinese growth figures from "GDP Growth (Annual %)," The World Bank.

219 *During this period, China consumed an average:* These figures are for total petroleum consumption. "Total Petroleum Consumption 2014," U.S. Energy Information Administration, 2014, www.eia.gov/cfapps/ipdbproject /IEDIndex3.cfm?tid=5&pid=5&aid=2.

219 *One study estimated that, in the absence:* ICF International, "U.S. Oil Impacts: The Impacts of Horizontal Multi-stage Hydraulic Fracturing Technologies on Historical Oil Production, International Oil Costs, and Consumer Petroleum Product Costs," presentation, The American Petroleum Institute, Washington, DC, October 30, 2014, www.api.org /~/media/Files/Policy/Hydraulic_Fracturing/ICF-Hydraulic-Fracturing-Oil-Impacts.pdf; "Europe Brent Spot Price FOB," U.S. Energy Information Administration, November 9, 2016, www.eia.gov/dnav/pet/hist /LeafHandler.ashx?n=pet&s=rbrte&f=a.

219 *If we simplify things to assume:* Calculations which should only be considered notional suggest $235 million a day and $86 billion a year. These numbers were produced by multiplying the number of barrels imported a day (5.5 million) from 2011 to 2014 by the difference in the actual

price (Brent price average was $107.60 from 2011 to 2014) from the price that above study calculated would have prevailed in the absence of the tight oil boom ($150 per barrel). If, for simplicity's sake, we attribute the whole price plunge to the production of unconventional resources, we can then estimate that in 2011, 2012, 2013, and 2014, China was saving $86 billion a year on its oil imports thanks to this boom. ICF International, "U.S. Oil Impacts: The Impacts of Horizontal Multistage Hydraulic Fracturing Technologies on Historical Oil Production, International Oil Costs, and Consumer Petroleum Product Costs"; "Europe Brent Spot Price FOB," U.S. Energy Information Administration, https://www.eia.gov/dnav/pet/hist/LeafHandler.ashx?n=pet&s=rb rte&f=a; "JODI-Oil: Joint Organizations Data Initiative–Primary (all data)," Joint Organizations Data Initiative, http://www.jodidb.org/.

219 *Moreover, the actual price plunge:* There is of course some relationship between China's weakening growth and lower oil prices, although China is not the biggest driver of the plunge by far. For instance, Andreas Economou, Bassam Fattouh, Paolo Agnolucci, and Vincenzo De Lipsi attribute $11 of the $50 price drop in oil between June 2014 and January 2015 to "the weakening of the global economy" (without specifying China's weight in that phenomenon). See Andreas Economou et al., "Oil Price Paths in 2017: Is a Sustained Recovery of the Oil Price Looming?" (Energy Insight: 1, Oxford Institute for Energy Studies, University of Oxford, Oxford. U.K., January 2017), 3, https://www.oxfordenergy.org/wpcms /wp-content/uploads/2017/01/Oil-Price-Paths-in-2017-Is-a-Sustained -Recovery-of-the-Oil-Price-Looming-OIES-Energy-Insight.pdf.

219 *The United States, for instance, took decades:* U.S. household final consumption expenditure as a percentage of GDP increased 7 percent (from 61 percent to 68 percent) over fifty-three years, from 1960 to 2013. "Household Final Consumption Expenditure, etc. (%ofGDP)," The World Bank, 2016, http://data.worldbank.org/indicator/NE.CON.PETC .ZS?locations=US.

220 *While consumer demand in China:* "China: Annual Data and Forecast," Economist Intelligence Unit, April 1, 2017.

220 *For every one-dollar drop:* Calculations here are derived from "JODI-Oil: Joint Organizations Data Initiative–Primary (all data)," Joint Organizations Data Initiative, http://www.jodidb.org/.

220 *Put another way, given that the price:* At $111.8 a barrel in June 2014, the

import of 6.8 million barrels of oil cost $760 million, whereas the same amount imported at $53 a barrel in December 2016 cost nearly $400 million less. "Europe Brent Spot Price Fob," U.S. Energy Information Administration, https://www.eia.gov/dnav/pet/hist/LeafHandler.ashx?n =pet&s=rbrte&f=m; "JODI-Oil: Joint Organizations Data Initiative– Primary (all data)," Joint Organizations Data Initiative, http://www .jodidb.org/.

220 *However, despite such benefits:* Antoine Halff, "The Outlook for Asia's Oil Market in a Lower Price Environment," in *Asia's Energy Security Amid Global Market Change,* ed. Muhamad Izham Abd. Shukor et al. (Seattle: National Bureau of Asian Research, 2016), http://nbr.org/pub lications/specialreport/pdf/Free/02172017/SR63_AsiasEnergySe curity_December2016.pdf?utm_source=Center+on+Global+Ener gy+Policy+Mailing+List&utm_campaign=1514364ae0-EMAIL_CAM PAIGN_2016_12_17&utm_medium=email&utm_term=0_0773077aac -1514364ae0-102087193.

221 *In February 2015, a video titled:* "Under the Dome (English subtitle, Complete) by Chai Jing: Air pollution in China," YouTube video, 1:43:57, posted by "Jiahua Guo," March 8, 2015, https://www.youtube.com/ watch?v=V5bHb3ljjbc&feature=youtube.

221 *In 2012, in Shifang:* See "Quiet Returns to Once-Restive-Shifang," *Wall Street Journal,* July 4, 2012, https://blogs.wsj.com/chinarealtime/2012/07 /04/quiet-returns-to-once-restive-shifang.

222 *Yang Chaofei, the vice chairman:* Jennifer Duggan, "Kunming Pollution Protest Is Tip of Rising Chinese Environmental Activism," *Guardian,* May 16, 2013, www.theguardian.com/environment/chinas-choice/2013/ may/16/kunming-pollution-protest-chinese-environmental-activism.

222 *A 2013 survey conducted:* Wang Hongyi, "Govt Environmental Transparency in Doubt," *China Daily: Europe,* May 9, 2013, http://europe.china daily.com.cn/china/2013-05/09/content_16486401.htm.

222 *This dirtiest of the fossil fuels:* Joseph Ayoub, "China Produces and Consumes Almost as Much Coal as the Rest of the World Combined," U.S. Energy Information Administration, May 14, 2014, www.eia.gov/today inenergy/detail.cfm?id=16271. In 2015, according to China's Bureau of Statistics, this number was 64 percent; in 2016, it was 62 percent. Michael Holz, "China's Coal Consumption Drops Again, Boosting its Leadership on Climate Change," *Christian Science Monitor,* March 3, 2017, www

.csmonitor.com/World/Asia-Pacific/2017/0303/China-s-coal-consump
tion-drops-again-boosting-its-leadership-on-climatechange.

222 *According to the U.S. Energy Information Administration:* Ayoub, "China
Produces and Consumes Almost as Much Coal."

222 *the scale of this challenge:* BP has Chinese demand for all fuels growing
robustly out to 2035. Demand for oil rises 63 percent, that for natural gas
increases 193 percent, and that for coal goes up 5 percent. Demand for
renewables in the power sector increases dramatically by 593 percent,
whereas nuclear power and hydropower go up by 827 and 43 percent respectively. "Country Insights: China," BP Energy Outlook, www.bp.com
/content/dam/bp/pdf/energy-economics/energy-outlook-2016/bp-ener
gy-outlook-2016-country-insights-china.pdf.

222 *Multiple obstacles have existed:* BP p.l.c., *BP Statistical Review of World
Energy 2016* (BP, June 2016), www.bp.com/content/dam/bp/excel/energy
-economics/statistical-review-2016/bp-statistical-review-of-world
energy-2016-workbook.xlsx.

223 *Its goal is for gas to account:* "Goals set for nuclear energy development
in next five years," *China Daily*, January 18, 2017, http://www.chinadaily
.com.cn/business/2017-01/18/content_27988526.htm.

223 *This may sound like a small amount:* Mark Dwortzan, "Enabling China
to Shift from Coal to Natural Gas," *MIT News*, November 18, 2016,
http://news.mit.edu/2016/enabling-china-to-shiftfrom-coal-to-natural
-gas-1118.

223 *But China's targets would equal:* This calculation is made by taking the
2014 projection for how much natural gas China will be using in 2020
and comparing it to Gazprom's total natural gas exports in 2014. "UP-
DATE 1-Russia's Gazprom Sees Gas Production at All-Time Low," Reuters, December 24, 2014, www.reuters.com/article/2014/12/24/russia
-gazprom-output-idUSL6N0U81HA20141224.

223 *To keep the country on track:* For instance, the government lowered by a
quarter the price local distributors pay to pipeline operators. See Brian
Spegele, "China Cuts Natural-Gas Prices to Spur Demand," *Wall Street
Journal*, November 18, 2015, www.wsj.com/articles/china-cuts-natural
-gas-prices-to-spur-demand-1447865810.

223 *Partially in response to such efforts:* Demand growth for natural gas had
dropped dramatically to approximately 4 percent in 2015, alarming Chinese policymakers. (In the years between 2009 and 2014, natural gas

demand had risen on average by 14 percent.) BP, *BP Statistical Review of World Energy 2017*. This flagging demand growth was one of the impetuses behind the price liberalization. Osamu Tsukimori, "Chinese Gas Demand to Rise; Will Help to Ease Glut: IEA," Reuters, November 24, 2016, http://www.reuters.com/article/us-lng-japan-iea-idUSKBN13J0OU.

223 *The global glut in natural gas:* "China's Cheaper Coal Seen Slowing Switch to Cleaner Natural Gas," Bloomberg, May 19, 2016, https://www.bloomberg.com/news/articles/2016-05-19/china-s-cheaper-coal-seen-slowing-switch-to-cleaner-natural-gas.

223 *However, if MIT researchers Sergey Paltsev:* Danwei Zhang and Sergey Paltsev, "The Future of Natural Gas in China: Effects of Pricing Reform and Climate Policy," *Climate Change Economics* 7, no. 4 (2016), https://globalchange.mit.edu/sites/default/files/MITJPSPGC_Reprint_16-19.pdf.

224 *China spent $90 billion dollars:* IEA, *World Energy Outlook 2016*, 406.

224 *this amount was more than:* Ibid.

224 *According to the IEA, electricity generated:* By IEA's assessment, electricity generation from solar photovoltaics (electricity generated by utilities rather than rooftop panels), onshore wind, geothermal, and hydroelectricity are cheaper than that generated by natural gas turbines in China for projects completed in 2015. Combined cycle gas turbines, however, are cheaper than all other options besides coal and geothermal. Ibid., 451.

225 *Concerned about the possibility:* "Police Deployed at South China Gas Stations," Reuters, August 18, 2005, www.freerepublic.com/focus/f-news/1466304/posts; the shortage seemed poised to move beyond Guangdong to Shanghai, with certain grades of gasoline selling out in China's largest city. "Fuel Shortage Shuts Half Shenzhen's Pumps," *South China Morning Post*, August 16, 2005, www.scmp.com/article/512395/fuel-shortage-shuts-half-shenzhens-pumps.

225 *As Zheng Bijian, a senior advisor:* Zheng Bijian, "China's 'Peaceful Rise' to Great-Power Status," *Foreign Affairs*, September/October, 2005, 19, https://www.foreignaffairs.com/articles/asia/2005-09-01/chinas-peaceful-rise-great-power-status.

225 *Shortly after assuming office:* Willy Lam, "Beijing's Energy Obsession," *Wall Street Journal*, April 2, 2004, https://www.wsj.com/articles/SB108086187874972230.

225 *In 2004, Li Junru, the vice president:* Quoted in Erica Downs, "The Energy

Security Series: China," Brookings, December 2006, 13, https://www
.brookings.edu/wp-content/uploads/2016/06/12china.pdf.

225 *And one year later, Zheng Bijian:* Bijian, "China's 'Peaceful Rise' to Great-
Power Status."

226 *True, the prospects for greater:* See "World Shale Resource Assessments,"
U.S. Energy Information Administration, September 24, 2015, https://
www.eia.gov/analysis/studies/worldshalegas/. The EIA published this
study on global shale resources first in 2011, and updated it in 2013.

226 *But difficulties in extracting this gas:* At least initially, the Chinese government had great ambitions to tap this resource to feed domestic markets.
In 2012, the government announced its goal of producing 2.1 tcf to 3.5
tcf of shale gas by 2020. It launched many initiatives to support this goal,
including opening up the sector to some foreign investment, creating incentives for shale development, and supporting technological research
related to shale. Yet, only two years later in 2014, China's State Council—
the country's highest administrative body—publicly scaled back the 2020
target to the production of only 1.1 tcf of shale gas in a rare downward
revision of the earlier goal. Speaking to Reuters in 2014, a Chinese government source admitted, "The previous targets were more of a vague
prospect, a hope. Thirty bcm is a more realistic goal." A year later, the
government announced that it would scale back the subsidy it provides to
shale gas developers, even as it continues to encourage the production of
the resource. See Zhongmin Wang, "China's Elusive Shale Gas Boom," The
Paulson Institute, March 2015, 6–9, www.paulsoninstitute.org/wp-con
tent/uploads/2015/04/PPEE_China-Shale-Gas_English.pdf; Chen Aizhu
and Judy Hua, "China Finds Shale Gas Challenging, Halves 2020 Output
Target," Reuters, August 7, 2014, www.reuters.com/article/2014/08/07/us
-china-shale-target-idUSKBN0G71FX20140807; Aibing Guo, "China Cuts
Subsidies for Shale Gas Developers Through to 2020," Bloomberg, April 29,
2015, www.bloomberg.com/news/articles/2015-04-29/china-cuts-subsidies
-for-shale-gas-developers-through-to-2020;

226 *While China may well become:* The U.S. Energy Information Administration assesses that "Chinese shale basins are tectonically complex with
numerous faults—some seismically active—which is not conducive to
shale development." See U.S. Energy Information Administration, "Technically Recoverable Shale Oil and Shale Gas Resources: China" (Washington, D.C.: U.S. Department of Energy, September 2015), xx-8, www

.eia.gov/analysis/studies/worldshalegas/pdf/China_2013.pdf. David Sandalow et al. predict that, after 2020, either high or low production scenarios could be realized for China's shale development. David Sandalow et al., "Meeting China's Shale Gas Goals" (working draft for public release, Center on Global Energy Policy, School of International and Public Affairs, Columbia University, New York, NY, September 2014), http:// energypolicy.columbia.edu/sites/default/files/energy/China%20 Shale%20Gas_WORKING%20DRAFT_Sept%2011.pdf.

226 *China's shale oil resources:* The EIA study ranks China as having the third largest shale *oil* reserves in the world, after Russia and the United States. "World Shale Resource Assessments," Analysis and Projections, U.S. Energy Information Administration, September 24, 2015, https://www.eia .gov/analysis/studies/worldshalegas/.

227 *In this positive economic environment:* China's LNG imports rose from 158 bcf in 2008 to 864 bcf in 2013. Anthony Fensom, "China: the Next Shale-Gas Superpower?" *The National Interest,* October 9, 2014, http://nation alinterest.org/feature/china-the-next-shale-gas-superpower-11432.

227 *The oil price plunge of 2014–2015:* LNG spot prices to Asia fell by 60 percent from 2014 to 2015. Longer-term contract prices were also affected, as most were indexed to the oil price. "Asian LNG Price Faces Steep Fall," Reuters, August 31, 2015, www.cnbc.com/2015/08/31/asian-lng-price-to -plunge-as-local-demand-wanes-supply-jumps.html; "Platts: August LNG Spot Prices to Asia Slide," LNG World News, July 23, 2015, https://www .lngworldnews.com/platts-august-lng-spot-prices-to-asia-slide/.

227 *Australia and the United States together:* From 2015 to 2021, global export LNG capacity is expected to increase by 45 percent, nearly all (90 percent) of which is anticipated to come from the United States and Australia. Stephen Letts, "LNG Glut Will Continue for Years as Demand Falls and Supply Surges: IEA," ABC News, June 8, 2016, http://www.abc.net.au/news/2016 -06-09/lng-glut-will-continue-as-demand-falls-and-supply-surges/7494850.

227 *Looking out two more decades:* International Energy Agency, *World Energy Outlook 2015* (Paris: OECD Publishing, 2015), 219–20, http://www .worldenergyoutlook.org/weo2015/.

227 *Lower than expected increases:* See "Asian LNG Price Faces Steep Fall," Reuters.

228 *After detailing the situation there:* See *Situation in Sudan, Before the Senate Foreign Relations Committee, 108th Congress.* (September 9, 2004,

testimony of Colin Powell, secretary of state of the United States); C-SPAN video, 04:28:07, https://www.c-span.org/video/?183435-1/situation-sudan.

228 *This was the first time the executive branch:* See Rebecca Hamilton, "Inside Colin Powell's Decision to Declare Genocide in Darfur," *Atlantic*, August 17, 2011, www.theatlantic.com/international/archive/2011/08/inside-colin-powells-decision-to-declare-genocide-in-darfur/243560/.

228 *The title of the U.N. press release:* Security Council, "Security Council Declares Intention to Consider Sanctions to Obtain Sudan's Full Compliance with Security, Disarmament Obligations on Darfur," United Nations, September 18, 2004, www.un.org/press/en/2004/sc8191.doc.htm.

228 *China's position toward Sudan:* For a detailed account of China's involvement with Sudan, see Luke Patey, *The New Kings of Crude: China, India, and the Global Struggle for Oil in Sudan and South Sudan* (London: Hurst Publishers, 2014).

228 *Equally important, China aimed to protect:* For an account of the experience of the Canadian company Talisman in Sudan, see Stephen J. Kobrin, "Oil and Politics: Talisman Energy and Sudan," *International Law and Politics* 36 (2004): 425–56, https://faculty.wharton.upenn.edu/wp-content/uploads/2012/05/nyujilp.pdf.

228 *By 2004, China had invested:* Chinese imports of Sudanese oil accounted for nearly half (47 percent) Sudan's export revenues exports in 2004. "UN Comtrade: International Trade Statistics," United Nations, https://comtrade.un.org/data/.

228 *"Business is business," he stated:* Quoted in Howard W. French, "China in Africa: All Trade, with No Political Baggage," *New York Times*, August 8, 2004, www.nytimes.com/2004/08/08/international/asia/08china.html.

229 *While China's approach to Sudan:* See Lydia Polgreen, "China, in New Role, Presses Sudan on Darfur," *New York Times*, February 23, 2008, www.nytimes.com/2008/02/23/world/africa/23darfur.html?n=Top%2FReference%2FTimes%20Topics%2FSubjects%2FO%2FOlympic%20Games; Daniel Hemel, "Darfur's Other Culprits," *Forbes*, April 27, 2008, www.forbes.com/2008/04/26/olympics-darfur-boycott-oped-cx_dhe_0427darfur.html.

229 *First articulated in 1997:* For an extensive analysis of China's "going out" policy as it relates to oil, see Bo Kong, *China's International Petroleum Policy* (Westport, Conn.: Praeger, 2009). Also see Xiaojie Xu, "Chinese NOCs' Overseas Strategies: Background, Comparison and Remarks," The James A. Baker III Institute for Public Policy, Rice University, March

2007, http://bakerinstitute.org/media/files/page/94235e0c/noc_chinese nocs_xu.pdf.

229 *The Chinese government and China's:* China National Petroleum Corporation (CNPC), China Petroleum & Chemical Corporation (Sinopec), and China National Offshore Oil Corporation (CNOOC) have played leading roles in this neo-mercantilist strategy. Headed by senior officials of the Chinese Communist Party, but also listed on the stock exchange, the NOCs are neither entirely independent of the government nor beholden to it. For a detailed history of China's oil industry, and the unfolding of the "going out" strategy, see Michal Meidan, "The Structure of China's Oil Industry: Past Trends and Future Prospects" (OIES Paper: WPM 66, Oxford Institute for Energy Studies, University of Oxford, Oxford, U.K., May 2016), https://www.oxfordenergy.org/wpcms/wp-content /uploads/2016/05/The-structure-of-Chinasoil-industry-past-trends -and-future-prospects-WPM-66.pdf.

229 *In 2010 alone, Chinese NOCs:* Julie Jiang and Jonathan Sinton, *Overseas Investments by Chinese National Oil Companies: Assessing the Drivers and the Impacts* (Paris: International Energy Agency, 2011), 38, www.iea.org /publications/freepublications/publication/PartnerCountrySeriesUpdate onOverseasInvestmentsbyChinasNationalOilCompanies.pdf. Also see ibid., Annex 2, p. 41.

229 *In the decade following 2004:* Ibid., 35–39.

230 *China seemed to believe:* For more on the objectives Chinese NOCS cite as motivating their overseas investment, see Julie Jiang and Jonathan Sinton, "Overseas Investment by Chinese National Oil Companies," International Energy Agency, February 2011, 12, https://www.iea.org/pub lications/freepublications/publication/overseas_china.pdf.

230 *Anticipating an ever-more-competitive:* Michal Meidan points out that this was an "untested theory." Meidan, "The Structure of China's Oil Industry," 54.

230 *In late 2004, Venezuelan president:* Juan Forero, "China's Oil Diplomacy in Latin America," *New York Times*, March 1, 2005, www.nytimes .com/2005/03/01/business/worldbusiness/chinas-oil-diplomacy-in-latin -america.html. Similarly, although Zimbabwe's wealth comes from diamonds and minerals, not oil, the country's regime has been able to withstand political and economic pressure in part as a result of Chinese unconditional backing. When Western powers and other African

countries were pressing for talks between Zimbabwean president Robert Mugabe and Zimbabwe's opposition in August 2005 amid a devastating economic crisis, Mugabe was clear about his sources of support. "I am aware there are shrill calls from many quarters [for talks]," he said in a televised address. "I am happy to announce that our Look East policies are beginning to assume a concrete form and yield quantifiable economic results for our nation. My recent state visit to China was most beneficial and is set to transform our economy in a fundamental way." "Mugabe Rejects Call for Dialogue; Looks to China for Help," Voice of America, October 30, 2009, http://m.voanews.com/a/a-13-2005-08-08 -voa32/395061.html.

231 *Perhaps the most oft-cited example:* Henry Lee and Dan Shalmon, however, have a different view, concluding that China is only hampering the promotion of good governance "on the margin." See Henry Lee and Dan Shalmon, "Searching for Oil: China's Oil Strategies in Africa," in Robert I. Rotberg, ed., *China into Africa: Trade, Aid, and Influence* (Washington, DC: Brookings Institution Press, 2008), 124, http://belfercenter.ksg .harvard.edu/files/Lee%20China%20into%20Africa.pdf.

231 *In 2002, Angola emerged:* Adam Taylor, "A 27-Year Civil War, for No Reason At All," *Washington Post*, October 14, 2015, https://www.washing tonpost.com/news/worldviews/wp/2015/10/14/a-27-year-civil-war-for -no-reason-at-all/.

231 *Once an exporter of food:* James Brooke, "War Status in Angola Is Status Quo," *New York Times*, December 27, 1987, www.nytimes .com/1987/12/27/world/war-status-in-angola-is-status-quo.html.

231 *Angola was already a significant:* Lee and Shalmon, "Searching for Oil: China's Oil Strategies in Africa," 119.

231 *Angola would repay the loan:* Ibid., 119–120.

231 *Around the same time, negotiations:* For more on this story, see Tom Burgis, Demetri Sevastopulo, and Cynthia O'Murchu, "China in Africa: How Sam Pa Became the Middleman," *Financial Times*, August 8, 2014, https://www.ft.com/content/308a133a-1db8-11e4-b927-00144feabdc0.

231 *Two years later, in June 2006:* Stephanie Hanson, "Angola's Political and Economic Development," Council on Foreign Relations, July 21, 2008, http://www.cfr.org/world/angolas-political-economic-development /p16820.

231 *Soon thereafter, China's EximBank:* Ibid.

231 *As of 2017, Angola was China's:* Sharon Cho, "Russia Reemerges as China's Top Oil Supplier Before OPEC Meet," Bloomberg, April 25, 2017, https://www.bloomberg.com/news/articles/2017-04-25/russia-reemerges-as-chinas-top-oil-supplier-before-opec-meet.

231 *By far the largest financer:* "China in Africa," *Ide-Jetro*, October 2009, www.ide.go.jp/English/Data/Africa_file/Manualreport/cia_10.html; Vivien Foster et al., "Building Bridges: China's Growing Role as Infrastructure Financier for Sub-Saharan Africa: Executive Summary," The World Bank, July 2008, http://siteresources.worldbank.org/INTAFRICA/Resources/BB_Final_Exec_summary_English_July08_Wo-Embg.pdf.

232 *a 2014 Pew poll found that:* "Chapter 2: China's Image," Pew Research Center, July 14, 2014, http://www.pewglobal.org/2014/07/14/chapter-2-chinas-image/.

232 *Writing in 2013, Nigeria's central bank governor:* Lamido Sanusi, "Africa Must Get Real About Chinese Ties," *Financial Times*, March 11, 2013, www.ft.com/intl/cms/s/0/562692b0-898c-11e2-ad3f-00144feabdc0.html#axzz3tRD2NLg2.

232 *African civil society groups:* See Carola McGiffert, ed., *Chinese Soft Power and Its Implications for the United States: Competition and Cooperation in the Developing World* (Washington, D.C.: Center for Strategic and International Studies, March 2009), 42, http://www.voltairenet.org/IMG/pdf/Chinese_Soft_Power.pdf.

232 *The more than one million:* "One Among Many," *Economist*, January 15, 2015, www.economist.com/news/middle-east-and-africa/21639554-china-has-become-big-africa-nowbacklash-one-among-many.

232 *Sata stirred up crowds:* Howard W. French, "In Africa, an Election Reveals Skepticism of Chinese Involvement," *Atlantic*, September 29, 2011, www.theatlantic.com/international/archive/2011/09/in-africa-an-election-reveals-skepticism-of-chinese-involvement/245832/.

232 *He reviled the Chinese for:* Ibid.

232 *Rumors circulated that his opponent's campaign:* Kristin Palitza, "Why Zambia's Elections Will Be All About China," *Time*, September 19, 2011, http://content.time.com/time/world/article/0,8599,2093381,00.html.

232 *Speaking in Beijing in 2012:* Leslie Hook, "Zuma Warns on Africa's Ties to China," *Financial Times*, July 19, 2012, https://next.ft.com/content/33686fc4-d171-11e1-bbbc-00144feabdc0?siteedition=uk&_i_location=http%3A%2F%2Fwww.ft.com%2Fcms%2Fs%2F0%2F2

F33686fc4-d171-11e1-bbbc-00144feabdc0.html%3Fsiteedition%3
Duk&_i_referer=&classification=conditional_standard&iab=barrier
-app#axzz3sgg1SWSH.

233 *Increasingly, observers concluded that NOCs:* See Bo Kong, "China's Quest for Oil in Africa Revisited" (Working Papers in African Studies, Paul H. Nitze School of Advanced International Studies, Johns Hopkins University, Washington, D.C., January 2011), http://www.sais-jhu.edu/sites /default/files/BoKongWP1-11_1.pdf .

233 *One industry insider shared with me:* Former CNOOC official, in-person conversation with author, Beijing, September 21, 2014.

233 *While the equity oil investments:* Jiang and Sinton report that "No evidence suggests that the Chinese government currently imposes a quota on the NOCs regarding the amount of their equity oil that they must ship to China." Julie Jiang and Jonathan Sinton, "Overseas Investments by Chinese National Oil Companies," 7.

233 *Other critics have focused:* See "The Chinese Oil Sector: Beijing's Latest Anti-Corruption Target," *Stratfor*, September 10, 2013, https://www.stratfor .com/analysis/chinese-oil-sector-beijings-latest-anti-corruption-target.

233 *For example, Chinese NOC Sinopec:* Meidan, "The Structure of China's Oil Industry," 55.

233 *Zhou had been an aggressive promoter:* See Patey, *The New Kings of Crude*, a book that traces Zhou's support for CNPC investments in Sudan.

233 *Sentenced in 2015 to life in prison:* Michael Forsythe, "Zhou Yongkang, Ex-Security Chief in China, Gets Life Sentence for Graft," *New York Times*, June 11, 2015, www.nytimes.com/2015/06/12/world/asia/zhou -yongkang-former-security-chief-in-china-gets-life-sentence-for-cor ruption.html.

234 *Chinese consumers especially chafe:* "The Chinese Oil Sector: Beijing's Latest Anti-Corruption Target," *Stratfor*.

234 *The loans-for-oil arrangements:* In 2013, Venezuela reportedly shipped more than 600,000 barrels of oil a day to China, receiving little or no payment for close to half the volume. Peitro D. Pitts and Nathan Crooks, "Venezuela Oil Sales to U.S. at 1985-Low Shows China Cost," Bloomberg, January 31, 2014, www.bloomberg.com/news /articles/2014-01-30/venezuela-oil-salesto-u-s-at-1985-low-shows -china-cost; Jonathan Kaiman, "China Agrees to Invest $20bn in Venezuela to Help Offset Effects of Oil Price Slump," *Guardian*,

January 8, 2015, www.theguardian.com/world/2015/jan/08/china-ven ezuela-20bn-loans-fancing-nicolas-maduro-beijing.

234 *Dr. Xue Li from the Chinese Academy of Sciences:* Xue Li and Xu Yanzhuo, "Why China Shouldn't Get Too Invested in Latin America," *The Diplomat,* March 31, 2015, http://thediplomat.com/2015/03/why-china -shouldnt-get-too-invested-in-latin-america/. Also see Michal Meidan, "China's Loans for Oil: Asset or Liability?" (OIES Paper: WPM 70, Oxford Institute for Energy Studies, University of Oxford, Oxford, U.K., December 2016), 11–14, https://www.oxfordenergy.org/wpcms/wp-content /uploads/2016/12/Chinas-loans-for-oil-WPM-70.pdf.

235 *Chinese officials insist Xi Jinping's:* Xie Tao, "Is China's 'Belt and Road' a Strategy?," *The Diplomat,* December 16, 2015, http://thediplomat .com/2015/12/is-chinas-belt-and-road-a-strategy/

235 *The "road" loosely retraces the maritime:* See Frank Viviano, "China's Great Armada," *National Geographic,* July 2005, http://ngm.nationalgeo graphic.com/ngm/0507/feature2/.

235 *Although there is considerable uncertainty:* The *Economist* anticipates more than \$1 trillion in government money will be spent on this initiative. "The New Silk Road," *Economist,* September 12, 2015, www.economist.com /news/special-report/21663326-chinas-latest-wave-globalisers-will-en rich-their-countryand-world-new-silk-road.

236 *Rather than being limited to tangible infrastructure:* See National Development and Reform Commission, Ministry of Foreign Affairs, *Vision and Actions on Jointly Building Silk Road Economic Belt and 21st-Century Maritime Silk Road* (Beijing, People's Republic of China, March 28, 2015), http://en.ndrc.gov.cn/newsrelease/201503/t20150330_669367.html.

236 *Booming infrastructure projects:* While many scholars see the One Belt, One Road initiative as partially motivated by the need to absorb overcapacity, others are skeptical that the initiative can provide nearly enough stimulus. See David Dollar, "China's Rise as a Regional and Global Power: The AIIB and the 'One Belt, One Road,'" Brookings Institution, July 15, 2015, www .brookings.edu/research/papers/2015/07/china-regional-global-power -dollar; Bert Hofman, "China's One Belt One Road Initiative: What We Know So Far," The World Bank, December 4, 2015, http://blogs.world bank.org/eastasiapacific/china-one-belt-one-road-initiative-what-we -know-thus-far.

237 *Chinese imports from Africa:* Valentina Romei, "China and Africa: Trade

Relationship Evolves," *Financial Times*, December 3, 2015, https://next.ft.com/content/c53e7f68-9844-11e5-9228-87e603d47bdc#axzz3tS8PqNxN.

237 *But what is most remarkable:* See Yun Sun, "Xi and the 6th Forum on China-Africa Cooperation: Major Commitments, but with Questions," Brookings Institution, December 7, 2015, www.brookings.edu/blogs/africa-in-focus/posts/2015/12/07-china-africa-focac-investment-economy-sun.

237 *Because, rather than acquiring equity oil:* Daniel Yergin, *The Quest: Energy, Security, and the Remaking of the Modern World* (London: Penguin Books, 2012), 193.

237 *"Industrial capacity cooperation":* Ibid.

237 *After decades of insisting:* See "Chasing the Chinese Dream," *Economist*, May 4, 2013, https://www.economist.com/news/briefing/21577063-chinas-new-leader-has-been-quick-consolidate-his-power-what-does-he-now-want-his.

238 *Graham Allison, a renowned professor:* See Graham Allison, *Destined for War: Can America and China Escape Thucydides's Trap* (Boston: Houghton Miller Harcourt Publishing, 2017). During his September 2015 visit to the United States, President Xi Jinping dismissed the idea of a "Thucydides trap"; during a speech in Seattle, he said, "There is no such thing as the so-called Thucydides trap in the world. But should major countries time and again make the mistakes of strategic miscalculation, they might create such traps for themselves." "Full Text of Xi Jinping's Speech on China-U.S. Relations in Seattle," *Xinhuanet*, September 24, 2015, http://news.xinhuanet.com/english/2015-09/24/c_134653326.htm.

238 *Energy will not be the only factor:* For other reflections on China and global energy governance, see Bo Kong, "Governing China's Energy in the Context of Global Governance," *Global Policy* 2, special issue (September 2011): 51–65.

239 *In April 2005, Mikkal Herberg:* See David Zweig and Bi Jianhai, "China's Global Hunt for Energy," *Foreign Affairs* (September/October 2005), https://www.foreignaffairs.com/articles/asia/2005-09-01/chinas-global-hunt-energy.

240 *In 2006, the Pentagon's annual report:* The report states, "In the near term, China's military build-up appears focused on preparing for Taiwan Strait contingencies, including the possibility of U.S. intervention. However, analysis of China's military acquisitions suggests it is also generating

capabilities that could apply to other regional contingencies, such as conflicts over resources or territory." Office of the Secretary of Defense, *Annual Report to Congress: Military Power of the People's Republic of China* (Washington, D.C.: Department of Defense, 2006), 1, http://web .archive.org/web/20150929063154/http://fas.org/nuke/guide/china/dod -2006.pdf.

240 *Two years later, a new version of:* Office of the Secretary of Defense, *Annual Report to Congress: Military Power of the People's Republic of China* (Washington, D.C.: Department of Defense, 2008), 13, http://www .mcsstw.org/www/download/China_Military_Power_Report_2008.pdf. Similarly, a 2008 presentation explaining the rationale for the establishment of AFRICOM—an overseas joint command similar to the Pentagon's CENTCOM, but for Africa, not the Middle East—highlighted the importance of Africa's oil and gas to the United States and China's growing influence over countries with such resources. Lauren Ploch, "Africa Command: U.S. Strategic Interests and the Role of the U.S. Military in Africa," *Congressional Research Service*, July 22, 2011, https://fas.org/sgp /crs/natsec/RL34003.pdf.

241 *During a Senate Committee hearing:* The *LNG Permitting Certainty and Transparency Act: Full Committee Hearing on S. Hrg. 114–9, Before the Senate Energy and Natural Resources Committee*, 114th Congress (January 29, 2015), C-SPAN video, 01:56:53 https://www.c-span.org /video/?324072-1/hearing-liquified-natural-gas-permitting%20.

241 *Michael Smith, the CEO of Freeport:* "U.S. Government Discouraged Chinese Investment in LNG Exports," Reuters, May 14, 2015, www.reuters .com/article/us-lng-china-idUSKBN0NZ2F720150514.

241 *As of early 2017, China had received:* "LNG Monthly 2017," U.S. Department of Energy, 2, https://energy.gov/fe/downloads/lng-monthly-2017.

242 *Finally, members of Congress:* The Obama Administration did seek to make this clear through statements issued after strategic and economic dialogues held between the United States and China. However, more recently, developments such as the Apple patent dispute with China and Uber's sale of its China business to a local competitor have heightened sensitivity around the need for reciprocal investment opportunities in China. A 2017 survey conducted by the American Chamber of Commerce in China found that close to three-quarters of Americans surveyed felt foreign businesses were less welcome in China than they did two years ago. See "Joint

U.S.-China Press Statements at the Conclusion of the Strategic & Economic Dialogue," U.S. Department of State, July 10, 2014, https://20092017.state .gov/secretary/remarks/2014/07/228999.htm; "Remarks by Vice President Biden and Chinese Vice President Xi at a U.S.-China Business Roundtable," The White House, Office of the Vice President, The White House President Barack Obama, August 19, 2011, https://obamawhitehouse .archives.gov/the-press-office/2011/08/19/remarks-vice-president-biden -and-chinese-vice-president-xi-us-china-busi; "China Business Climate Survey Report," AmChamChina, 2017, https://www.amchamchina.org /policy-advocacy/business-climate-survey/.

242 *While Chinese investment has quietly flowed:* A 2015 report by the Rhodium Group revealed that there has been significant Chinese investment in U.S. energy from 2000 to 2015. The study points to 113 energy deals valued cumulatively at $13.4 billion as of early 2017. "Chinese Investment Monitor," Rhodium Group, http://rhg.com/interactive/china-invest ment-monitor. However, in 2012, the U.S. government—through a White House order—forced the reversal of a wind farm investment in Oregon by a private Chinese firm. The wind farm was close to a military base. This investment was the first to be blocked on such grounds in twenty-two years. "US Blocks Chinese Firm's Investment in Wind Farms," *BBC News*, September 28, 2012, www.bbc.com/news/world-us-canada-19766965.

242 *Recognizing Beijing and building a relationship:* Henry Kissinger, *On China* (London: Penguin Books, 2012), 213–15.

242 *For China, the opening brought:* See Denny Roy, *China's Foreign Relations* (Lanham, MD: Rowman & Littlefield, 1998), 29–30.

242 *On November 11, 2014:* See Cary Huang, "Xi Jinping and Barack Obama Walk Down the Path of History," *South China Morning Post*, November 15, 2014, www.scmp.com/news/china/article/1640193/xi-jinping-and -barackobama-walk-down-path-history; Carol E. Lee, "U.S., China Reach New Climate, Military Deals," *Wall Street Journal*, November 12, 2014, www.wsj.com/articles/u-s-china-ready-deals-to-avert-military-confronta tions-1415721451.

243 *According to the Chinese press:* "Xi, Obama meeting: a lively history lesson," *China Daily*, November 15, 2015, http://www.chinadaily.com.cn /china/2014-11/15/content_18920403.htm.

243 *As discussed in Chapter Seven:* These goals are slightly less ambitious than the ones made by President Obama in 2009, when he announced

the target of reducing emissions by 83 percent by 2050 compared to 2005 levels. Doing so would entail a 17 percent reduction in 2020, a 30 percent reduction in 2025, and a 42 percent reduction in 2030. "President to Attend Copenhagen Climate Talks," The White House, November 25, 2009, https://obamawhitehouse.archives.gov/the-press-office/president-at tend-copenhagen-climate-talks.

243 *China, for its part, announced:* John Podesta and John Holdren, "The U.S. and China Just Announced Important New Actions to Reduce Carbon Pollution," The White House, President Barack Obama, November 12, 2014, https://obamawhitehouse.archives.gov/blog/2014/11/12 /us-and-china-just-announced-importantnew-actions-reduce-car bon-pollution.

244 *But doing so will create challenges:* For more on the politics behind better environmental regulations, see Jost Wübbeke, "The Three-year Battle for China," *China Dialogue,* April 25, 2014, https://www.chinadialogue.net /article/show/single/en/6938-The-three-year-battle-for-China-s-new -environmental-law.

244 *China's trade with the Middle East:* "The Great Well of China," *Economist,* June 18, 2015, http://www.economist.com/news/middle-east-and-africa /21654655-oil-bringing-china-and-arabworld-closer-economically-poli tics-will.

245 *In 2040, the U.S. EIA anticipates:* International Energy Agency, *World Energy Outlook 2014* (Paris: OECD Publishing, 2014), 81, http://www .worldenergyoutlook.org/weo2014.

245 *The 2017 decision of Saudi Aramco:* "Saudi Arabia's Aramco to Invest $7 bn in Malaysia Oil Refinery," *Yahoo! News,* February 27, 2017, https://www.yahoo.com/news/saudi-arabias-aramcoinvest-7-bn-malay sia-oil-100804170.html.

245 *China's vision for closer economic and political:* See "Xi Jinping Attends Opening Ceremony of Sixth Ministerial Conference of China–Arab States," Ministry of Foreign Affairs of the People's Republic of China, June 5, 2014, http://www.fmprc.gov.cn/mfa_eng/zxxx_662805/t1163554 .shtml.

245 *Harking back to historical cooperation:* "Conference Proceedings: China in the Middle East," *Georgetown Security Studies Review,* June 2015, http:// georgetownsecuritystudiesreview.org/wp-content/uploads/2015/07 /GSSR-Asia-Conference.pdf.

245 *such trade had stood at less:* Piero Formica, *Stories of Innovation for the Millennial Generation: The Lynceus Long View* (London: Palgrave Macmillan, 2013), 131.

245 *Beijing kept quiet when:* Yitzhak Shichor, "Iran Keeps China in a Chokehold," *Asia Times Online,* accessed September 26, 2008, www.atimes.com/atimes/China/JI26Ad01.html.

246 *Prime Minister Wen Jiabao:* Michael Wines, "China Leader Warns Iran Not to Make Nuclear Arms," *New York Times,* January 20, 2012, www.nytimes.com/2012/01/21/world/asia/chinese-leader-wen-criticizes-iran-on-nuclear-program.html?_r=0.

246 *In speaking with journalist Tom Friedman:* Barack Obama, "The Obama Interviews: China as a Free Rider," *New York Times Video,* August 9, 2014, video recording, www.nytimes.com/video/opinion/100000003047788/china-as-a-free-rider.html.

247 *When asked about Chinese views:* Alexa Olesen, "China Sees Islamic State Inching Closer to Home," *Foreign Policy,* August 11, 2014, http://foreignpolicy.com/2014/08/11/china-sees-islamic-state-inching-closer-to-home/.

247 *Both powers, in short, are in need:* For some ideas of ways in which the United States and China can begin to increase their cooperation, see Christopher Yung and Wang Dong, "U.S.-China Relations in the Maritime Security Domain," in *U.S.-China Relations in Strategic Domains,* ed. Travis Tanner and Wang Dong (Seattle: The National Bureau of Asian Research, April 2016), http://www.nbr.org/publications/element.aspx?id=889.

Eleven: The Middle East—Trying to Make the Most of a Tough Situation

249 *Sir Mark Sykes slid his finger:* For a vivid account of this history, and the Anglo-French rivalry that underpinned it, see James Barr, *A Line in the Sand: The Anglo-French Struggle for the Middle East, 1914–1948* (New York: W. W. Norton, 2013). Quote in text can be found on page 7.

250 *Enraged, House wrote:* Ibid., 30.

250 *Walid Jumblatt, the leader of the Druze:* Jim Muir, "Sykes-Picot: The Map That Spawned a Century of Resentment," *BBC News,* May 16, 2016, www.bbc.com/news/world-middle-east-36300224.

250 *Barham Salih, a Kurdish leader:* Robin Wright, "How the Curse of

Sykes-Picot Still Haunts the Middle East," *New Yorker*, April 30, 2016, www.newyorker.com/news/news-desk/how-the-curse-of-sykes-picot -still-haunts-the-middle-east.

250 *Even ISIS leader Abu Bakr al-Baghdadi:* Ibid.

251 *Turkey saw its GDP per capita:* "GDP per capita, PPP (current international $), Turkey," The World Bank, http://data.worldbank.org/indicator /NY.GDP.PCAP.PP.CD?locations=TR.

251 *Oil and gas are the economic lifeblood:* As mentioned later in the chapter, from 2011 to 2014, Saudi Arabia alone reportedly provided nearly $23 billion in aid to Bahrain, Egypt, Jordan, Oman, Yemen, Palestine, Morocco, Sudan, and Djibouti. "KSA [Kingdom of Saudi Arabia] Has Allocated SR252bn in Foreign Aid Since 1990," *Arab News*, September 29, 2014, www.arabnews.com/economy/news/637176.

252 *With dozens of American sailors:* "Ibn Saud Greets President Franklin D. Roosevelt on USS Quincy in the Suez Canal," YouTube, 3:49, posted by "CriticalPast," March 15, 2014, www.youtube.com/watch?v=FH1cU z7O7mg.

252 *He slept outside on the deck:* Thomas W. Lippman, "The Day FDR Met Saudi Arabia's Ibn Saud," *The Link* 38, no. 2 (April/May 2005): 5, www .ameu.org/getattachment/51ee4866-95c1-4603-b0dd-e16d2d49fcbc /The-Day-FDR-Met-Saudi-Arabia-Ibn-Saud.aspx.

253 *Sitting close together, the leaders:* Ibid., 7.

253 *Over the course of the twentieth century:* See Bruce R. Kuniholm, "The Carter Doctrine, the Reagan Corollary, and Prospects for United States Policy in Southwest Asia," *International Journal* 41, no. 2 (1986).

253 *A few years earlier, the CIA:* See "Intelligence Memorandum: The Impending Soviet Oil Crisis," U.S. Central Intelligence Agency, ER 77-10147, CIA Historical Review Program, March 1977, www.cia.gov/library/reading room/docs/DOC_0000498607.pdf.

253 *In his last State of the Union:* Jimmy Carter, "The State of the Union Address Delivered Before a Joint Session of the Congress," House of Representatives, Washington, DC, January 23, 1980, The American Presidency Project, www.presidency.ucsb.edu/ws/?pid=33079.

254 *His words were interrupted:* See "1980 State of the Union Address," C-SPAN, https://www.c-span.org/video/?124054-1/1980-state-union-address.

254 *Just one year later, newly elected President:* Ronald Reagan, "The President's News Conference," East Room, White House, Washington, DC,

October 1, 1981, The American Presidency Project, www.presidency
.ucsb.edu/ws/?pid=44327.

254 *In 1990, Iraqi forces overran:* Oil was, of course, not the only decisive factor in the U.S. response to Iraq's invasion of Kuwait in 1990.

254 *"The economic lifeline of the industrial world":* Thomas L. Friedman, "U.S.
Jobs at Stake in Gulf, Baker Says," *New York Times*, November 14, 1990,
www.nytimes.com/1990/11/14/world/mideast-tensions-us-jobs-at
-stake-in-gulf-baker-says.html. Also see a PBS interview with Baker,
in which he responds to the question of what was the impetus for war
by saying, "The fundamental reason was that this was very much in
the vital national interest of the United States. It had been seen to be
in the vital national interests of the United States through both Republican and Democratic administrations going all the way back to
Roosevelt. That is secure access to the energy resources of the Persian Gulf. That, plus the fact that we had here an outrageous case of
unprovoked aggression by a large country against its small neighbor.
We had the potential of a dictator who was in the process of developing nuclear weapons or trying to develop nuclear weapons, sitting
astride the economic lifeline of the West. We had a situation where
if he had been successful it would have adversely impacted the economies of all of the West. It would have impacted jobs in the United
States. All of those reasons were valid reasons for fighting this war.
My suggestion that it boiled down to jobs got a lot of attention and
flak but the fact of the matter is it would have boiled down to jobs if
Saddam Hussein had been able to control the flow of oil from the Persian Gulf or to, by controlling his own oil and Kuwait's oil act in a way
to influence prices." "Oral History: James Baker," PBS, accessed October 17, 2016, www.pbs.org/wgbh/pages/frontline/gulf/oral/baker/1
.html.

254 *As late as 1989, the United States:* Tim Kane, "Global U.S. Troop Deployment, 1950–2003," Heritage Foundation, October 27, 2004, http://www
.heritage.org/defense/Report/Global-US-Troop-Deployment-1950-2003.

254 *Yet, even a decade after Iraq:* Joshua Rovner and Caitlin Talmadge, "Less
Is More: The Future of the U.S. Military in the Persian Gulf," *The Washington Quarterly* 37, no. 3 (Fall 2014), https://twq.elliott.gwu.edu/less
-more-future-us-military-persian-gulf.

255 *According to economist Anders Aslund:* "A Symposium of Views: The

Geopolitics of U.S. Energy Independence," *The International Economy* (Summer 2012): 23–24, www.international-economy.com/TIE_Su12 _GeopoliticsEnergySymp.pdf.

255 *Middle East policy expert:* Ibid.

255 *In 2012, President Obama voiced:* Benjamin Alter and Edward Fishman, "The Dark Side of Energy Independence," *New York Times*, April 27, 2013, www.nytimes.com/2013/04/28/opinion/sunday/the-dark-side-of-energy -independence.html.

255 *Obama's Republican opponent in the 2012:* Ibid.

255 *In 2016, Donald Trump's campaign:* Markham Hislop, "So Long, OPEC—You're at the Top of Donald Trump's Hit List," *Financial Post*, November 14, 2016, http://business.financialpost.com/fp-comment/so -long-opec-youre-at-the-top-of-donald-trumps-hit-list.

255 *Then, beginning with the economic slowdown:* "U.S. Imports by Country of Origin," U.S. Energy Information Administration, https://www.eia .gov/dnav/pet/pet_move_impcus_a2_nus_epc0_im0_mbblpd_a.htm.

255 *U.S.-Saudi crude trade reflected:* The peak of 2.2 mnb/d in May 2003 turned into a trough of 0.8 mnb/d by December 2014. Ibid.

256 *The IEA and some oil and gas experts:* For instance, see Leonardo Maugeri, "Global Oil Production is Surging: Implications for Prices, Geopolitics, and the Environment," Geopolitics of Energy Project, Belfer Center for Science and International Affairs, Kennedy School of Government, Harvard University, June 2012, 2, http://www.belfercenter.org/sites/default /files/legacy/files/maugeri_policybrief.pdf; International Energy Agency, *World Energy Outlook 2016*, 142.

256 *If history is any indication:* See Blake Clayton and Michael Levi, "The Surprising Sources of Oil's Influence," *Survival* 54, no. 6 (2012).

256 *They were approximately 1.4 mnb/d:* "U.S. Crude Oil Imports," U.S. Energy Information Administration, http://www.eia.gov/dnav/pet/pet_move _impcus_a2_nus_epc0_im0_mbblpd_m.htm; Quote can be found in David Ottaway, *The King's Messenger: Prince Bandar bin Sultan and America's Tangled Relationship with Saudi Arabia* (New York: Walker, 2008), 86.

256 *And they were more or less:* Ottaway, *The King's Messenger*, 167.

257 *They estimate the costs:* Mark A. Delucchi and James J. Murphy, "US Military Expenditures to Protect the Use of Persian Gulf Oil for Motor Vehicles," *Energy Policy* 36 (2008): 2259, https://escholarship.org/uc /item/0j9561zd#page-12.

258 *By 2040, one of every two:* International Energy Agency, *World Energy Outlook 2016,* 143.

258 *America's energy boom may allow:* President Eisenhower initially played hardball with France, the U.K., and Israel, urging them to withdraw from the Suez before orchestrating the Oil Lift. See Daniel Yergin, *The Prize: The Epic Quest for Oil, Money and Power* (New York: Simon & Schuster, 2009), 466–77.

258 *But, at least for the coming decade:* See International Energy Agency, *World Energy Outlook 2016,* 128–38.

258 *For instance, growing Iranian influence:* As of 2016, together, Iran and Iraq have 18 percent of world oil reserves and 9 percent of global oil production. "International Energy Statistics," U.S. Energy Information Administration, https://www.eia.gov/beta/international/data/browser/#/?pa=00000000000000000000800000000000g&c=0000000000000000000003&ct=0&tl_id=5-A&vs=INTL.53-1-IRN-TBPD.A&cy=2016&vo=0&v=H&start=2010.

259 *As Joe Nye of Harvard mused:* "A Symposium of Views: The Geopolitics of U.S. Energy Independence," 23.

259 *Being responsive has required:* The EIA defines spare capacity as "the volume of production that can be brought on within 30 days and sustained for at least 90 days." "What Drives Crude Oil Prices: Supply OPEC," U.S. Energy Information Administration, https://www.eia.gov/finance/markets/crudeoil/supply-opec.php.

259 *If anything, we may see:* In a 1986 press conference, with oil at $10 a barrel, Vice President Bush described what he intended to say in an upcoming trip to Riyadh, "My plea will be for stability in the marketplace. . . . [Prices cannot continue a] free fall like a parachutist jumping out without a parachute." He explained to his Washington audience that low prices were harming U.S. interests; a former oilman, Bush considered the beating the U.S. oil industry was taking from low prices to be a threat to American national security. "A Plea for Stability, not Price-Setting, on the Oil Market," *New York Times,* April 6, 1986, www.nytimes.com/1986/04/06/weekinreview/a-plea-for-stability-not-price-setting-on-the-oil-market.html. Vice President Bush found himself in the political spotlight after these comments, which appeared at odds with the views of President Reagan, who felt the market alone should determine the price of oil and believed low prices were an enormous boon to

American consumers. See Gerald M. Boyd, "Bush Seeks to End Confusion Stirred by Oil Price Views," *New York Times*, April 8, 1986, www.nytimes .com/1986/04/08/business/bush-seeks-to-end-confusion-stirred-by-oil -price-views.html.

260 *As a result, there will be:* Oil has long been a central point of diplomatic conversations. For multiple declassified transcripts and meeting notes of such encounters, see *Foreign Relations of the United States, 1969–1976*, ed. Steven G. Galpern and Edward C. Keefer, vol. 37, *Energy Crisis, 1974– 1980* (Washington, DC: U.S. Government Printing Office, 2012), https:// history.state.gov/historicaldocuments/frus1969-76v37.

260 *He said "Leadership is needed.":* Senior Turkish diplomat, in-person conversation with author, Ankara, Turkey, August 26, 2016.

261 *It provided billions of dollars:* See Mariam Fam and Nadeem Hamid, "Saudi King Calls for Egypt Aid Drive After El-Sisi Win," Bloomberg, June 4, 2014, www.bloomberg.com/news/articles/2014-06-03/saudi-king -seeks-aid-drive-for-egypt-after-el-sisi-win.

261 *And it finalized deals to buy:* Gopal Ratnam, "U.S. Seeks $10.8 Billion Weapons Sale to U.A.E., Saudis," Bloomberg, October 15, 2013, www .bloomberg.com/news/articles/2013-10-15/u-s-seeks-10-8-billion -weapons-sale-to-u-a-e-saudis. See also "China, Saudi Arabia Vow to Strengthen Military Ties," *China Military Online*, November 18, 2014, http://english.chinamil.com.cn/news-channels/china-military-news /2014-11/18/content_6229730.htm.

261 *It will take time for these efforts:* For now, Europe has little economic heft and few security assets to undergird a more robust political engagement. China, while gradually appreciating the need to engage more broadly in the Middle East, is definitively a reluctant player, having few if any of the capabilities needed to play a meaningful role alongside the United States, despite its deep energy interests there. See Chapter Ten.

262 *According to Ottaway, Prince Bandar:* Ottaway, *The King's Messenger: Prince Bandar bin Sultan and America's Tangled Relationship with Saudi Arabia,* 1.

262 *Certainly, neither Saudi Arabia:* As noted earlier in the chapter, given the fluidity of the global oil market, merely stopping exports to one destination would not have a huge impact in any case; truly destabilizing markets require taking production *off* the market.

262 *The IEA created a "Low Oil Price Scenario":* See the Low Oil Price Scenario

in International Energy Agency, *World Energy Outlook 2015* (Paris: OECD Publishing, 2015), 154. The U.S. EIA also has a low-price scenario, where the Brent price of crude oil is only $38 in 2020 and $43 in 2025. See "Annual Energy Outlook 2016," U.S. Energy Information Administration, www.eia.gov/forecasts/aeo/data/browser/#/?id=19-AEO2016.

262 *In this scenario, low-cost producers:* International Energy Agency, *World Energy Outlook 2015*, 164–66.

262 *According to the IEA, by 2040:* International Energy Agency, *World Energy Outlook 2015*, 188, http://www.worldenergyoutlook.org/weo2015/.

262 *Asia in particular becomes more vulnerable:* Ibid., 178. It is important to distinguish between this number—which is indicative of Asia's increasing reliance on Middle Eastern oil for supply—with the 90 percent number cited earlier in the book, which indicates growing Middle East dependency on Asia for demand.

263 *Later dubbing his theory:* Thomas L. Friedman, "The First Law of Petropolitics," *Foreign Policy*, October 16, 2009, http://foreignpolicy.com/2009/10/16/the-first-law-of-petropolitics/.

263 *Pointing out that Bahrain:* "Friedman on Petropolitics and Global Corruption," NPR, March 4, 2006, www.npr.org/templates/story/story.php?storyId=5383613.

263 *Friedman came under a barrage:* See, for example, Romain Wacziarg, "The First Law of Petropolitics," *Economica* 79, no. 316 (October 2012): doi:10.1111/j.1468-0335.2011.00902.x.

264 *countries whose economies are heavily reliant:* For more on the resource curse, see Andrew Bauer and Juan Carlos Quiroz, "Resource Governance," *The Handbook of Global Energy Policy*, ed. Andreas Goldthau (Malden, MA: John Wiley & Sons, 2013), 244–64; Shah M. Tarzi and Nathan Schackow, "Oil And Political Freedom In Third World Petro States: Do Oil Prices and Dependence On Petroleum Exports Foster Authoritarianism?," *Journal of Third World Studies* 29, no. 2 (Fall 2012): 231–50.

264 *Even though it was the thirteenth largest:* "Production of Crude Oil, NGPL, and Other Liquids 2012," U.S. Energy Information Administration: Beta, https://www.eia.gov/beta/international/rankings/#?cy=2012&pid=55&tl_id=5-A; "Nigerians Living in Poverty Rise to Nearly 61%," *BBC News*, February 13, 2012, www.bbc.com/news/world-africa-17015873.

265 *In response, the Saudi government increased:* See "World Report 2012:

Saudi Arabia, Events of 2011," Human Rights Watch, https://www.hrw
.org/world-report/2012/country-chapters/saudi-arabia.

265 *the total price of these perks:* Neil MacFarquhar, "In Saudi Arabia, Royal
Funds Buy Peace for Now," *New York Times*, June 8, 2011, http://www
.nytimes.com/2011/06/09/world/middleeast/09saudi.html.

266 *Shortly after acceding to the throne:* Ben Hubbard, "Saudi King Unleashes
a Torrent of Money as Bonuses Flow to the Masses," *New York Times*,
February 19, 2015, www.nytimes.com/2015/02/20/world/middleeast
/saudi-king-unleashes-a-torrent-as-bonuses-flow-to-the-masses.html
?action=click&contentCollection=Middle East&region=Footer&module
=MoreInSection&pgtype=article; "Budget: Research for the People," National Institutes of Health, https://www.nih.gov/about-nih/what-we-do
/budget.

266 *Appreciative Saudis created a new:* Hubbard, "Saudi King Unleashes a
Torrent of Money."

266 *The yawning gap between:* This fiscal breakeven price almost always exceeds a different threshold price—the actual cost of producing a single
barrel of oil. On average, the cost of finding and lifting a barrel of crude
oil or natural gas equivalent in the Middle East is less than $17, com-
pared with $45 in Africa and $34 for conventional resources in the United
States (in 2009 dollars). "How much does it cost to produce crude oil and
natural gas?," U.S. Energy Information Administration, January 15, 2014,
http://web.archive.org/web/20150211220237/www.eia.gov/tools/faqs
/faq.cfm?id=367&t=6.

267 *In Saudi Arabia, the breakeven price:* According to the IMF, Saudi Arabia's
fiscal breakeven price in 2008 was $38. Six years later, in 2014, it was
$106. "Breakeven Fiscal Oil Price, US Dollars Per Barrel: Saudi Arabia,
Dataset: MCD Regional Economic Outlook October 2016," International
Monetary Fund, http://data.imf.org/?sk=388DFA60-1D26-4ADE-B505
-A05A558D9A42&ss=1479331931186.

267 *At the end of 2014, soon after:* "Breakeven fiscal oil price, US dollars
per barrel, Dataset: MCD Regional Economic Outlook October 2016,"
International Monetary Fund, http://data.imf.org/?sk=388DFA60-1D26
-4ADE-B505-A05A558D9A42&sId=1479331931186.

267 *Saudi Arabia, Bahrain, and* Oman: Ibid.

267 *In late 2014, only Kuwait:* Kuwait's breakeven price in 2014 was $56 and
UAE's was $79. Ibid.

267 *In 2014, Saudi Arabia, for example:* The United Nations Economic and Social Commission for Western Asia estimated this figure for reconstruction in September 2014. "UN: Rebuilding Syria, Iraq and Gaza Will Cost $750bn," *Middle East Monitor*, September 16, 2014, www.middleeast monitor.com/news/americas/14162-un-rebuilding-syria-iraq-and-gaza -will-cost-750bn; "Total reserves (includes gold, current US$)," The World Bank, data.worldbank.org/indicator/FI.RES.TOTL.CD?end=2015&loca tions=SA&start=2011; "GDP (current US$)," The World Bank, data .worldbank.org/indicator/NY.GDP.MKTP.CD?end=2015&loca tions=SA&start=2011.

267 *The reserves of the emirates:* "Investment Corporation of Dubai," Sovereign Wealth Fund Institute, http://www.swfinstitute.org/swfs/investment -corporation-of-dubai/; "Abu Dhabi Investment Authority," Sovereign Wealth Fund Institute, www.swfinstitute.org/swfs/abu-dhabi-investment -authority/; "How Much is a Trillion dollars? What a Trillion Can Buy," *Fox Business*, April 30, 2015, www.foxbusiness.com/features/2015/04/30 /how-much-is-trillion-dollars-what-trillion-can-buy.html.

267 *or roughly twice the annual:* Kevin McCormally, "14 Ways to Spend $1 Trillion," Kiplinger.com, January 28, 2011, www.kiplinger.com/article /business/T043-C000-S001-14-ways-to-spend-1-trillion.html.

267 *For a small country with more expatriates:* AFP, "Kuwait Fiscal Reserves Hit Record $592 bn: Report," *Yahoo News*, July 9, 2015, https://www.yahoo .com/news/kuwait-fiscal-reserves-hit-record-592-bn-report-092133093 .html.

267 *Even the few nations that did:* As of November 24, 2014, Bahrain was expected to have a deficit of $2.2 billion in 2015, while its reserves were estimated at just $6.4 billion. As of January 15, 2015, Oman expected a deficit of $2.7 billion for 2015, with 2014 reserves of $19.6 billion. Toby Illes, ed., *Country Report: Bahrain*, Economist Intelligence Unit, November 24, 2014, 10; Robert Powell, ed., *Country Report: Oman*, Economist Intelligence Unit, January 15, 2015, 9.

267 *In the three months between October 2014:* "Breakeven fiscal oil price, US dollars per barrel," International Monetary Fund.

268 *Bahrain, Kuwait, Oman:* International Monetary Fund, "Statistical Appendix," *Regional Economic Outlook Update: Middle East and Central Asia: Learning to Live with Cheaper Oil Amid Weaker Demand*, Washington, D.C., January 21, 2015. In some cases, canceled projects go beyond

the borders of a country, such as the decision by Kuwait Petroleum International to terminate its investment in a refinery in the Netherlands. "UPDATE 1—Kuwait's KPI Cancels Refinery Investment, Considers Sale," Reuters, October 6, 2014, http://uk.reuters.com/article/2014/10/06/kpi-refinery-europoort-idUKL6N0S11P020141006.

268 *Of all the Gulf monarchies:* In contrast to Saudi Arabia, which relies on oil to provide between 80 and 90 percent of fiscal earnings, oil revenues comprise 4 percent of the UAE's federal revenue. Kuwait relies on oil for a similarly high percentage of its budget to Saudi Arabia, but its financial reserves per capita dwarf those of Saudi Arabia. Robert Powell, ed., *Country Report: Saudi Arabia*, Economist Intelligence Unit, December 18, 2014, 5; Frank Kane, "Budget Shows Oil Is Not the Lifeblood of Dubai's Growth," *The National*, January 6, 2015, www.thenational.ae/business/economy/budget-shows-oil-is-not-the-lifeblood-of-dubais-growth; "Kuwait Launches Another Development Plan," Economist Intelligence Unit, n.d.

268 *From 2014 to the end of 2016:* "Saudi Arabia: Annual Data and Forecast," Economist Intelligence Unit, April 3, 2017, http://country.eiu.com.ezp-prod1.hul.harvard.edu/article.aspx?articleid=235353607&Country=Saudi%20Arabia&topic=Economy&subtopic=Charts+and+tables&subsubtopic=Annual+data+and+forecast&aid=1&oid=235353607.

268 *At more than thirty million people:* Approximately two-thirds of this thirty million are Saudi citizens, whereas one-third are expatriates living in the kingdom. "Trends in International Migrant Stock," The United Nations Population Division, 2015, www.un.org/en/development/desa/population/migration/data/estimates2/data/UN_MigrantStockTotal_2015.xlsx.

268 *Birth rates in the kingdom:* Department of Economic and Social Affairs, Population Division, World Population Prospects: The 2017 Revision, DVD Edition (New York: The United Nations, 2017), https://esa.un.org/unpd/wpp/DVD/Files/1_Indicators%20(Standard)/EXCEL_FILES/1_Population/WPP2017_POP_F01_1_TOTAL_POPULATION_BOTH_SEXES.xlsx.

268 *But in Saudi Arabia:* Ben Hubbard, "Young Saudis See Cushy Jobs Vanish Along With Nation's Oil Wealth," *New York Times*, February 16, 2016, https://www.nytimes.com/2016/02/17/world/middleeast/young-saudis-see-cushy-jobs-vanish-along-with-nations-oil-wealth.html.

268 *More than a quarter of a million:* General Authority for Statistics, Kingdom of Saudi Arabia, https://stats.gov.sa/en/node.

268 *In 2012, a report by Alkhabeer Capital:* "Analysis of Saudi Unemployment," Alkhabeer Capital, March 10, 2014, http://jef.org.sa/files/analysis-of-saudi-unemployment.pdf.

269 *Providing health care and education:* Fahad M. Alturki, Asad Khan, and Rakan Alsheikh, "Saudi Arabia's 2015 Fiscal Budget" (Riyadh: Jadwa Investment, December 28, 2014), www.jadwa.com/en/download/2015-budget/2015-saudi-budget.

269 *In 2015, gasoline was just 45:* Kevin Sullivan, "If You Think Gas Is Cheap These Days, Look What It Costs in Saudi Arabia," *Washington Post*, February 8, 2015, www.washingtonpost.com/world/middle_east/if-you-think-gas-is-cheap-these-days-look-what-it-costs-in-saudi-arabia/2015/02/07/889536ef-fb15-4453-b99b-eb99622dcf4e_story.html.

269 *In 2013, electricity prices:* Christopher Segar, "Saudi Energy Mix: Renewables Augment Gas," International Energy Agency, November 3, 2014, web.archive.org/web/20141128053419/www.iea.org/ieaenergy/issue7/saudi-energy-mix-renewables-augment-gas.html.

269 *As a result, domestic demand:* As of 2014, natural gas fuels 43 percent of Saudi electricity, with fuel oil and diesel providing the balance. Ibid.

269 *In 2009, overall Saudi energy demand:* Khalid A. Al-Falih, "Saudi Aramco and Its Role in Saudi Arabia's Present and Future," *Saudi Arabia Oil & Gas* speech, MIT Club of Saudi Arabia, Riyadh, Saudi Arabia, April 19, 2010, http://saudiarabiaoilandgas.com/index.php?option=com_content&view=article&id=61:saudi-aramco-and-its-role-in-saudi-arabias-present-and-future-issue14&catid=43:current-issue&Itemid=55.

269 *Extrapolating that year's demand:* Ibid.

269 *Aid to Bahrain, Egypt, Jordan:* "KSA [Kingdom of Saudi Arabia] Has Allocated SR252bn in Foreign Aid Since 1990," *Arab News*, September 29, 2014, www.arabnews.com/economy/news/637176.

269 *Moreover, during a visit of Saudi:* "Egypt Will Receive Oil Shipments from KSA Late March: El Molla," *Daily News Egypt*, March 17, 2017, http://www.dailynewsegypt.com/2017/03/16/egypt-will-receive-oil-shipments-ksa-late-march-el-molla/.

269 *Rather than simply hoping for a revival:* For more on energy reforms in the Gulf, see Bassam Fattouh, Anupama Sen, and Tom Moerenhout, "Striking the Right Balance? GCC Energy Reforms in a Low Price

Environment" (*Oxford Energy Comment*, The Oxford Institute for Energy Studies, University of Oxford, Oxford, U.K., May 2016), https://www.oxfordenergy.org/publications/striking-right-balance-gcc-energy-reforms-low-price-environment/.

270 *The countries of the Gulf Cooperation Council:* "Oil Exporters Respond to Price Slump," Economist Intelligence Unit, January 1, 2016.

270 *The favorite son of the elderly:* Karen Elliott House, "Uneasy Lies the Head That Wears a Crown: The House of Saud Confronts Its Challenges," Senior Fellow paper, Belfer Center for Science and International Affairs, Kennedy School of Government, Harvard University, Cambridge, MA, March 2016, 3, http://belfercenter.ksg.harvard.edu/files/Saudi%20Paper%20web.pdf.

270 *MbS has anchored his own personal ambitions:* See Kingdom of Saudi Arabia, *Saudi Vision 2030*, April 25, 2016, http://vision2030.gov.sa/en.

270 *The proceeds will provide added capital:* See Ibid.; Stefania Bianchi, "The Key Questions Asked About Saudi Arabia's $2 Trillion Fund," Bloomberg, May 25, 2016, https://www.bloomberg.com/news/articles/2016-05-25/key-questions-raised-by-the-2-trillion-saudi-wealth-fund-plan.

271 *The young prince declared:* See Adam Taylor, "Saudi Arabia Announces Plan to End Its 'Addiction' to Oil," *Washington Post*, April 25, 2016, www.washingtonpost.com/news/worldviews/wp/2016/04/25/saudi-arabia-announces-plan-to-end-its-addiction-to-oil/.

271 *An even greater obstacle:* See Laura El-Katiri, "Saudi Arabia's Labor Market Challenge," *Harvard Business Review*, July 6, 2016, https://hbr.org/2016/07/saudi-arabias-labor-market-challenge.

271 *The energy boom in the United States:* The unconventional boom could frustrate the progress that Saudi Arabia has made in building an extensive petrochemical sector by opening up the possibility of either more chemical plants in the United States or the export of significant amounts of U.S. ethane. Either outcome would constitute a disruption to Saudi Arabia and could challenge its markets in Asia, Europe, and Latin America. See Figure 2, "Shale Gas: Reshaping the US Chemicals Industry," PricewaterhouseCoopers, October 2012, 5, https://web.archive.org/web/20160302155537/http://www.pwc.com/us/en/industrial-products/publications/assets/pwc-shale-gas-chemicals-industry-potential.pdf. Nick Butler, "Ethane—the Next Challenge for the Energy Market," *Financial Times*, January 18, 2015, http://blogs.ft.com/nick-butler/2015/01/18/ethane-the-next-challenge-for-the-energy-market/; Alex Chamberlin,

"Ethane Production: The Prospects and Possibilities in the Market," *Market Realist*, April 14, 2014, http://marketrealist.com/2014/04/ethane-production-prospects-possibilities-market/.

273 *But also critical were new arrangements:* These arrangements were first enshrined in an interim constitution known as the Transitional Administrative Law and later became part of Iraq's permanent constitution. The 17 percent was supposed to be roughly in accord with the percentage of the Iraqi population that was Kurdish. Certain sovereign expenditures were first to be deducted from the overall revenues; 17 percent of the remaining amount was to then flow to the Kurdish region. See Coalition Provisional Authority, "Law Of Administration For The State Of Iraq For The Transitional Period," March 8, 2004, www.au.af.mil/au/awc/awcgate/iraq/tal.htm; "Full Text of Iraqi Constitution," *Washington Post*, October 12, 2005, www.washingtonpost.com/wp-dyn/content/article/2005/10/12/AR2005101201450.html.

273 *While assurances of a significant portion:* These fears appeared borne out when Baghdad decided to discontinue these transfers to Erbil once it started to export oil; the logic was that the revenues of these exports should be transferred to Baghdad for general redistribution as was the case with other Iraqi revenues, but on the whole, they were not. For more on recent events, see "Kurdish divisions spill over into federal politics," Economist Intelligence Unit, September 27, 2016; Yaroslav Trofimov, "Battle for Mosul Resets Ties Between Kurds and Baghdad," *Wall Street Journal*, November 3, 2016, https://www.wsj.com/articles/battle-for-mosul-resets-ties-between-kurds-and-baghdad-1478165405.

273 *In spring of 2014, sipping a glass:* Ashti Hawrami, in-person conversation with author, Erbil, Iraq, March 30, 2014.

274 *With oil prices north of $100:* David and Marina Ottaway posited in 2014—before the price plunge—"by the time Kurdish oil exports reach 450,000 barrels a day, perhaps as soon as the end of this year, Kurdistan will be earning enough to replace what it receives from Baghdad, which was $12 billion last year [2013]." Marina Ottaway and David Ottaway, "How the Kurds Got Their Way: Economic Cooperation and the Middle East's New Borders," *Foreign Affairs*, May 1, 2014, www.foreignaffairs.com/articles/141216/marina-ottaway-and-david-ottaway/how-the-kurds-got-their-way.

274 *The Iraqi government even enlisted:* Laurel Brubaker Calkins, "Iraq

Allowed to Sue Kurds over Texas Oil Tanker in U.S.," Bloomberg, January 8, 2015,www.bloomberg.com/news/articles/2015-01-09/iraq-allowed-to -sue-kurds-over-texas-oil-tanker-in-u-s-1-.

274 *Iraq could no longer send:* The pipeline itself became inoperable in March 2014 due to repeated attacks. "Iraq Faces Up to Export Undershoot," *Middle East Economic Survey* 57, no. 44 (October 31, 2014): 14.

275 *The push for Kurdish independence from Baghdad:* See Mahmut Bozarslan, "Will Kurds finally vote for an independent Kurdistan?" *Al-Monitor*, April 7, 2017, http://www.al-monitor.com/pulse/originals/2017/04/tur-key-iraqi-kurdistan-independence-voting.html#ixzz4gzrReqx8; Omar Sattar, "Could Kurds hold independence referendum this year?" *Al-Monitor*, April 10, 2017.

275 *Together, the two fields are believed:* Tamar and Leviathan are believed to hold 10 tcf and 22 tcf respectively. The comparison with the U.S. National Petroleum Reserves is based on the amount of reserves that would be extracted when the price of natural gas is close to $8 per mnbtu. "USGS Sees 18–32 Tcf Recoverable in NPR-Alaska," *Oil & Gas Journal*, May 5, 2011, www.ogj .com/articles/2011/05/usgs-sees-18-32-tcf.html; "Noble Energy Sells Three Percent Interest in Tamar Field, Offshore Israel, for $369 Million," Noble Energy, July 5, 2016, investors.nblenergy.com/releasedetail .cfm?releaseid=978097.

276 *Despite threats from Hezbollah:* Associated Press, "Hezbollah Warns Israel Against 'Stealing' Gas from Lebanon," *Haaretz*, July 27, 2011, http:// www.haaretz.com/israel-news/hezbollah-warns-israel-against-steal ing-gas-from-lebanon-1.375518.

276 *After working through a range:* "Israel: Leviathan FID Set for 4Q 2016. Development Concept Revealed," OffshoreEnergyToday.com, February 25, 2016, www.offshoreenergytoday.com/israel-leviathan-fid-set -for-4q-2016-development-concept-revealed/.

276 *Turkey was once Israel's closest ally:* Gulsen Solaker and Jonny Hogg, "Turkish PM Erdogan Says Israel 'Surpasses Hitler in Barbarism,'" Reuters, July 19, 2014, http://uk.reuters.com/article/uk-israel-turkey-travel-idUKKBN0 FO0XD20140719.

276 *In a telephone call orchestrated:* Jodi Rudoren and Mark Landler, "With Obama as Broker, Israelis and Turkey End Dispute," *New York Times*, March 22, 2013, www.nytimes.com/2013/03/23/world/middleeast/pres ident-obama-israel.html.

276 *It was not until a 2016 meeting:* Oddly, while the elements of the agreement were finalized in Rome, the agreement was actually signed separately and announced by the two prime ministers in two different locations. See Barak Ravid, "Israel and Turkey Reach Reconciliation Deal; Formal Announcement Postponed Until Monday," *Haaretz*, June 26, 2016, http://www.haaretz.com/israel-news/.premium-1.727205.

276 *Relations between the two capitals began to thaw:* "Russia Halts Turkish Stream Project over Downed Jet," *RT*, December 3, 2015, www.rt.com/business/324230-gazprom-turkish-stream-cancellation/.

277 *At the time of the diplomatic agreement:* Prime Minister Benjamin Netanyahu, "PM Netanyahu's Statement at His Press Conference in Rome," Prime Minister's Office, Rome, Italy, June 27, 2016, www.pmo.gov.il/English/MediaCenter/Speeches/Pages/speechTurkeytreaty270616.aspx.

277 *In September 2016, Jordan's state-owned:* "Jordan Agrees Gas Purchase Deal with Israel," Economist Intelligence Unit, September 29, 2016, http://country.eiu.com/article.aspx?articleid=1894657373.

277 *Israel had similarly sought to sell:* In 2014, the owners of Leviathan had signed a preliminary, non-binding $30bn deal with BG to sell natural gas to supply its LNG terminal in Idku, Egypt. John Reed, "Israel's Leviathan Partners Target $30bn Supply Deal with BG," *Financial Times*, June 29, 2014, https://www.ft.com/content/7a51810a-ff6e-11e3-9a4a-00144feab7de.

277 *Late in the summer of 2015:* Christopher Adams, "Eni Discovers 'Supergiant' Gasfield Near Egypt," *Financial Times*, August 30, 2015, https://www.ft.com/content/899031ec-4f0f-11e5-b029-b9d50a74fd14.

278 *Its absolute population has declined:* "Population, Total," The World Bank, data.worldbank.org/indicator/SP.POP.TOTL?locations=SY.

278 *A video clip filmed by a drone:* "Aerial: Drone Footage Shows Total Devastation in Homs, Syria (EXCLUSIVE)," YouTube, 1:55, posted by "RT," January 20, 2016, www.youtube.com/watch?v=DoRdCbDd50o.

278 *The group had commandeered oil fields:* For more details, see Erika Solomon, Robin Kwong, and Steven Bernard, "Inside Isis Inc: The journey of a barrel of oil," *Financial Times*, February 29, 2016, ig.ft.com/sites/2015/isis-oil/.

278 *Although Syria itself was never:* Before the conflict started, Syrian oil production was down to about 380,000 barrels a day. Ninety-five percent of Syria's exports went to the European Union; the revenue accounted for 25 percent of overall Syrian revenue.

279　*Russia and Saudi Arabia are the third:* These budgets are denominated in 2015 U.S. dollars. *SIPRI Military Expenditure Database* (Stockholm: Stockholm International Peace Research Institute, 2017), https://www .sipri.org/sites/default/files/Milex-constant-2015-USD.pdf.

279　*Iran's official military budget is smaller:* Ibid.

279　*The $500 million that the United States:* The U.S. defense budget for both years was close to $600 billion (2015 US$), Ibid. For more on the Syrian program, see Michael D. Shear, Helene Cooper, and Eric Schmidt, "Obama Administration Ends Effort to Train Syrians to Combat ISIS," *New York Times*, October 9, 2015, https://www.nytimes.com/2015/10/10 /world/middleeast/pentagon-program-islamic-state-syria.html?_r=0 .

279　*In speaking to the press:* Mark Mazzetti, Eric Schmitt, and David D. Kirkpatrick, "Saudi Oil Is Seen as Lever to Pry Russian Support from Syria's Assad," *New York Times*, February 3, 2015, www.nytimes.com/2015/02/04 /world/middleeast/saudi-arabia-is-said-to-use-oil-to-lure-russia-away -from-syrias-assad.html.

279　*He blamed the rumors on fake news:* "No Saudi-Russian Talks to Bump Up Oil Price in Return for Disowning Assad—Moscow," *RT*, February 4, 2015, http://rt.com/news/229183-saudi-russia-oil-assad/.

279　*Speaking anonymously, a Saudi diplomat:* Mazzetti, Schmitt, and Kirkpatrick, "Saudi Oil Is Seen as Lever to Pry Russian Support from Syria's Assad."

280　*But central government revenues:* "Iran's Economic Outlook, October 2016," The World Bank, 2016, http://pubdocs.worldbank.org/en/2065 81475460660337/Iran-MEM-Fall-2016-ENG.pdf.

280　*For this reason, many Iranians feel:* Jonas Siegel, "UMD Poll Reveals That As Benefits From Nuclear Deal Fall Short of Iranian Public's Expectations, Ahmadinejad Closes In On Rouhani," University of Maryland, July 15, 2016, https://umdrightnow.umd.edu/news/umd-poll-reveals-benefits -nuclear-deal-fall-short-iranian-publics-expectations-ahmadinejad.

280　*In the words of Ebrahim Mohseni:* Barbara Slavin, "New Poll Underlines Iranian Disappointment with US, Nuclear Deal," *Al-Monitor*, July 11, 2016, www.al-monitor.com/pulse/originals/2016/07/iran-new -poll-disappointment-nuclear-deal-jcpoa.html#ixzz4HdpPuuKH.

281　*Whereas some international companies:* To offset the effects of these challenges and incentivize foreign oil companies to invest in Iran after the sanctions are lifted, the Iranian oil ministry created draft contracts that replaced the traditional buy-back provisions with more relaxed terms, going as far as allowing joint ventures with the Iranian National Oil Company, thus providing higher returns and lower investment risks than those the international oil companies would get from Iran's rivals. See Parisa Hafezi and Jonathan Saul, "Exclusive—Iran Sweetens Oil Contracts to Counter Sanctions and Price Plunge," Reuters, February 3, 2015, http://uk.reuters.com/article/2015/02/03/uk-iran-oil-sanctions-exclusive-idUKKBN0L70G620150203.

Conclusion: From Serendipity to Strategy

284 *Even the Marshall Plan:* See David S. Painter, "Oil and the Marshall Plan," *The Business History Review* 58, no. 3 (Autumn 1984): 359–83.

284 *In 1990, after years of accusing:* Thomas C. Hayes, "Confrontation in the Gulf; The Oilfield Lying Below the Iraq-Kuwait Dispute," *New York Times*, September 3, 1990, www.nytimes.com/1990/09/03/world/confrontation-in-the-gulf-the-oilfield-lying-below-the-iraq-kuwait-dispute.html.

288 *In 2013, National Security Advisor:* "Remarks by Tom Donilon, National Security Advisor to the President, at the Launch of Columbia University's Center on Global Energy Policy," The White House, April 24, 2013, https://www.whitehouse.gov/the-press-office/2013/04/24/remarks-tom-donilon-national-security-advisor-president-launch-columbia-.

288 *Similarly, the National Security Strategy:* National Security Strategy, The White House, February 2015, 16, https://www.whitehouse.gov/sites/default/files/docs/2015_national_security_strategy.pdf.

289 *For example, the Bipartisan:* "Geopolitics of Lifting the Crude Oil Export Ban," Bipartisan Policy Center, September 2015, http://cdn.bipartisanpolicy.org/wp-content/uploads/2015/09/BPC-Energy-Crude-Oil-Export-Ban-Geopolitics.pdf.

289 *In 2007, the U.S. government:* "Annual Energy Outlook 2007," U.S. Energy Information Administration, 200, www.eia.gov/oiaf/archive/aeo07/pdf/0383(2007).pdf.

289 *Absent the new energy abundance:* According to the EIA, the all-time high of U.S. net crude imports was 11.6 mnb/d in 2006. "International Energy Statistics," U.S. Energy Information Administration, https://www.eia.gov/beta/international/data/browser/#/?pa=0000000000000400000 000000000000000vg0000000000c&tl_id=5-A&c=00000000000000000

000000000000000000000000000000002&ct=0&f=A&s=&cy=2015&start
=1980&end=2015&vs=INTL.55-1-USA-TBPD.A~INTL.57-3-USA-TB
PD.A&ug=g&tl_type=p&v=T.

290 *In his book,* A World: Richard N. Haass, *A World in Disarray* (New York: Penguin Press, 2017).

290 *Imagine, therefore, how much harder:* Robert B. Zoellick has written and spoken extensively about the fundamental link between economic and national security. See, for example, "Economics & Security in American Foreign Policy: Back to the Future?," an address at the John F. Kennedy Jr. Forum, Kennedy School of Government, Harvard University, October 2, 2012.

295 *As this program fizzles out:* Even before the price plunge, the Obama administration launched a task force to examine how the United States could help Caribbean states increase their energy security and sustainability; these efforts can and should be accelerated and augmented. See David L. Goldwyn and Cory R. Gill, "The Waning of Petrocaribe? Central American and Caribbean Energy in Transition," The Atlantic Council's Adrienne Arsht Latin America Center, May 2016, http://publications .atlanticcouncil.org/Petrocaribe/petrocaribe.pdf.

298 *For the decade ahead:* For more on the Thucydides trap, see Graham T. Allison, *Destined for War: Can America and China Escape Thucydides's Trap?* (Boston: Houghton Mifflin Harcourt, 2017).

301 *The days that Steve Coll:* See Steve Coll, *Private Empire: ExxonMobil and American Power* (New York: Penguin Press, 2012).

302 *This is the case as:* See "China's Cheaper Coal Seen Slowing Switch to Cleaner Natural Gas," Bloomberg, May 19, 2016, www.bloomberg.com /news/articles/2016-05-19/china-s-cheaper-coal-seen-slowing-switch -to-cleaner-natural-gas.

304 *As of April 2017:* "World Economic Outlook, April 2017: Gaining Momentum?" International Monetary Fund, April 2017, 15, www.imf.org/~ /media/Files/Publications/WEO/2017/April/pdf/text.ashx?la=en.

304 *BP's* Energy Outlook 2035: BP, *BP Energy Outlook: 2016 edition,* 75, https://www.bp.com/content/dam/bp/pdf/energy-economics/energy -outlook-2016/bp-energy-outlook-2016.pdf.

305 *In 2000, George Mitchell:* Joseph W. Kutchin, *How Mitchell Energy & Development Corp. Got Its Start and How It Grew: An Oral History and Narrative Overview* (Boca Raton, FL: Universal Publishers, 2001), 86–87.

Index

About the Author

Meghan L. O'Sullivan is the Jeane Kirkpatrick Professor of the Practice of International Affairs at Harvard University's Kennedy School of Governent. She is also the Director of the Geopolitics of Energy Project, which explores the complex interaction between energy markets and international politics. Between 2004 and 2007, she was special assistant to President George W. Bush, a time which included two years as Deputy National Security Advisor for Iraq and Afghanistan. She lives in Cambridge, Massachusetts. *Windfall* is her third book.